Digital Television Systems

Digital television is a multibillion-dollar industry with commercial systems now being deployed worldwide. In this concise yet detailed guide, you will learn about the standards that apply to fixed-line and mobile digital television, as well as the underlying principles involved, such as signal analysis, modulation techniques, and source and channel coding.

The digital television standards, including the MPEG family, ATSC, DVB, ISDTV, DTMB, and ISDB, are presented to aid understanding of new systems in the market and reveal the variations between different systems used throughout the world. Discussions of source and channel coding then provide the essential knowledge needed for designing reliable new systems. Throughout the book the theory is supported by over 200 figures and tables, whilst an extensive glossary defines practical terminology. Additional background features, including Fourier analysis, probability and stochastic processes, tables of Fourier and Hilbert transforms, and radiofrequency tables, are presented in the book's useful appendices.

This is an ideal reference for practitioners in the field of digital television. It will also appeal to graduate students and researchers in electrical engineering and computer science, and can be used as a textbook for graduate courses on digital television systems.

Marcelo S. Alencar is Chair Professor in the Department of Electrical Engineering, Federal University of Campina Grande, Brazil. With over 29 years of teaching and research experience, he has published eight technical books and more than 200 scientific papers. He is Founder and President of the Institute for Advanced Studies in Communications (Iecom) and has consulted for several companies and R&D agencies.

Digital Television Systems

MARCELO SAMPAIO DE ALENCAR

Federal University of Campina Grande

CAMBRIDGE UNIVERSITY PRESS
Cambridge, New York, Melbourne, Madrid, Cape Town, Singapore, São Paulo, Delhi

Cambridge University Press
The Edinburgh Building, Cambridge CB2 8RU, UK

Published in the United States of America by Cambridge University Press, New York

www.cambridge.org
Information on this title: www.cambridge.org/9780521896023

First published 2009

Printed in the United Kingdom at the University Press, Cambridge

A catalog record for this publication is available from the British Library

ISBN 978-0-521-89602-3 hardback

For
Silvana, Thiago, Raphael and Marcella

Contents

Contents

Figures

Tables

Contributors

Professor Valdemar Cardoso da Rocha Jr.
Federal University of Pernambuco.

Professor José Ewerton Pombo de Farias
Federal University of Campina Grande.

Professor Mylène Christine Queiroz de Farias
Federal University of São Paulo.

Professor Marcelo Menezes de Carvalho
Federal University of São Paulo.

Jean Felipe Fonseca de Oliveira
Federal University of Campina Grande.

Jeronimo Silva Rocha
Federal University of Campina Grande.

Paulo Ribeiro Lins Júnior
Federal University of Campina Grande.

Mozart Grizi Correia Pontes
Federal University of Campina Grande.

Preface

The book presents the historical evolution of television. It also introduces the basic concepts of digital television, including signal analysis, modulation techniques, source coding, probability and channel coding. The digital television standards, including the MPEG family, the ATSC, DVB, ISDTV, DTMB and ISDB standards are discussed. Several appendices, with topics including Fourier analysis, probability and stochastic processes, tables of Fourier and Hilbert transforms, tables of radiofrequency and a glossary complement the book. Many illustrations and graphics help the reader understand the theory.

Digital television is a new topic in the educational market; it evolved from the amalgamation of different areas, such as source coding, modulation, transmission techniques, channel coding, signal analysis and digital signal processing.

In commercial terms, digital television is a driving force of the economy and its deployment throughout the world generates a huge market as the analog television sets are replaced by the new HDTV devices.

It is important to realize that the media industry, which deals with enormous sums of money each year, relies on equipment to produce and distribute its series, movies and shows. The employment market in this area requires information technology professionals and engineers.

Few books have been published covering all the subjects needed to understand the subsystems that form the digital television network. This book is aimed at senior undergraduate students, graduate students, engineers and information technology professionals in the areas of electrical engineering and computer science. It can be used as a textbook for a course on digital television. The reader is expected to have a background in calculus and signal analysis.

Chapter 1 presents the fundamentals of digital television. It is self-contained, and concisely describes the most important television standards, in terms of percentage of the world population affected.

An overview of audio and video coding standards is presented in Chapter 2, which covers the basics of the subjects. In Chapter 3 the coding standards for compression of audio and video are discussed, with emphasis on the MPEG series of standards.

Channel coding for digital television is the subject of Chapter 4, which presents the most important algorithms used to protect the television signal from noise and other disturbances during transmission.

Digital and analog modulation techniques are presented in Chapter 5. The ATSC standard, used in the USA and other countries, is the subject of Chapter 6. Chapter 7 introduces the DVB standard, which was developed in Europe and now has been adopted by more than a hundred countries.

Chapter 8 discusses the Japanese digital television standard, ISDB. The Brazilian standard, ISDBT, is the subject of Chapter 9, and Chapter 10 presents the DTMB standard, developed in China.

The book includes three appendices. Appendix A relates, in a concise manner, the evolution of television since its inception. The basics of signal analysis, which is needed to understand the book, are presented in Appendix B. Random signals and noise are the subjects of Appendix C. The book also has a glossary of the terms frequently encountered in the sphere of digital television.

Acknowledgments

The author would like to acknowledge the help of Dr. Phil Meyler, Publishing Director, Engineering, Mathematical and Physical Sciences of Cambridge University Press, who agreed to publish the book and helped with all phases of the work, and Miss Sarah Matthews, Assistant Editor (Engineering), and Eleanor Collins, Production Editor, of Cambridge University Press who saw the book through all its production stages and gave precise and helpful suggestions, along with corrections for the manuscript.

The author also wishes to thank Mr. Marcus André Matos, who helped with the translation of selected parts of the book and revised the manuscript, and Jeronimo Silva Rocha, Paulo Ribeiro Lins Júnior, Jean Felipe Fonseca de Oliveira, Carlos Danilo Miranda Regis, Pedro Leonardo Falcão Costa, Daniel Cardoso de Morais, Raíssa Bezerra Rocha, Gilney Christierny Barros dos Anjos, Roana d'Ávila Souza Monteiro and Antonio Ricardo Zaninelli do Nascimento, for the preparation of graphs, figures, tables and diagrams for the book.

1 Fundamentals of digital television

1.1 Digital television

Digital television appeared as a natural evolution of analog television. Previously, the phases that constituted the production of a TV show (shooting the scenes, editing, finalizing and storing videos), broadcasting (generating the video composite, modulation, amplification, radio transmitting) and reception (the capture of the signal by the antenna, the demodulation of the television set receiver and the presentation of the image and sound to the viewer) of the signal by the user were all analog, i.e. the signals that represented the image and the sound generated in the studio were all analog, as well as the signals transmitted to the TV receiver (Carvalho, 2006).

Nowadays, the information is generated digitally in the studio. These signals are converted into analog signals and transmitted to analog television receivers. With digital television, all of the processes are digital; thus the image, the sound and all the additional information are generated, transmitted and received as digital signals. This gives the best definition of image and sound: the image is wider than the original one (panoramic screen), with a higher degree of resolution (high resolution) and stereo sound (Graciosa, 2006, Zuffo, 2006).

A digital television system is made up of a set of standards, as presented in Figure 1.1, which identifies the basic components: video and audio represent the services that are essential to the broadcasting of digital television; interactivity and the new services (e-commerce, Internet access) are added to the system by the middleware (Herbster *et al.*, 2005). These new services, introduced by digital television, originated from data transmission with video and audio. They may be used to offer new concepts in the broadcasting of TV programs to the users, or even to send data for applications that do not have a direct connection with television programming (Crinon *et al.*, 2006).

With digital television, the viewers will be renamed users, as they participate in interaction with the TV stations and the companies that supply services (Manhaes and Shieh, 2005, Valdestilhas *et al.*, 2005).

1.2 High-definition television

High-definition television (HDTV) is a digital television system that presents better image quality than the traditional system. HDTV allows better, more detailed image

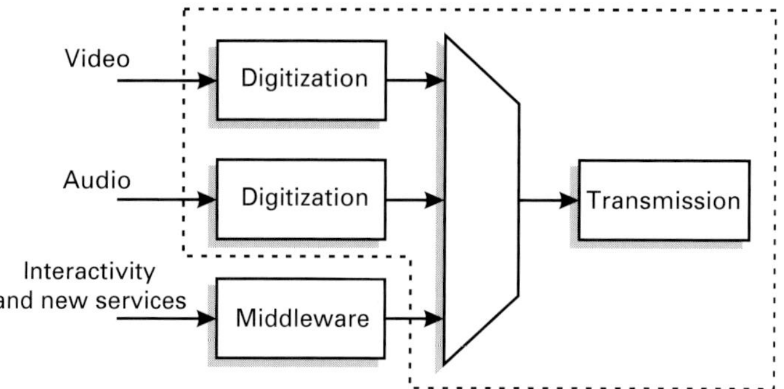

Figure 1.1   Set of standards in a digital television system for land broadcasting (Graciosa, 2006)

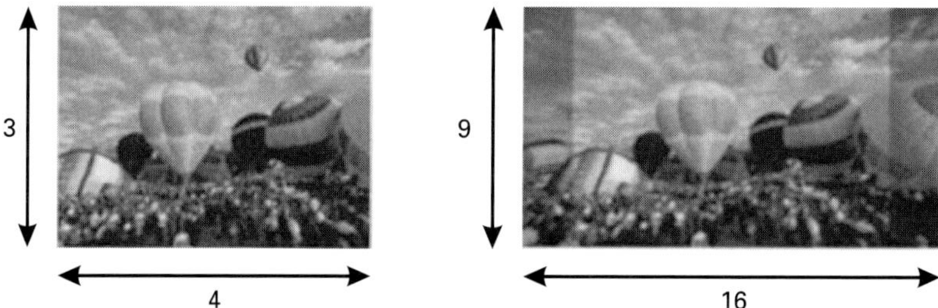

Figure 1.2   Comparison of aspect ratios 4:3 and 16:9

transmission, a wider picture (aspect ratio of 16:9) and stereo sound with up to six channels, which makes it possible to use several languages, among other services (Jones *et al.*, 2006).

Figure 1.2 presents comparison between two television sets with aspect ratios of 4:3 and 16:9. The usual aspect ratios for the presentation of films in movie theaters are 1.85:1 and 2.39:1 The most appropriate comparison between conventional television and HDTV is, however, not based upon the aspect ratio, but on image detail (HDTV makes it possible to see the image from a much wider angle) (Poynton, 2003a).

Currently, the most popular systems of HDTV are:

- The system with 750 lines/picture, 60 pictures/second, progressive scanning of 60 fields/second (non-interlaced) and 720 active lines per picture;
- The system with 1125 lines/picture, 30 pictures/second and alternated scanning of 60 fields/second and 1080 active lines per picture.

In interlaced scanning, only half the picture is on the screen at any given moment: while one frame shows only the odd numbered lines (1,3,5,...), the next frame shows only the even numbered ones (2,4,6,...). This happens so fast that the human eye perceives it as only one image. Progressive scanning shows each complete frame one at a time. Instead of alternating the lines, each is shown as lines 1, 2, 3, and so forth. The final result is a clearer image (HDTV.NET, 2006).

The HDTV signals are broadcast in the 720p or 1080i format, respectively: 720p means that there are 720 horizontal lines which are scanned progressively, and 1080i shows that there are 1080 horizontal lines which are scanned alternately. Despite the fact that there is a significant difference between the number of scanned horizontal lines, the images obtained by means of the 720p and 1080i systems are very similar (Poynton, 2003).

A television channel can broadcast HDTV programs as well as those of standard-definition television (SDTV), or even both simultaneously. The number of programs depends on the allotted bandwidth. Many countries still broadcast their digital television programs in the SDTV format (Jones *et al.*, 2006). SDTV is a system with a spatial resolution of 480 lines, with 640 picture elements (pixels) per line, and a timing resolution of 60 pictures per second in interleaved mode. A pixel is the smallest information element of an image; it has a unique set of attributes, such as color and luminance. The image quality of SDTV is higher than that received by open analog television stations, as it does not present problems such as the crossing over of colors and static that occur in the domestic reception of analog signals.

Currently, most of the transmissions are made in 4:3 format, though there is a trend to move to the 16:9 format (widescreen). Comparatively, the bit rate corresponding to one program on HDTV allows the broadcasting of four SDTV programs.

As well as HDTV and SDTV, there is also (HDTV.NET, 2006):

- Enhanced-definition television (EDTV): EDTV is of intermediate quality and, despite not having the same resolution as HDTV; it has better image quality than SDTV. Typically, it uses wide screen format (16:9) and a resolution of 480 lines, 720 pixels per line, and progressive mode scanning. The audio is stereo (5.1), as in HDTV.
- Low-definition television (LDTV): LDTV has a resolution quality lower than SDTV. A typical example is the system with 240 lines, 320 pixels per line and progressive scanning. A large amount of software and many microcomputer capture circuits currently run on images at this resolution level. Another typical example is the home-use VHS, which gives a resolution of 480 interleaved lines and an average 330 pixels per line (besides a clear decay in the chromatic resolution, which does not happen on the LDTV).

1.3   The digital programming platform

Middleware is the software layer, or programming platform, between the system and its applications, and permits interactive services on digital TV. Its main objective is to offer a set of tools that make possible interoperability of video transmission systems with various kinds of transmission media, including satellites, cables, land networks and microwaves.

At its most basic level, middleware has a software that has access to the flow of video, audio and data, routing them to an output element (television screen) or a storage element. The middleware receives input from the viewer's input gadgets (remote control

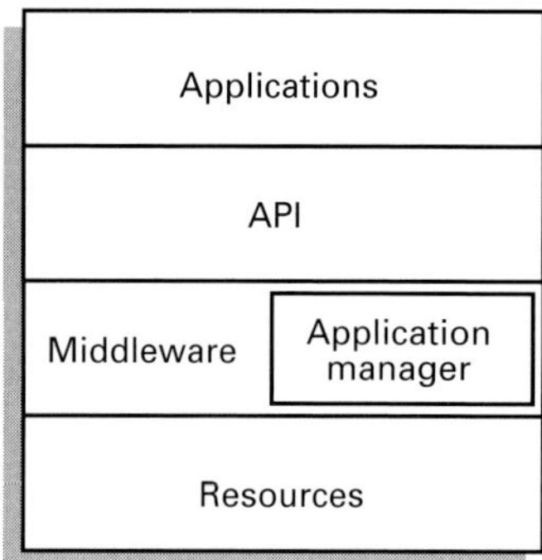

Figure 1.3   Basic structure of the elements of middleware (MC/MCT/FINEP/FUNTTEL, 2004)

or keyboard), and sends out information to the television screen and speakers, and also provides communication with remote entities by means of a remote channel.

The basic organizational structure of the elements of middleware, as shown in Figure 1.3, may be described as follows (MC/MCT/FINEP/FUNTTEL, 2004):

- Resources: the lower layer represents the hardware and software resources of the platform, whose elements (motherboards, microprocessors, subsystems, and operational systems in real time (RTOS)) vary according to the manufacturer. The middleware visualizes the resources in an abstract manner, in such a way that they may be charted in one or more distinct hardware entities.
- Middleware: the applications do not have direct access to the resources, and the middleware provides them with an abstract view of the resources. It isolates the hardware application, making its portability possible. In addition, middleware is responsible for managing all applications, including the resident ones.
- Applications programming interface (API): the API provides the services associated with the applications. In practice, there are several APIs that implement different interfaces. The middleware implements the API, presenting an abstract model of:
  - streams of audio and video executed from different sources and channels to carry them;
  - commands and events;
  - records and files;
  - hardware resources.
- Applications: these implement interactive services in the form of software to be executed in one or more hardware entities. The middleware API is the view that the applications have of the systems software.

Currently, there are four middleware standards in use: DASE, from the American digital television standard (ATSC), MHP, from the European standard (DVB), ARIB, from the Japanese standard (ISDB), and Ginga, from the Brazilian standard (ISDTV). Apart from these, other standards have been developed to support interactive video, such as MHEG and MPEG-4.

1.4   Interactivity

In an interactive digital television system, local storage of information is needed. Independently of the existence of the interactivity channel, user interaction is basically provided by the locally stored information processing. Thus, there must be local storage of information or a return channel to provide interactivity (Moreira, 2006). Some digital television sets include transcoders which make interactivity possible, although conventional sets are able to receive the content of digital TV and achieve interactivity by means of a device called a set-top box as shown in Figure 1.4.

The set-top box appears in the current scenario as an alternative to the costly digital television set. It is a decoder that receives the digital television content and converts it into analog format in such a way that the user may access digital technology. It also makes it possible to browse the web, using a return channel. And from such an initial contact with digital technology, the user may decide to change to a digital television device (Valdestilhas *et al.*, 2005).

Figure 1.5 shows a model of an interactive digital television system. This model shows the broadcasting of television programs by the stations to users spread all over the country.

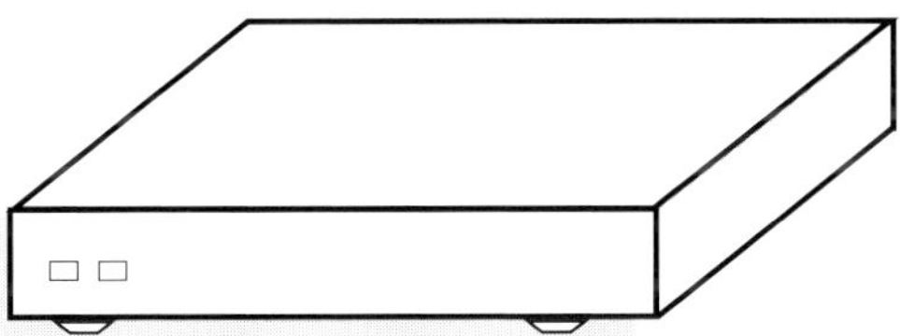

Figure 1.4   Set-top box model

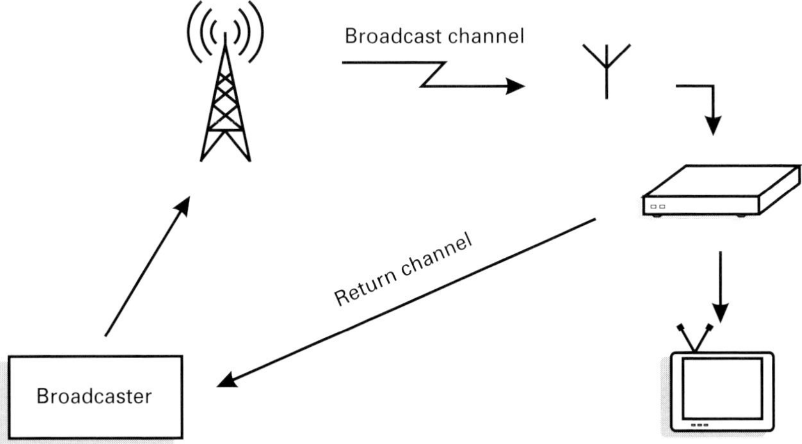

Figure 1.5   Model of interactive digital television system

The user receives the digital programs that are converted by the set-top box, which makes it possible to watch these programs on analog devices. The information coming from the station/broadcasting transmitter is sent through a broadcast channel, whereas the interactive information may be broadcast through the interactivity channel or even by the broadcasting channel. In addition, the user information is broadcast through the interactivity channel (Moreira, 2006).

The most basic kind of interactivity is called local interactivity, and uses the user's device or the set-top box. The data from certain interactive services are broadcast and stored on the device. This device may react to the user's queries without requiring data exchange throughout the network.

When it is wished to enable the user to respond to the interactive service in any way, in order for the broadcaster or network operator to capture such response and react to it, an interactive channel must be provided throughout the broadcasting network. This interactivity may be simple, as it is when one votes for a contestant on a television show. For this to happen, only one directional return channel from the viewer to the broadcaster is necessary.

If a higher level of interactivity is necessary, when the user needs a response (for example, in on-line shopping in which the buyer sends credit card data and receives confirmation of the purchase), a direct channel between the supplier and consumer must be implemented. The broadcast channel alone will not be sufficient if there is not an individual address, since such information is confidential and must be addressed solely to the relevant client.

If the information expected or requested by the interactive service user is more complex or demands a high broadcasting capacity, another level of interactivity must be achieved. This is the case, for example, when a potential consumer that, upon seeing an advertisement of a given manufacturer's product, asks for additional information. In that case, the direct interactivity channel must be broadcast. The interactive service is then akin to a two-way communication service with similar demands for the capacity and quality of broadcasting, in both the direct and reverse directions.

The addition of interactivity to the infrastructure of digital television requires the installation of the system be extended to include the components that promote communication between the end-user and the supplier of the interactive service. The high bit rate of the digital television broadcast channel may be used in the distribution of information to the user of the interactive service at rates of up to 20 Mbit/s per channel on land broadcasting networks, and up to 38 Mbit/s per channel on satellite or cable broadcasting networks. The transmission capacity of the interactivity channel will depend upon the type of network used for the transmission (Reimers, 2005b).

1.4.1   Interactive services

The term interactive service may describe a number of different types of services that call for a variable level of interaction between the user and the supplier of the service or network operator. Some of these services that are already available, or that will soon be, include (Resende, 2004, Crinon *et al.*, 2006):

- Electronic programming guide (EPG): this may be the oldest mode of interactivity on television, and allows the user to keep up with the programming of hundreds of channels, making the choice of a desired program easier.
- Enhanced television: this is an evolution of the television programs that already use interactivity; the difference lies in the format through which the user interacts with the station, which is no longer via the Internet (through a computer) or via a telephone, but through the digital television set itself.
- Individualized television: in this type of service, users will have at their disposal a level of interactivity similar to that of a DVD player (it will be possible to set cameras, sound, and subtitles at will).
- Internet television: this service makes it possible to access the Internet on the TV screen.
- Video on demand (VOD): this is an interactive application for which there has been great demand in recent years; it offers users a selection of films or TV shows that are available at that moment. VOD differs from EPG by allowing the user to search for a program in a databank with thousands of alternatives, to see if the program is showing and if so on which channel.
- Commercials: these can be incremented with the option of providing details if the user is interested in the product announced. There are also applications in which the user makes direct contact with the salesperson, with the possibility of purchasing the product through the television. A similar application is already available on shopping channels.
- Purchase of MP3 files, films or educational products that can be downloaded from a supplier immediately after the transaction.

Besides the interactive services, other services are available on digital television (CPqD, 2005):

- Monoprogramming: the showing of a program, with associated video and audio content, on a frequency exclusively designated for the station/programmer. This option, not obligatory in the environment of land digital television, is being used in some countries for the transmission of HDTV.
- Multiprogramming: this offers a multitude of simultaneous programs through only one frequency channel. Thanks to the encoding of video/audio and data signals, it is possible to broadcast between four and six simultaneous programs in SDTV.
- Mobility/portability: this allows the reception of digital television signals by the user in different movement conditions (stationary, moving, or even within a speeding vehicle). The reception may be by means of television sets on vehicles, television receivers integrated in cell phones or palmtops (Faria *et al.*, 2006a, Itoh and Tsuchida, 2006).
- Multiservices: these simultaneously combine several broadcasting and telecommunication services on the same digital television platform.

1.4.2   Television on mobile phones

Some mobile phone operators have launched handsets with digital television reception. Although the system is still of low resolution, it may become another success story for a product associated with the mobile phone, just as it was with the transmission of messages and the digital camera.

The Scandinavians and the Koreans are already using television on mobile phones. The Nordic operator Telia Sonera believes that this will be the next big trend in the mobile world. The giant Nokia launched its activities in this area with a major event in Singapore in the second quarter of 2005.

North American Strategy Analytics estimates that the mobile phone networks will have close to 51 million television users around the world by 2009, generating a total income of US\$ 6.6 billion.

Korean SK Telecom launched, in May 2005, a satellite TV service on mobile phones which offers 12 channels of video and audio. The Korean operator has also integrated TV and mobile Internet services on mobile phones. South Korean LG Electronics has a display that processes 30 image frames per second, an evolution of the 20 frames per second on earlier handsets.

There is competition in relation to two standards, Digital Video Broadcasting Handsets (DVB-H), and Digital Multimedia Broadcasting (DMB). LG prefers the DMB standard, which transmits twice as much data as DVB-H and does not discharge batteries so quickly. The Japanese, the Koreans, and Sweden's Ericsson support the DMB standard. Taiwan's Samsung has a DMB handset which was used in the satellite TV service on mobile phones from SK Telecom. Nokia is interested in the DVB-H standard.

In Europe, the telephone operators Vodafone and O2 are implementing plans in this direction. In the USA, a country in which the process is more advanced, Cingular has launched the MobiTV service with 22 channels, and Verizon Wireless has launched the VCAST service with CDMA 1xEV-DO technology. Qualcomm launched the MediaFLO service on its own frequency band of 700 MHz which can accommodate from 50 to 100 broadcasting channels with up to 15 streaming channels.

In Brazil, Bandeirantes Television Network (Band) has an agreement with the telephone operators for broadcasting news. Globo TV Network is awaiting the definition of the market and legislation before deciding what action to take. It does not want to give up control of the content. However, the possibility of broadcasting soap opera compacts and news is certainly tempting (Alencar, 2005a).

A study shows a world panorama of the experiences of the introduction and exploration of land digital television, in some of the countries pioneering its use. The data are presented in Table 1.1. The table relates to other services besides the interactive ones: monoprogramming, multi-programming, services based upon mobility and portability, and in multi-service environments (CPqD, 2005).

The mobility of the handset is an undeniable advantage of the DVB and ISDB systems when compared to the American ATSC. Consumers have already become used to the portable telephone and may be interested in a TV set that can be carried to the stadium, used in automobiles, or even in city and intercity bus fleets. A new market niche may

Table 1.1. Interactive services and business models for digital TV in some countries (CPqD, 2005)

	Germany	Australia	South Korea	Spain	USA	Finland	Netherlands	Italy	Japan	United Kingdom	Sweden
Services											
Monoprogramming		•	•		•				•		
Multiprogramming	•	•		•	•	•	•	•	•	•	•
Interactive:											
(without return channel)	•	•	•	•	•	•	•	•	•	•	•
(with return channel)			•			•		•	•		
Mobility/portability			•			•			•		
Multiservice	•	•		•	•	•	•	•	•	•	•
Business models											
Open TV	•	•	•	•	•	•		•	•	•	•
Cable TV				•	•	•	•			•	•
Pay-per-view								•			

appear, surprisingly not predicted by the Americans, who have invested so much in mobile communications in the last three decades (Alencar, 2002b).

1.5 Return channel for digital television

The interactivity channel is seen as a subsystem that makes it possible for each user, individually and independently from the others, to interact by forwarding or receiving information and requests from the networks (Manhaes and Shieh, 2005). That is, the return channel is the medium through which the networks and the television announcers directly reach the viewers. It is through this channel that a direct connection is established between the user and the seller of a given product. The subscriber may choose a program that is different from the one showing, or change the angle of the camera broadcasting the show.

In addition to having the benefits of the regular display of the network's programs as well as the major interaction with subscribers, viewers might decide to follow a court case through an e-government service, to check if their salaries have been credited to their bank accounts through home banking, or even call upon their children for an on-line French class. All these possibilities might seem intangible in the short run, although such advances are possible by means of the interactivity channel of digital TV.

The model shown in Figure 1.6 illustrates a generic system of interactive services. The downlink establishes communication between the network and the viewers; it can happen through broadcasting (open and available to all users/subscribers) or be individualized. The user is enabled to interact with the network by means of an uplink or return channel, expressing preferences and opinions. The return channel may be implemented

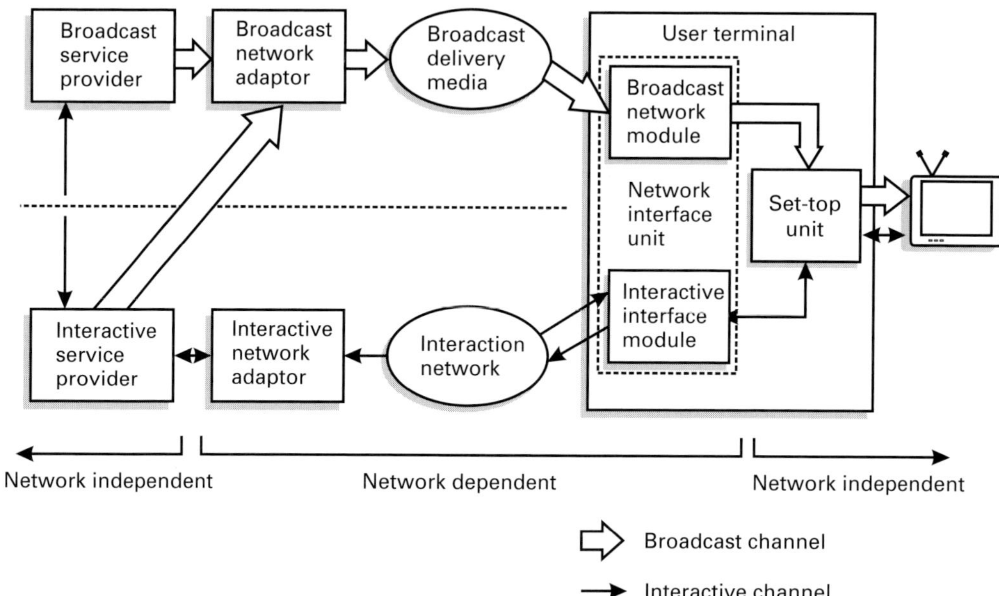

Figure 1.6   Model for a generic system of interactive services (Reimers, 2005b)

by any communication technology which establishes a connection from the users to the networks (Manhaes and Shieh, 2005).

Figure 1.6 shows that two channels are established between the users and the supplier of the service. The network adapter provides the connectivity between the supplier of the service and the network, while the interface unit connects the network to the user. The supplier of the broadcast service distributes the MPEG-2 stream of transport from the unidirectional broadcast channel to the user's set-top box. The supplier of the interactive service offers an interactive channel for the bidirectional communication which is divided between the interaction route in the direct direction, to the downstream, and the return route in the reverse direction, to the upstream.

In order to offer high-speed services to the user, the supplier of the interactive service can choose the broadcasting link to fit the data on the MPEG-2 transport stream. In that case, the broadcasting channel may contain data application or communication control, in such a way as to connect to the distribution network. The user may, for example, use a cable modem instead of a set-top box. A bidirectional control application and a communication channel will also be required from the different services suppliers, with the intention of obtaining synchronization (Reimers, 2005b).

It is worth mentioning that the interactivity channel may be designed to send data through the broadcasting channel networks or by means of a specific channel.

The introduction of a return channel, however, is not an easy task. Table 1.1 shows that the interactivity channel is well established in only four countries: South Korea, Finland, Italy, and Japan (CPqD, 2005). In the UK, despite being available since 2003, the interactivity channel did not appeal to the population and so did not reach the desired number of households. The Brazilian standard has a recommendation for the adoption of the intraband channel, which uses the IEEE WiMax technology for the return channel.

- In South Korea, the land digital broadcasting standard adopted is the ATSC, as a result of the country's decision to choose the high-definition image quality (HDTV) and to choose the same standard adopted by the North American market (considered important for South Korean electronic industry exports). The return channel is implemented by means of ADSL technology (which has great acceptance in the country), and has been used, since 2003, in applications such as question-and-answer sessions on variety television shows with an audience, on educational programs, and in opinion polls on TV news.
- Finland, just like the other European countries, uses the DVB-T digital broadcasting standard. Some applications of interactive services are offered by means of a return channel, implemented by the shared land telephone system (SLTS) or by the short message system (SMS) of the cell phone operators.
- Italy also uses the DVB-T standard for broadcasting digital television, and has already implemented its return channel through SCTS.
- In Japan, in order to promote the expansion of its electro-electronic industry and to remain a world leader in the television sector (besides favoring high-definition image quality), the ISDB-T standard was developed, in which the return channel is implemented by the ADSL technology.

1.6   Digital television standards

There are five main digital television systems in operation in the world: the American Advanced Television Systems Committee (ATSC), the European Digital Video Broadcasting Terrestrial (DVB-T), the Japanese Integrated Services Digital Broadcasting Terrestrial (ISDB-T), the Brazilian International Standard for Digital Television (ISDTV or ISDB-Tb) and the Chinese Standard for Digital Television (DTMB) (Farias *et al.*, 2008; Resende, 2004). In this section, these five established systems are analyzed.

The similarities between the systems are that they maintain the same frequency band used today, improve the vertical and horizontal special resolutions, improve the representation of colors, present an aspect rate of 16:9 to approximate the format to that of a movie theater (the analog set uses the aspect ratio 4:3), support multichannel sound of high fidelity and data transmission. There are also standards for digital television via cable or satellite. The specifications for terrestrial broadcasting of signals are shown on Table 1.2 (Tude, 2006).

For a digital television standard, the modulation technique used to broadcast the signal is the main feature. There are two methods generally used: the single-carrier modulation (SCM), and the multiple-carrier modulation (MCM); each model causes different behaviors of the signal in the communication channel, in addition to using distinct methods of encoding (Drury *et al.*, 2001).

Table 1.2. Specifications for terrestrial broadcasting (TELECO, 2006)

	ATSC	DVB-T	ISDB-T	ISDTV	DTMB
Video digitization	MPEG-2	MPEG-2	MPEG-2	H.264	MPEG-2
Audio digitization	Dolby AC-3	MPEG-2 ACC	MPEG-2 AAC	H.264	MPEG-2
Multiplexing	MPEG	MPEG	MPEG	MPEG	MPEG
Signal transmission	8-VSB modulation	Multiplex COFDM	Multiplex COFDM	Multiplex COFDM	SCM and MCM
Middleware	DASE	MHP	ARIB	Ginga	IMP

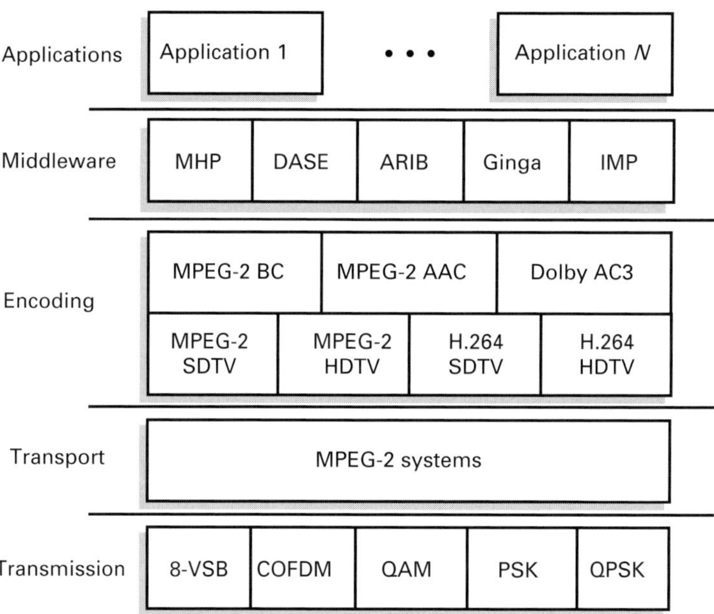

Figure 1.7   Standard options for digital television

The American system (ATSC) uses the SCM method, with the modulation schemes 8-VSB (8-level vestigial side band), and OQAM (offset quadrature amplitude modulation) respectively; whereas the European (DVB-T), the Brazilian (ISDTV) and the Japanese (ISDB-T) systems make use of the MCM method and work with coded orthogonal frequency division multiplexing (COFDM) (Resende, 2004). The Chinese standard uses both SCM and MCM.

Frequency division multiplexing (FDM) is not a new technique, and appears on various standards for digital radio, television and transmission systems. They all use digital orthogonal signals, i.e. without interference between them. Figure 1.7 illustrates the usual standard options for digital television.

1.6.1 The DVB-T standard

The European digital television standard, Digital Video Broadcasting (DVB), was started in 1993 by a consortium of more than 300 members. Among them are equipment manufacturers, network operators, software developers and the regulation departments of 35 countries. The standard consists of a set of documents related to broadcasting, transporting, encoding, and middleware. Currently, DVB is adopted within the European Union, Australia, New Zealand, Malaysia, Hong Kong, Singapore, India and South Africa, and over 100 other countries (DVB, 2006).

DVB-T was developed to meet the several different needs of various countries, and, for that reason, it is a flexible standard in relation to the setting modes (from a total of 126 possible settings). The broadcasting system operates in channels of 6, 7, or 8 MHz, with COFDM multiplexing, with 1705 carriers (2K system) or 6817 carriers (8K system), and its broadcasting rates may vary between 5 and 31.7 Mbit/s. SDTV broadcasting on DVB-T allows the simultaneous broadcasting of up to six programs on the same terrestrial bandwidth. The channel encoding is done to reduce the effect of the channel on the broadcast signal, thus reducing the number of errors.

For protection against errors, the DVB standard uses the Reed–Solomon code combined with a convolutional code of the type used in mobile communications, such as the CdmaOne system (IS-95) manufactured by Qualcomm, with some suppressed bits. The use of guarding intervals between the symbols of various carriers guarantees more robustness in relation to intersymbol interference (Alencar, 2002b).

Still in terms of modulation, Digital Video Broadcasting Cable (DVB-C) uses 64-QAM modulation, with six bits of data per symbol; Digital Video Broadcasting Satellite (DVB-S) uses QPSK modulation; Digital Video Broadcasting Microwave operating on frequencies of up to 10 GHz (DVB-MC) makes use of the Multipoint Multichannel Distribution System (MMDS) with 16.32 or 64-QAM; and Digital Video Broadcasting Microwave operating at frequencies above 10 GHz (DVB-MS) uses the Local Multipoint Distribution Service (LMDS) with QPSK (Alencar, 2007a, Resende, 2004, Fernandes *et al.*, 2004, MHP, 2006).

Specifications for the DVB standard

The DVB European consortium has developed a set of specifications for interactive services, describing solutions for a variety of possible network settings, including the broadcasting specifications of its standard, as well as interactive networks capable of providing the digital television systems with the return channel. Figure 1.8 depicts the DVB standard architecture and Figure 1.9 illustrates the standard scheme. Table 1.3 summarizes the specified technical areas and the respective acronyms.

Although this set of specifications has been defined for the European standard of digital television, only a few countries, including Finland, Italy and England, currently use the return channel on their digital television broadcasting (CPqD, 2005).

Multimedia home platform (MHP)

MHP is the open middleware system designed for the European digital television standard. It defines a generic interface between interactive digital applications and the

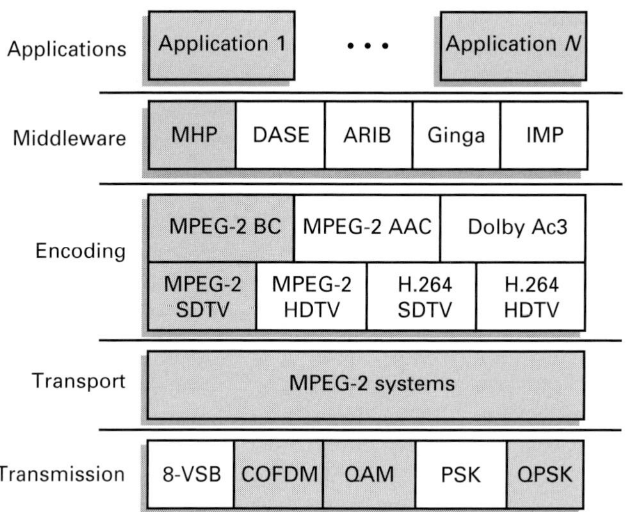

Figure 1.8   DVB standard architecture

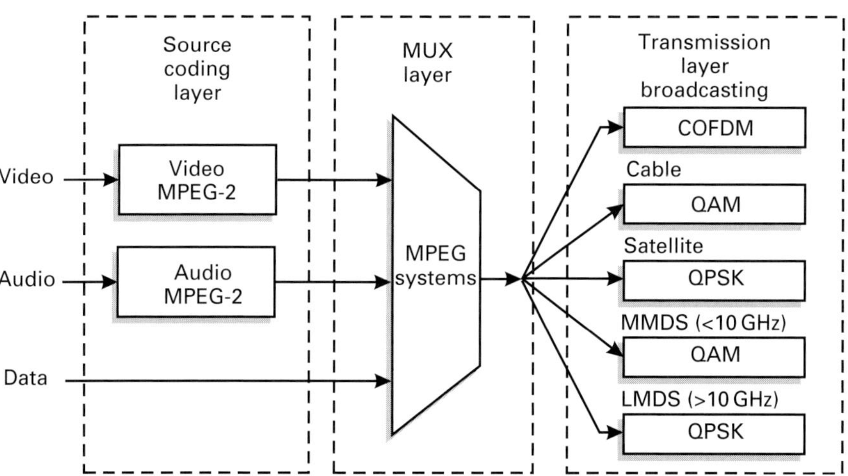

Figure 1.9   DVB standard scheme (TELECO, 2006)

terminals where the applications are executed. This interface buffers the vendors from the specific hardware or software used by the terminal. The MHP extends the existing DVB standards to the interactive and broadcasting services in all the broadcasting networks, including terrestrial, satellite, cable and microwave systems. MHP is based on a platform denominated DVB-J, which includes a virtual machine defined in compliance with specifications of Java Virtual Machine, from Sun Microsystems.

Several software packages provide generic APIs for the large number of resources of the platform. An API is a source code interface that an operating system usually provides to support requests for services to be made by computer programs. The MHP applications

Table 1.3. Set of specifications of the DVB standard
for the interactivity channel (Reimers, 2005b)

Interactivity channel	DVB acronym
ISDN	DVB-RCP
DECT	DVB-RCD
GSM	DVB-RCG
CATV	DVB-RCC
LMDS	DVB-RCL
Satellite	DVB-RCS
SMATV	DVB-RCCS
Terrestrial	DVB-RCT

access a platform only from these specific APIs. The MHP has three different profiles that provide sets of resources and functions for local interactivity, interactivity with a return channel and Internet access. Besides the procedural applications the MHP also accepts declarative applications, using the DVB-HTML format (MC/MCT/FINEP/FUNTTEL, 2004, MHP, 2006, Piesing, 2006).

1.6.2 The ATSC standard

The American standard for digital television (ATSC), covers HDTV, SDTV, data transmission, audio with multi-channel sound, and direct-to-home broadcasting (ATSC, 1995b).

Started in 1982 and currently composed of approximately 130 members (equipment manufacturers, network operators, software developers, and regulation departments), the ATSC grand alliance has been functioning in the USA since November 1998, and the standard was adopted by a few other countries, including Canada, South Korea and Mexico. In 2006, it was estimated that over 28 million digital TV sets had been sold in the USA since 1998. ATSC is made up of a set of document which define the adopted standards, including those related to broadcasting, transportation, encoding, and middleware (Richer *et al.*, 2006).

For terrestrial broadcasting it operates with channels of 6, 7 or 8 MHz. The original information, with a 1 Gbit/s rate is compressed to 19.3 Mbit/s and then encoded, for protection against errors, with a Reed–Solomon encoder (the same code used on DVD) and a trellis encoder. The resulting signal is modulated in 8-VSB for broadcasting on a 6 MHz channel, using SCM.

The VSB modulation technique is used on PAL-M and NTSC due to the bandwidth economy in relation to AM (for video broadcasting) and because its generation demands less precise and cheaper equipment than those necessary for the single side band (SSB) (Alencar, 2007a). However, it shows problems in the reception by regular television set antennas and does not work well with mobile reception. Meanwhile, cable television uses 64-QAM modulation (similar to DVB), and satellite broadcasting uses QPSK modulation (also similar to DVB).

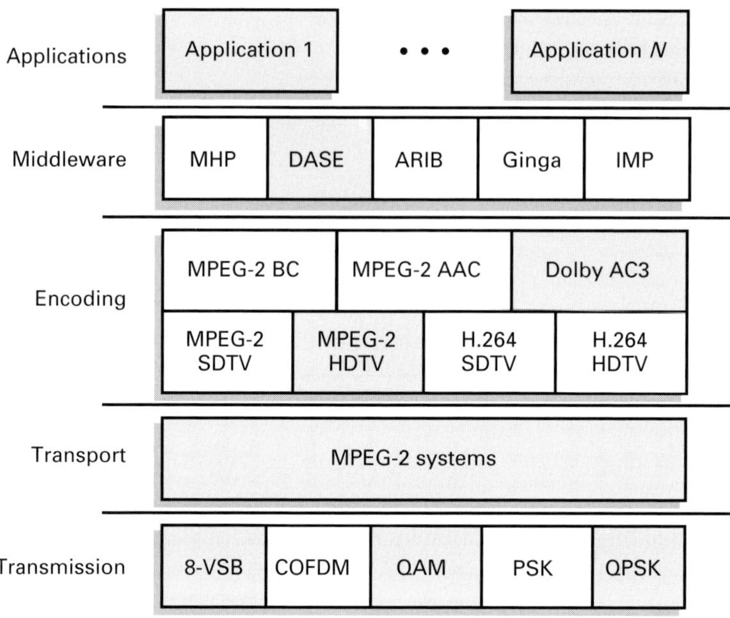

Figure 1.10   Architecture of the ATSC standard

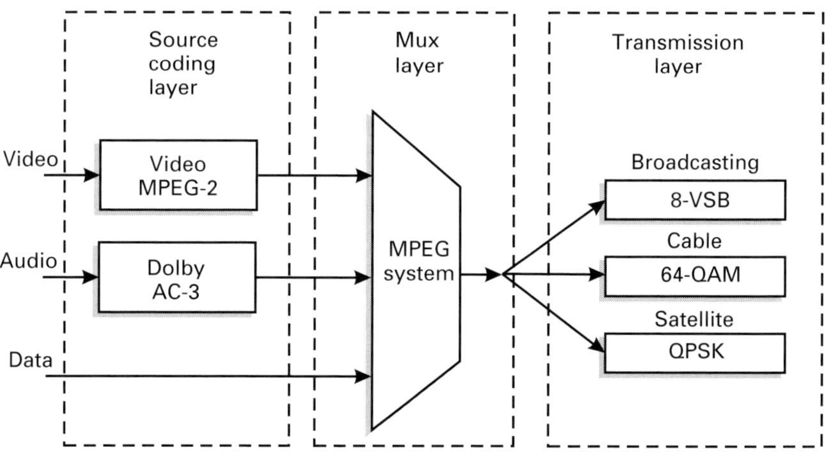

Figure 1.11   Scheme of the ATSC standard (TELECO, 2006)

The ATSC system, whose architecture is described in Figure 1.10, uses the 1920×1080 pixels format with interlaced scanning of 60 fields/s, or 1280×720, with progressive scanning of 30 pictures/s. The video uses the MPEG-2 encoding, created by the Moving Picture Experts Group (MPEG), and the audio follows the standard Dolby AC-3, version of the standard used in theaters and in high-quality sound equipment. Figure 1.11 illustrates the scheme of the ATSC standard.

The system allows for several qualities of image with 18 different video formats (SDTV, HDTV, or intermediate picture qualities achieved with different frame rates), besides making data transmission possible. This system was developed to operate on channels with several attenuating features, from white noise to multi-path to impulsive and phase noises. It was also designed to operate on overloaded bands with optimum spectral efficiency, without interference from the NTSC television (Alencar, 2007a, Resende, 2004, Fernandes *et al.*, 2004, DASE, 2003, Richer *et al.*, 2006).

DTV application software environment (DASE)

DASE is an ATSC standard which defines the platform for advanced functions in the receiver. It interacts with the platform services of the receiver, in that the receiver can accept inputs from the transmission carrier and from the final user and generate output of audio and graphs to the receiving systems. The receivers' platform offers essential services to the DASE, such as operating systems services, input and output services, and memory services.

It also has two application environments for backup to declarative and procedural applications: the declarative application environment (DAE); and the procedural application environment (PAE), which processes contents with active objects for a Java virtual machine. Besides DAE and PAE, DASE has content decoders which serve not only the procedural but also the declarative applications for decoding and presentation of common types of content, such as the PNG (portable font resource) and the JPEG formats (DASE, 2003, MC/MCT/FINEP/FUNTTEL, 2004).

1.6.3 The ISDB-T standard

The digital television standard ISDB-T was conceived to execute the digital broadcasting of the television signals, making it possible for HDTV to be accessible not only to users fixed receivers but also to those with mobile, wireless ones, with low image definition.

Specified in Japan in 1999, by the Digital Broadcasting Experts Group (DIBEG), the ISDB-T involved several enterprises and television operators from that country. To date, it has only been adopted in Japan; nevertheless, the ISDB is said to gather the greatest set of technical facilities among the three main digital television standards: high definition, data transmission, and mobile and portable reception (Asami and Sasaki, 2006, Takada and Saito, 2006).

The ISDB system, whose architecture is described in Figure 1.12, can broadcast video, sound, data, or a combination of all three, since it presents a high flexibility of settings, thanks to the way it was conceived. The way its bandwidth is segmented defines the transmission method, known as band segmented transmission OFDM (BST-OFDM). It has 13 distinct segments that can be set in three different modes; these modes, layers of the system, may be modulated in an independent way by means of multilevel modulation schemes, and transmitted through an MCM system, which is the OFDM. It may be seen as an improved variant of the European standard. Figure 1.13 shows the ISDB standard scheme.

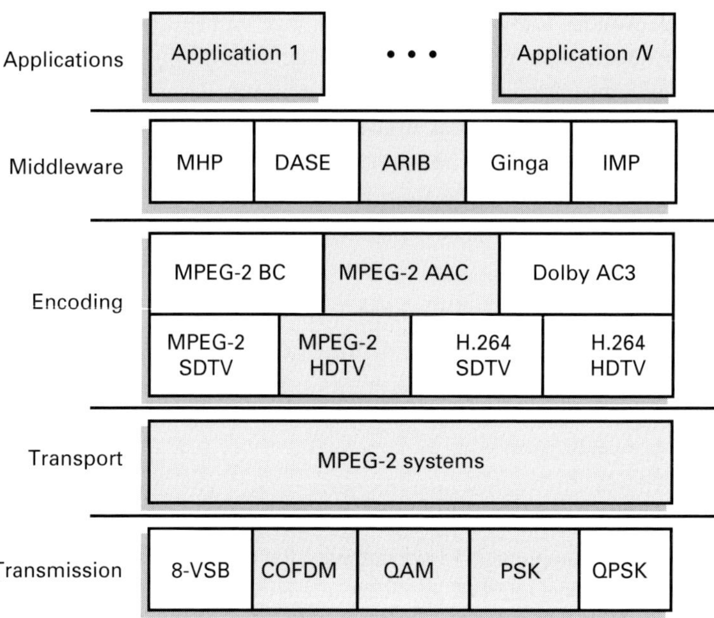

Figure 1.12   Architecture of the ISDB standard

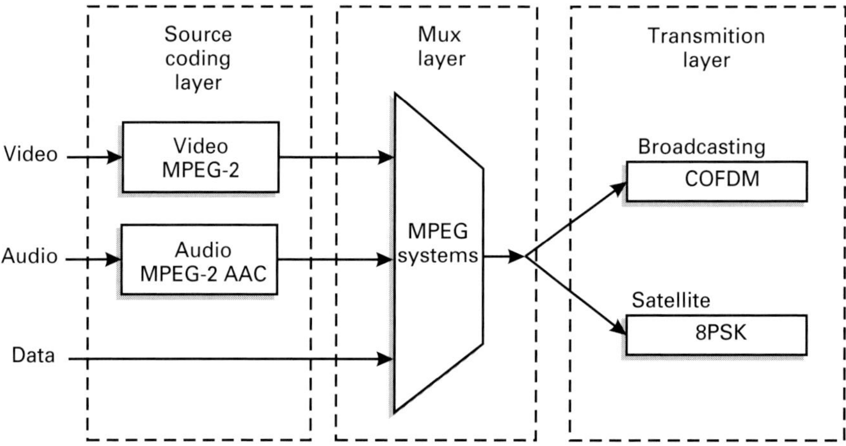

Figure 1.13   Scheme of the ISDB standard (TELECO, 2006)

The main characteristics of the ISDB-T are:

- transmission of HDTV, SDTV, LDTV;
- transmission of multiple programs;
- interactive services and high-quality multimedia for mobile and conventional receivers;
- hierarchic transmission, which accepts a singular setting for various receivers, partial reception inclusive.

ISDB-T operates with channels of 6, 7, 8 MHz, uses COFDM multiplexing with variations, and encoding of signal payload with MPEG-2. The protection against errors is provided by Reed–Solomon code combined with convolutional code; designed for hierarchic systems with multiple levels, ISDB-T reaches a transmission rate which varies between 3.65 and 23.23 Mbit/s. For cable television, it makes use of the 64-QAM modulation scheme, while satellite broadcasting uses the 8-PSK modulation (Alencar, 2007a, Resende, 2004, Fernandes *et al.*, 2004, ARIB, 2006, Asami and Sasaki, 2006, Takada and Saito, 2006).

Association of Radio Industries and Business (ARIB)

The Association of Radio Industries and Business (ARIB) established standards for transmission and encoding of data for digital broadcasting, which are based on an XML specification, consisting of three parts: encoding of a single medium (created to maintain the compatibility with the old multiplexed data transmission system that was already in use in Japan), encoding of multimedia (to establish compatibility with the network standards and methods of data transmission used in the European and American systems), and specification of data transmission.

There are two specifications for the execution of applications: the first one is based on the ARIB B24 standard with support to declarative applications, using broadcast markup language (BML); and the second one is based upon the ARIB B23, which defined an environment of procedural applications, based on the standards DVB/MHP and globally executable MHP (GEM) (ARIB, 2006, MC/MCT/FINEP/FUNTTEL, 2004).

1.6.4   The ISDTV standard

The first actions to implement digital television in Brazil started in the late 1990s with the work of the Brazilian Commission of Communications of the National Telecommunications Agency (Anatel). From November 1998 to May 2000, extensive field and laboratory tests were carried out with the digital television standards available at that time.

The project for the development of the Brazilian Digital Television System (SBTVD), which later became known as International System for Digital Television (ISDTV or ISDB-Tb), was launched in November, 2003. More than a hundred institutions were involved in the ISDTV project, including industry, universities, research centers, and broadcasting companies. In February 2006, the report containing the recommendations for the ISDTV standard was released (CPqD, 2006).

The ISDTV standard uses a technology that is similar to the Japanese standard ISDB-T for coding and modulation of digital television signals. The signals are transmitted using the band segmented transmission (BST) technique and orthogonal frequency-division multiplexing (OFDM). The architecture of the ISDTV standard is shown in Figure 1.14.

The ISDTV committee adopted H.264 as the video compression standard. H.264 is used to code both standard and high-definition video, as well as reduced resolution videos for mobile or portable receivers. The adoption of H.264 is a key innovation in relation to all other digital television standards.

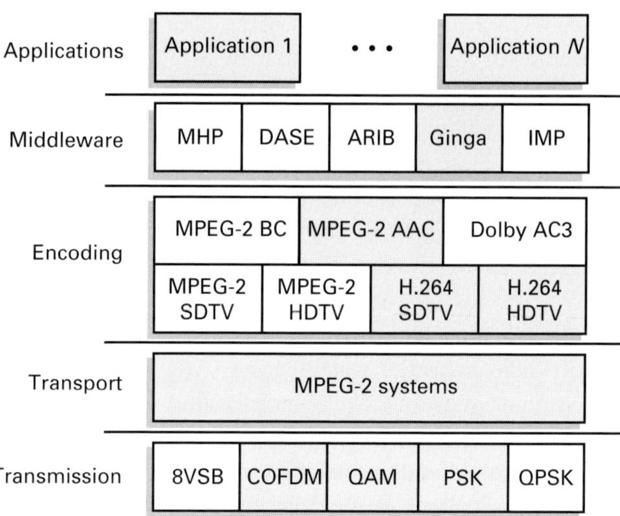

Figure 1.14 Architecture of the ISDTV standard

Middleware

Middleware uses the Ginga standard, which specifies a set of common functionalities that support the Ginga application environments. The Ginga-Core is composed of the common content decoders and procedures that can be used to prepare the data to be transported through the interactivity channel. The Ginga-Core also supports the conceptual display model of ISDTV. Ginga specifications for the architecture and applications are designed to work on terrestrial broadcasting digital television receivers, but may also be used for other systems, such as satellite or cable DTV systems.

A number of technologies have been studied for the return channel. The analysis favored WiMAX technology, in particular, WiMAX-700 technology, which is a new WiMAX profile (Meloni, 2007). The profile operates in the 400–900 MHz primary frequency band (UHF) and, optionally, from 54 MHz to 400 MHz as a secondary band (VHF).

ISDTV maintains the same main characteristics as analog television, including the use of the VHF/UHF band and the 6 MHz channel. ISDTV deployment plans require that all state capitals must be covered by the end of 2009, while all other Brazilian cities should be receiving the digital signal by the end of 2013.

For mobile reception, Brazil and Japan use the OneSeg standard, which includes H.264/MPEG-4 AVC for video and HE-ACC for audio, encapsulated in an MPEG-2 transport stream. The resolution is 320/240 pixels (QVGA) and the video transmission rate is 220–320 kbit/s. The audio transmission rate is 48–64 kbit/s. The modulation scheme is QPSK, with a rate 2/3 encoder and final transmission rate of 416 kbit/s.

1.6.5   The DTMB standard

China has more than 400 million television receivers, approximately 30% of the 1.4 billion television sets worldwide. The Chinese government started to develop the Chinese digital television standard in 1994 and completed the process in 2006, after stringent laboratory and field tests.

In 2001, the Chinese government published a call for proposals for the standard. The proposals were supposed to provide higher bandwidth efficiency larger coverage than the existing system, low-power consumption as well as high mobility, fixed and mobile and indoor and outdoor reception, and also support SDTV as well as HDTV. Several proposals were submitted.

In 2004, the competing proposals, Digital Multimedia Television Broadcasting Terrestrial (DMB-T), the multicarrier approach from Tsinghua University, Advanced Digital Television Broadcasting Terrestrial (ADBT-T), the single-carrier approach from Shanghai Jiaotong University, and Terrestrial Interactive Multiservice Infrastructure (TiMi), a multicarrier approach from the Academy of Broadcasting Science, started the merging process.

The system was named the Framing Structure, Channel Coding and Modulation for Digital Television Terrestrial Broadcasting System (Standard GB 20600-2006), but it is usually called Digital Terrestrial Television Multimedia Broadcasting (DTMB). It became mandatory on August 1, 2007 (Song *et al.*, 2007a). The middleware for the Chinese standard is called interactive Media Plataform (IMP). It is based on the MHP, DASE, and ARIB plataforms and can be described in three categories: Enhanced Service profile, T-Commerce profile and Multinetwork Service profile (Hong-pang and Shi-bao, 2004). The architecture of the DTMB standard is shown in Figure 1.15.

Coding and modulation

The forward error correction (FEC) subsystem uses the concatenation of two codes. An outer BCH(762, 752) code and an inner low-density parity-check (LDPC) code. The output binary sequence from the FEC is converted to an M-ary quadrature amplitude

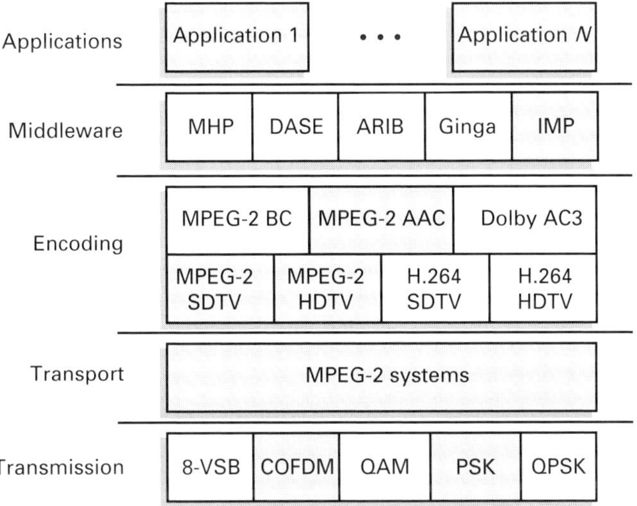

Figure 1.15 Architecture of the DTMB standard

modulation (MQAM) symbol stream. The standard supports 4-QAM, 4-QAM-NR (Nordstrom Robinson), 16-QAM, 32-QAM and 64-QAM constellations.

1.6.6   Ultrahigh-definition television

Digitization has its advantages, but it also creates problems for the future, mainly on the consumer side, for it excessively speeds up the development process of new products.

The Japan Broadcasting Corporation (Nippon Hoso Kyokai – NHK) is leading a research initiative for the creation of a new standard for digital television, named Super Hi-Vision. This ultrahigh-definition wide-screen system has 4000 scanning lines and is supposed to deliver images so real that viewers feel as if they are actually at the site of the broadcast (NHK, 2008).

The features of Super Hi-Vision, also called ultrahigh definition (UHDTV), include:

- The system uses a video format with 7680×4320 pixels, approximately 33 megapixels, which is 16 times higher than the HDTV standard, for a 1.77 aspect ratio (16×9).
- The scanning lines are not visually noticeable even when relatively close to the screen and a wider viewing angle conveys a stronger sense of reality.
- The new three-dimensional audio system with 24 loudspeakers (22.2 channels) enhances the feeling of being present.
- A codec has been developed by NHK for the efficient transmission and broadcast of Super Hi-Vision signals. The encoder compresses the video signal from approximately 24 Gbit/s down to 180–600 Mbit/s and the audio one from 28 Mbit/s to 7–28 Mbit/s.
- The standard offers a resolution of 32 million pixels compared to HDTV's two million pixels.
- The refresh rate is 60 frames per second.

The audio system, which has already been demonstrated by NHK in the USA, has 20 speakers and two special effects loudspeakers (22.2 channels). The cameras that capture the image have a resolution of 8 megapixels (Alencar, 2006).

The Japanese Ministry of Internal Affairs and Communications will start a public and private partnership to develop technology for UHDTV in the hope of setting an international standard for Super Hi-Vision. Broadcasting is not expected to begin before 2015, and it is proposed that it will be via fiber optic cable.

1.7   Television receivers

Most of the television sets manufactured in the world still use cathode ray tubes (CRTs). They are reliable, robust, and have good contrast and color display. However, they are big and heavy. Some manufacturers have invested in flatter, narrower tubes, which may give the CRT an extended life.

The technologies for manufacturing plasma and liquid crystal display (LCD) television sets have been known for decades. The first operational LCD dates back to 1963, while the plasma set was invented in 1964. The first TV based on plasma technology was developed in 1997.

The fundamental difference between plasma and LCD screens is that the plasma one emits light from each spot on the screen, with the use of cells that use neon and xenon gas, while the luminescence of an LCD screen depends on a source of light positioned behind the screen.

In both cases, each pixel is made up of three subpixels: red, green, and blue, which form the RGB system. On LCD the liquid crystal controls the passage of light to the sub-pixels. This luminescence may be reflective, when it uses the light of the environment around it, transmissive, when the light is behind the screen, or transflective, if it does both procedures.

Without considering the initial cost, the LCD screen is more economical because it uses less energy than the plasma screen. The difference may be up to 20% and varies according to the image shown: the LCD screen has constant consumption, while the plasma screen has higher consumption when showing brighter scenes. The plasma screen has some advantage in relation to the angle of viewing and the contrast. The LCD screen is brighter than the plasma one, which improves the visualization in lighter environments. The reproduction of colors on the plasma monitor is more ample and precise.

As far as the resolution goes, i.e., the definition of the image on the screen, LCD has the advantage because it usually contains more pixels. Plasma and LCD monitors are estimated to last between 50 and 60 thousand hours, but the plasma screen loses luminescence faster than the LCD one.

There are several monitors with rear projection that use CRT or microdisplay. A CRT monitor is very heavy, but produces a high-quality image. The microdisplay monitor may use an LCD screen, a Digital Light Processing (DLP) screen, or a Liquid Crystal on Silicon (LCoS) screen.

In the DLP projector, the light of its bulb is focused on the surface of an integrated circuit (chip), whose reflecting surface is composed of thousands of micromirrors, each of which modulates the behavior of each pixel that is projected on the screen. DLP one-chip systems are able to project 16.7 million colors (24 bits of color levels), while the DLP three-chip systems can project 35 million colors.

The LCoS technology creates images using a fixed mirror assembled on the surface of a chip, and uses a crystal liquid matrix to control the quantity of light reflected. However, manufacturing LCoS chips is more complex, which makes the television set more expensive (Wikipedia, 2007a).

Companies like Samsung Electronics are investing in high-definition televisions with light-emitting diodes (LEDs) and lasers, which produce a far wider range of colors than the bulbs.

LEDs produce the basic colors: red, green and blue. Beams are emitted in a narrow band of wavelengths very close to those of single, pure colors, giving off brilliant images. The three primary colors are beamed in varying intensities at the same spot on a television screen, and create a palette of hues in a wider range than in televisions without this technology.

As of now, LEDs within television sets are all in rear-projection models. The devices are expected to last the lifetime of the television, unlike the bulbs typically used in rear-projection TVs, which generally must be replaced every few years at a high cost.

Laser televisions, unlike LED models, are not yet available commercially, but several manufacturers have demonstrated them at trade shows. Mitsubishi Digital Electronics introduced a large-screen laser television at the Consumer Electronics Show in Las Vegas in January, 2008. The lasers produce extremely saturated colors. Lasers are able to give more than 90% of the color range that the human eye can perceive. Plasma displays and LCDs present a color gamut which reaches only 40% and 35%, respectively.

The distinctive range of colors produced by lasers and LEDs may provide a competitive edge for rear-projection TVs, which lost market share to plasma and liquid crystal display models. Rear-projection sets are typically lit by high-pressure white-light mercury lamps, which last for two years, at most. LEDs are typically used in sets with screens of 50 inches, 56 inches and 61 inches (Eisenberg, 2007).

2   Audio and video coding

2.1   Introduction

Digital television signals are usually generated by cameras, when captured from real-life footage, or are produced by computer animation. In either case, there is a need for coding the source signal. Analog to digital conversion is necessary in the first case, in order to reduce the entropy of the information source. The original analog signal has an infinite number of levels, which implies an infinite storage space in the camera or studio equipment. For the second case, coding is also needed to compress the original signal for transmission.

Digital television typically generates audio and video signals, as well as the data for controlling the broadcasting and for interactivity. The audio and video signals, after being captured by the camera, are processed by a piece of equipment generically known as a source encoder. The source encoder turns the analog signal captured into a digital signal either for storage in a mass memory device, such as the hard disk (HD) or digital video disk (DVD), or for direct broadcasting to the studio.

The analysis of any communication system, including television, reveals that, roughly speaking, the systems may be represented as shown in Figure 2.1, the blocks of which are defined as follows:

- Source: this generates the information to be transmitted. The following are examples of output sources: a computer terminal, a microphone, a television camera, and a remote sensor in a telemetry system. The source is commonly modeled by a stochastic signal or by a random data generator to be transmitted.
- Transmitter: this converts the output of the source into waveforms that are suitable for broadcasting on the channel. The function of the transmitter can be subdivided as follows:
  - (1) Source coder: in many cases, this consists only of an analog to digital converter. In more sophisticated applications, it perform the function of removing unnecessary details of information, such as the ones in the image processing.
  - (2) Channel coder: this adds controlled redundancy to the output of the source encoder, to counter the effects of noise. The most used ones in digital television systems are the Reed–Solomon coder, which encodes the information in blocks, and the Viterbi coder, which uses convolutional encoding.

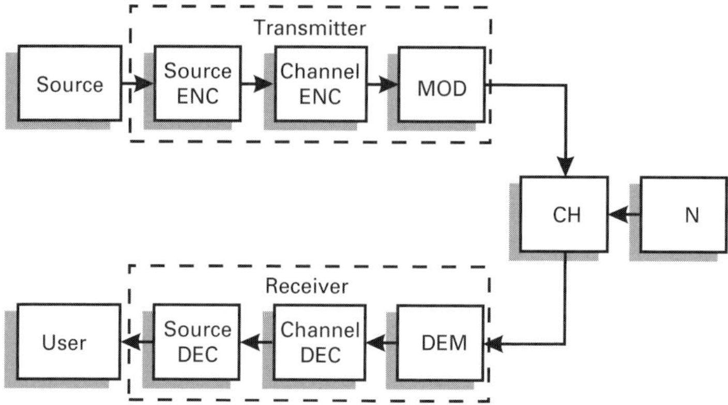

Figure 2.1   Generic communication system

(3) Modulator: this turns the output of the channel encoder into some waveform and frequency suitable for broadcasting through the channel. The modulator is generally followed by a power amplifier, which raises the level of the signal and matches the impedance with the waveguide that carries the signal to the antenna, when the broadcasting is on the air. The antenna is the element that sends out the electromagnetic waves to the broadcasting medium. To produce the maximum gathering of power, the antenna must have the same impedance as the broadcasting medium.

- Channel: this is the physical medium through which the information passes before reaching the receiver. A channel may consist of a pair of wires, an optical fiber in a cable television system, or a microwave link like multichannel multipoint distribution service (MMDS), for example. Upon being carried through the channel, the information is subject to noise in the form of undesirable and unpredictable disturbances. As a result of the influence of noise, part of the information may end up somewhat broken. In order to mathematically establish the performance of the system, it is necessary to characterize the noise by means of a stochastic process: i.e., the channel must be mathematically modeled.

- Receiver: it is up to the receiver to process the noisy output of the channel in order to determine the transmitted waveform. This is normally the most complex part of a communication system, and may be subdivided as follows:

(1) Demodulator: from the signal received from the channel, the demodulator emits an estimate of the type of waveform sent by the transmitter and delivers the corresponding digital version in the output. Due to the noise, this version is not always correct and, in this way, estimates containing errors might be passed to the channel decoder. It is usually preceded by a high-gain signal amplifier, known as a front end amplifier. For reception with a satellite dish, which is in fact a paraboloid of revolution, i.e., a surface obtained by means of the rotation of a parabola on its own axis, a low noise amplifier (LNA) is used at the focus of the parabola. This amplifier has a low noise value, i.e., the ratio between the power input and the power output of the device, due only to the thermal noise, is small. As there

are usually many serial amplifiers in the receiver, if the gain of the LNA is high enough, its noise value prevails for all the amplification circuit (Alencar, 2001).

(2) Channel decoder: with encoding techniques applied to the symbols provided by the demodulator, the channel decoder tries to correct possible errors and thus produces its estimates of the source encoder output digits.

(3) Source decoder: this processes the output of the channel decoder, restoring the redundancy that was removed in the source encoder, restructuring the message to be delivered to the final receiver. In the case of television, there is a decoder for the video signal and another one for the audio signal. At the final stage of the decoding process the signal is converted to analog, for the reproduction of colors and sound.

- Final receiver: this is the ultimate receiver of the information transmitted. It can be an individual at the other end of the line, a television viewer, or a computer, for example.

2.2 Source coding

Any information that needs to be stored on computer or transmitted through a digital communication system, including voice and video signals, time series, measurement results, and results of signals processing, must necessarily go through a quantization process (Wagdy, 1987). An immediate application is the quantization of voice, which makes up the encoding process (Flanagan *et al.*, 1979).

In fact, the quantization is done in such an automatic manner in day-to-day life that it is overlooked. Grading students at the end of each semester involves a rounding-up process, which presents, as a result, a quantization error.

The interest aroused by quantization can be measured by the number of articles published about it. The references at the end of this work, however limited, show how important this topic is. The scalar quantizers project is still an area of continuous research, with the focus on the applications of signal detection (Blum and Kassam, 1991), Markov sources (Goblirsch and Farvardin, 1992), combined encoding of source and channel (Zeger and Manzella, 1994), optimum quantizers (Trushkin, 1993, Wu and Zhang, 1993), dither quantization (Zamir and Feder, 1992, Gray, 1993), multiple stage vector quantizers (Phamdo *et al.*, 1993), fixed rate vector quantizers (Laroia and Fravardin, 1993), sigma-delta (He *et al.*, 1992) and trellis coding quantizers (Fischer and Wang, 1992).

Despite being very simple in its description and construction, the uniform quantizer has been surprisingly difficult to analyze, due to its inherent non-linearity. Estimating the spectrum of quantization noise and retrieving most of the signal, once this error is present, is one of the topics of this chapter. Only scalar quantization is dealt with, but other quantization schemes can be explored by means of a suitable mathematical treatment.

The following sections introduce digital encoding techniques, such as the pulse code modulation (PCM) scheme, which is the basic building block of most source coders. Quantization schemes are discussed, with emphasis on quantization noise effects. Section 2.8 presents an introduction to the MPEG family, used in audio, video and digital television broadcasting systems.

2.3   The source coding process

The source encoding processes (also called signal digitization) have, as a final objective, the reduction, in a controlled way, of the entropy of the signal generated by the data source. The entropy represents the average information provided by the source symbols and is defined for an alphabet X of symbols, such as

$$H(X) = \sum_X p(x) \log_2 \frac{1}{p(x)}, \tag{2.1}$$

in which $p(x)$ represents the probability of the symbol x. The entropy, as information measurement, is given in units of shannon (Sh), but the use of bit to refer to the content of information is common.

Claude E. Shannon borrowed the concept of entropy for the theory of information from the homonymous concept of thermodynamics. Apparently, the Hungarian mathematician John von Neumann suggested this term. It is believed that he had recommended it for two reasons: first, because there was the thermodynamics mathematical function with this name; second, and most important, because few people really knew what entropy was, and its use in a discussion would always constitute an initial advantage.

For instance, one of the techniques studied by Shannon, PCM, turns an analog signal into a series of binary pulses, which can be manipulated in an efficient manner by the communication system. However, this procedure results in an intrinsic error, or noise, caused by the attribution of quantum levels to the signal.

The sampling and coding stages do not cause significant distortion to the signal during the demodulation. Quantization error, or noise, however, is inevitable in the process of quantization, which is the conversion of the signal from analog into digital.

The development of the source coding systems can be summarized, as follows (Alencar, 1991a). Alec Reeves patented the PCM scheme in the USA, in 1942. Apparently, Reeves invented the system in England, in 1937, and obtained a patent in France, in 1939 (Bedrosian, 1958, Reeves, 1942). William R. Bennett produced the work on the quantization noise. He introduced the model for the characteristic function $f(x)$ of the quantizer, and calculated the spectrum of the noise for a Gaussian input signal (Bennett, 1948).

Claude Shannon established the mathematical basis of information theory, with the use of the concepts of entropy, mutual information, and channel capacity (Shannon, 1948). He demonstrated the usefulness of the PCM scheme, computing the capacity of this channel for a quantization noise with uniform distribution (Shannon, 1948b).

In a meeting of the Institute of Mathematical Statistics, in Atlantic City, USA, in 1957, Stuart P. Lloyd presented a seminal paper on PCM quantization (Lloyd, 1982). In 1960, Joel Max developed a method of quantizing for minimum distortion (Max, 1960). Edwin D. Banta introduced the Fourier series decomposition for the calculation of the self-correlation for a sum of a deterministic signal and a Gaussian noise (Banta, 1965). B. Lévine used the idea of Fourier series decomposition for the Gaussian signal in the calculation of the quantization noise spectrum (Lévine, 1973). Sripad and Snyder

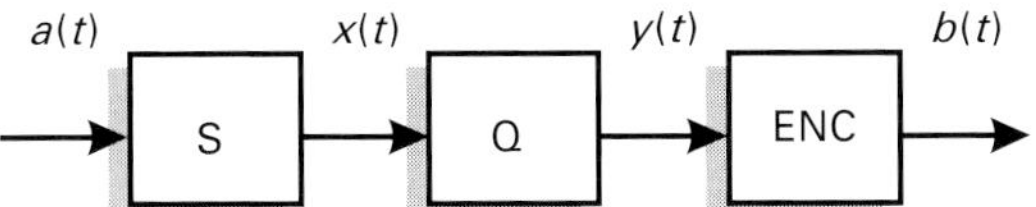

Figure 2.2   Generic model for a source encoding system

established the conditions for the quantization noise to be uniform and white (Sripad and Snyder, 1977).

The introduction of Woodward's theorem, along with the Fourier series analysis, for deterministic signals, was done by T. A. C. M. Claasen, in 1989. He managed to obtain the solution of the quantization noise spectrum (Claasen and Jongipier, 1981). This can be seen as the first attempt at using Woodward's theorem to tackle the problem, but the results were presented without a formal proof, leading to questions about the convergence of the model introduced. Marcelo S. Alencar deduced the general formula of the quantization noise spectrum for random signals in 1989, including the spectrum produced by non-uniform schemes (Alencar, 1988, 1989).

In 1990, Robert Gray developed a mathematical model for the self-correlation of the quantization noise for the uniform quantifier. The result is based on the bidimensional characteristic function of the input signal. The connection between the model and the practical results is not obvious (Gray, 1990).

Figure 2.2 illustrates a generic model for source encoding with the sampling, quantization, and encoding blocks separate. The system input signal, $a(t)$, is analog and the output signal, $b(t)$, is digital.

2.4   Signal sampling

In the sampling process, a voice signal that is continuous in time is turned into one that is discrete in time. In order for this to happen, samples of the signal are taken at regular intervals. The process of sampling and generating the pulse amplitude modulation (PAM) signal is illustrated in Figure 2.3.

For the original signal to be retrieved from the sampled signal, the sampling frequency has to comply with the Nyquist criterion: it has to be at least twice as much as the maximum frequency of the signal. For a signal with maximum frequency f_M, the Nyquist sampling rate, or frequency, is $f_A = 2f_M$ ($\omega_A = 2\omega_M$). The sampling frequency of the signal is determined by the signal maximum frequency and is given in samples per second.

For applications in telephony, the internationally adopted sampling frequency is $f_A = 8$ ksamples/s. Recordings on CD are made with a sampling rate of 22 ksamples/s. In the case of the MPEG-1, the audio sampling rates are 32, 44.1, or 48 ksamples/s. The television standard of 480 interlaced lines (480i) has a sampling rate of 13.5 Msamples/s. Figure 2.4 illustrates spectral features of the PAM signal.

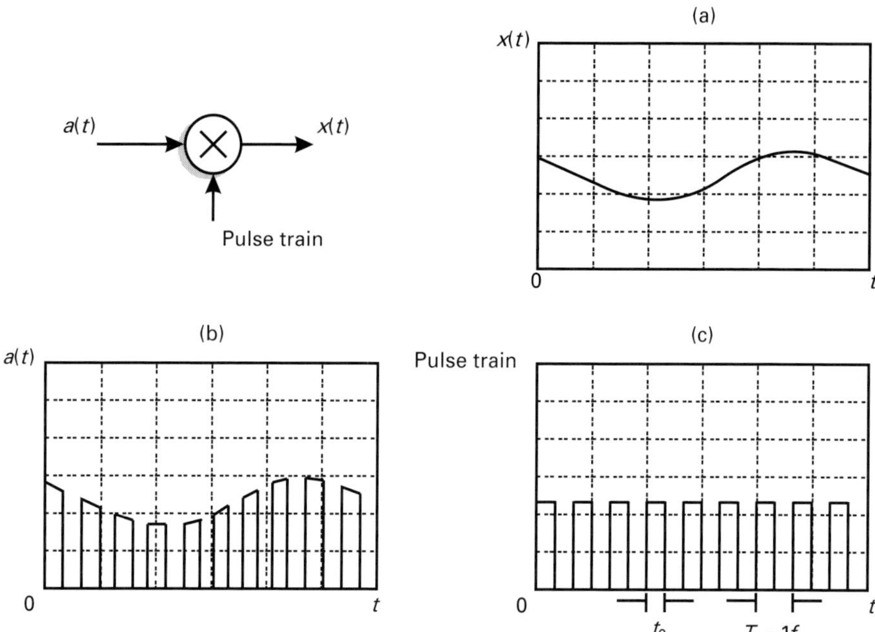

Figure 2.3   Pulse amplitude modulation: (a) original signal; (b) sampled signal; (c) pulse train

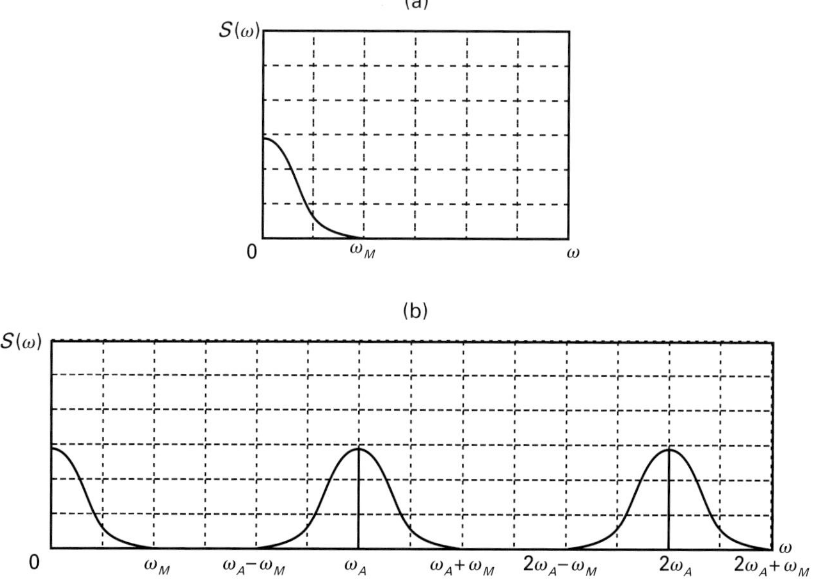

Figure 2.4   (a) Signal spectrum; (b) Spectrum of the PAM signal

If the sampling frequency is lower than the Nyquist frequency, the signal is not fully retrieved, as there is spectral superposition, resulting in distortion in higher frequencies. This phenomenon is known as aliasing.

2.4.1 Sampling of the video signal

In the video sampling process, the camera lenses project over the sensor's surface (charge coupled device – CCD), a segmented image in picture elements (pixels). The resolution of the image depends directly on the number of pixels. The sampling device analyzes the analog signal that comes from reading the pixels and digitizes it, generating the signal in digital format.

After the digitization, an image composed of a given number of pixels is generated, according to the format used. In the NTSC DV, for instance, there are 720 pixels across by 480 pixels high, i.e., 345 600 pixels in total. For the NTSC standard, which is the interlaced type, there are two fields (odd- and even-numbered lines) making up each frame of the image. In this way, half of these pixels represent the odd-numbered lines and the other half represent the even-numbered lines.

On editing programs, the smallest unit of image shown on the screen is a frame that contains both fields. In this way, as a frame is seen on the computer screen, all the lines that compose the image (both fields together) are visualized, i.e., all 345 600 pixels. This form of visualization combines with the display mode used on computer screens, which is the progressive scanning type, instead of interwoven, used on television sets.

The analog signal read from the CCD contains luminescence and color information, be it with the use of color lenses that cover the pixels in the shape of a mosaic (RGB), or with the use of three separate CCDs, one for each color. The generated RGB signal is converted into the YUV signal, which is then encoded.

In the digitization for the DV format, the luminescence component is sampled at the rate of 13.5 Msamples/s. The U and V components, which represent the color differences in relation to the luminescence signal Y, are sampled at a lower rate, 3.37 Msamples/s, since the color information is not as important in the composition of the image for the human eye as the luminescence. Thus, the video DV format sampling rate is 4:1:1, based on the sampling rate of the three components.

In the Digital Betacam, DVCPRO50, DVCPRO HD, and Digital-S formats, for example, this rate is 4:2:2, resulting in better image quality and better color resolution. The compression algorithm MPEG-2 can compress data using this rate. This algorithm, when employed for DVD-Video, and the standards MPEG-1, JPEG, MJPEG, as well as the DV format in the European PAL standard, and HD HDV format, uses the rate 4:2:2, alternating the components U and V at each line, which results in the notation 4:2:0.

If during the digitization the original RGB signal does not suffer reduction of the resolution of its color components, it is said that the signal is of the 4:4:4 type. However, direct comparison of formats using only these numbers, if not interpreted correctly, may lead to dubious results. Thus, for example, the HDCAM, a high-definition format, uses a sampling rate of 3:1:1, which means that the luminance component has a sampling rate three times higher than that of the U and V components. This format, however, gives a better image than the Digital Betacam format, with 4:2:2. What happens is that the digits used in the definition of the rate do not have an absolute value, but represent the ratio of the number of luminescence and color samples. In fact, while the luminescence sampling rate in the DV is 13.5 M samples/s, in the HDCAM it is 74.5 Msamples/s (Views, 2006).

2.5   The quantization process

From a certain point of view, the quantization process may be seen as the mapping of the signal from the continuous domain to a countable number of possible levels of output (Alencar, 1991b). The need to represent signals with a finite number of bits results in quantization noise in almost all systems for digital processing of signals. This incurs a quantization error, which is intrinsic to the analog to digital conversion process (Sripad and Snyder, 1977).

Figure 2.5 illustrates the quantization scheme for a uniform quantizer that has a quantization step d. The input signal suffers a non-linear transformation, just like its probability density function, which goes on to present, after the quantization, the shape of a series of impulses in the output of the quantizer.

Figure 2.6 shows that the signal to quantization noise ratio (SQNR) falls with the inverse of the signal amplitude $(1/X)$, i.e., smaller amplitude values suffer more from

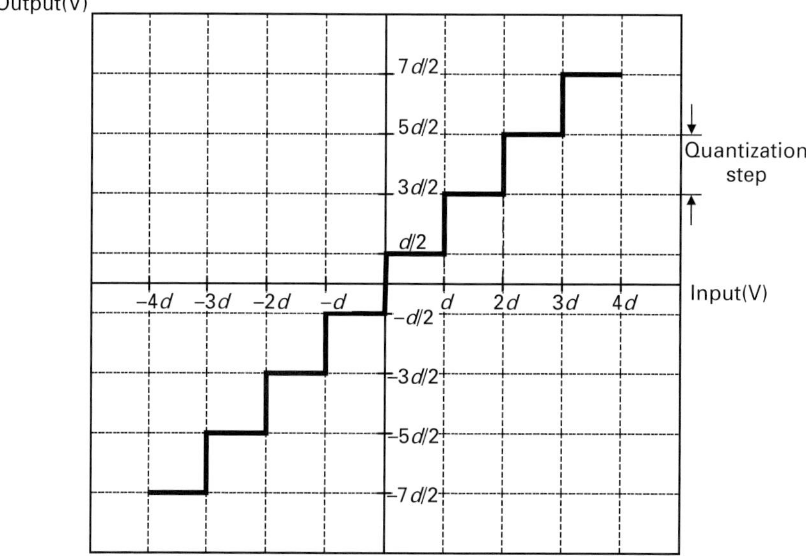

Figure 2.5   Quantization scheme

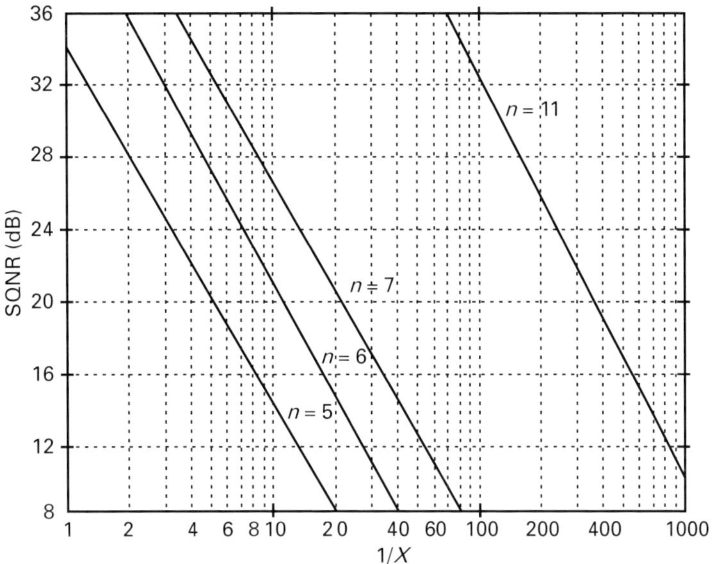

Figure 2.6   Signal to quantization noise ratio (SQNR)

the quantization noise. An increase of 1 bit, which is the same as doubling the number of levels of the quantized signal, increases the SQNR by 6 dB (Alencar, 2002a).

The quantization noise

The quantization error, or noise, is the difference between the input signal of the quantizer and the discrete output signal, $n(t) = y(t) - x(t)$, in which $y(t) = q(x(t))$ and $q(\cdot)$ represents the quantization function. The encoding system's performance, or signal processing, is limited by the quantization noise level. The channel capacity itself is limited, because of the noise.

Therefore, the most used figure of merit in comparative analysis is the SQNR (Paez and Glisson, 1972). The quadratic average error for a uniform quantizer is approximately $d^2/12$, assuming there is a uniform distribution for the quantization noise, in which d represents the quantization step (Bennett, 1948).

Non-uniform quantization

In a non-uniform quantizer the quantization step is not constant, as it was in the quantizer previously discussed, but is a function of the signal amplitude value. For lower signal levels the quantization step is smaller. In order for the error percentage to be constant, the quantization levels have logarithmic spacing. Figure 2.7 illustrates a non-uniform quantizer.

The first efforts to project optimum quantizers were made in the late 1950s and early 1960s, and were based on the maximization of the SQNR (Max, 1960). It is possible

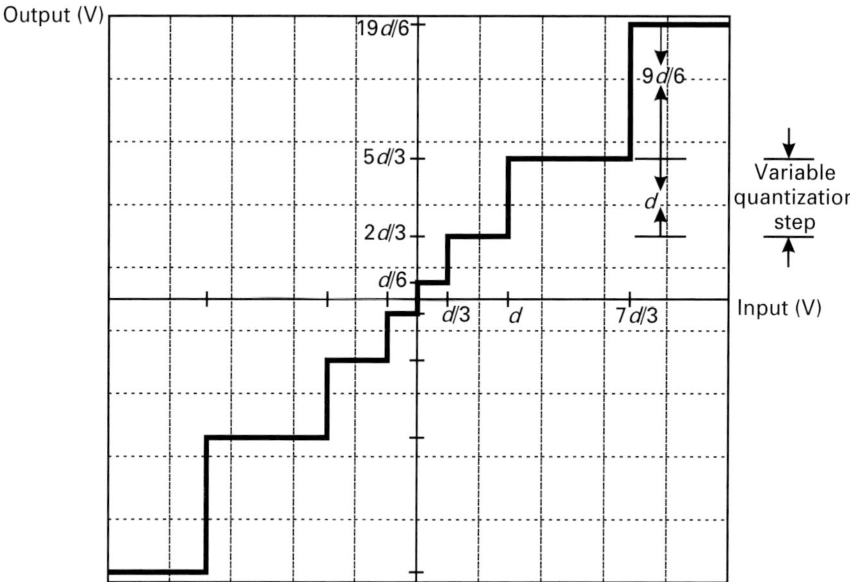

Figure 2.7   The non-uniform quantizer

to reach a solution which involves the probability density function, recognizing that the non-uniform quantizer is composed of a compression (curve) function (discussed later) and a uniform quantizer (analog to digital converter).

As the uniform quantizer has a minimal error for a signal with uniform distribution, it is enough to transform the probability density function of the signal to be quantified, by means of an appropriate circuit, to obtain the lowest quantization noise. The problem is then to find which function executes such a transformation and design the corresponding circuit (Alencar, 1995).

The word companding was coined to describe the processes of compression and expansion of the signal to be encoded. The compression is needed to increase the lowest levels of the signal, in comparison with the high levels, and in this way to make the encoder more robust. The expansion is done at the receiver as an inverse function of the compression. Figure 2.8 illustrates generic curves of compression and expansion.

The SQNR curve is more uniform than that for the inverse of the signal amplitude for non-linear quantizers. Thus, the lower levels of the signal are preserved.

The next section presents a general model for the calculation of the quantization noise spectrum for a random input signal. It must be noted that the uniform quantizer is ideal for signals with a uniform probability distribution. In practice, the signals of interest have probability density functions that are very different from the uniform one. The Gaussian and Laplacian distributions are generally used to represent voice signals, for example. For these distributions, non-uniform quantizers need to be designed.

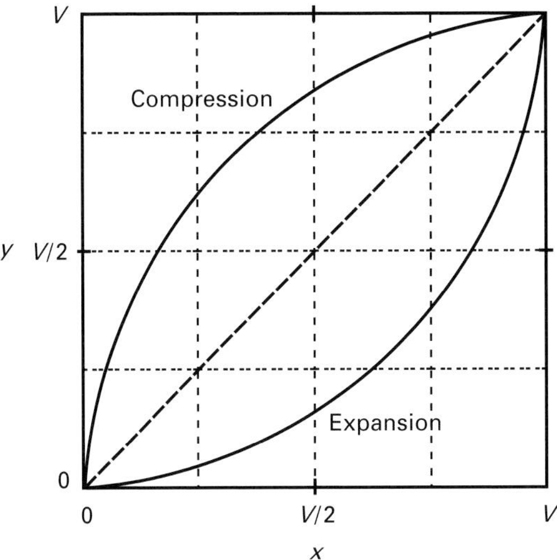

Figure 2.8   Curves of compression and expansion

2.6   Development of the model for the quantizer

The following development is based on the theorem proposed by P. M. Woodward (Woodward, 1952), whose most elaborate demonstration was developed and published years later by N. M. Blachman and G. A. McAlpine (Blachman and McAlpine, 1969). The strict demonstration of the theorem, using the linear estimator, was developed by Marcelo S. Alencar (Alencar, 1988). The application of the theorem to estimate the probability density function was an almost immediate consequence (Alencar, 1989).

Nonetheless, the correct estimate for the approximation error, made when the theorem is used in the analysis of signal quantization, was published only more recently (Alencar, 1993b). To introduce the model, the quantization noise is considered to be generated by the passage of the signal through a characteristic function $f(x)$, as shown in Figure 2.9 (Bennett, 1948).

$$f(x) = x - md,$$

$$\left(m - \frac{1}{2}\right)d < x \leq \left(m + \frac{1}{2}\right)d, \ m = 0, \pm 1, \pm 2, \ldots \tag{2.2}$$

The non-linear function $f(x)$ is periodic, with period d, and has a Fourier series representation given by (Alencar, 2007b)

$$f(x(t)) = \frac{d}{\pi} \sum_{n=1}^{\infty} \frac{(-1)^n}{n} \sin\left[\frac{2\pi n x(t)}{d}\right], \ |x(t)| \leq X_M, \tag{2.3}$$

in which d represents the quantization step, $x(t)$ is the input signal, and X_M is the limit for the dynamic range of the non-uniform quantizer (Lévine, 1973).

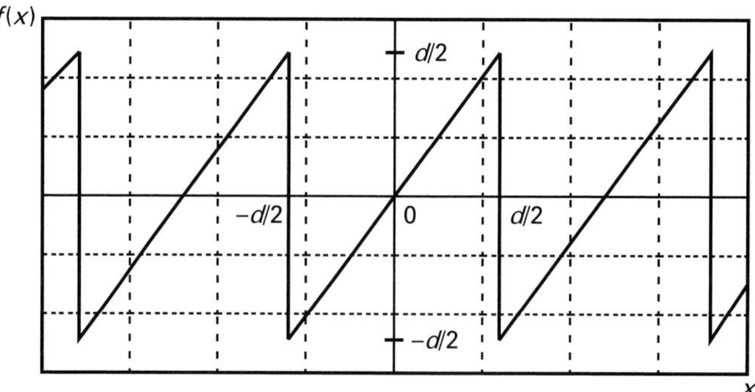

Figure 2.9   Characteristic function of the quantizer

The power spectral density of the quantization noise is closely related to the probability density function of the input signal derivative (Classen and Jongepier, 1981) as shown in the following.

Equation (2.3) represents the sum of numerous phase modulated signals, with phase deviation $F = 2\pi n/d$ and amplitude $d/\pi n$. The autocorrelation function of this set can be calculated, assuming that the phase modulated signals form an orthogonal set, which implies zero cross-correlation (Alencar, 1993a).

The linear mean square estimator is used, which results in a small estimation error for the case of a high-phase deviation. The phase deviation is inversely proportional to the quantization step, which implies this is an adequate model for a quantizer with a small quantization step (Alencar, 1998). This leads to the autocorrelation function

$$R_N(\tau) = \frac{d^2}{4\pi^2} \sum_{n=1}^{\infty} \frac{1}{n^2} E\left[e^{-j\frac{2\pi n}{d}\tau x'(t)} + e^{+j\frac{2\pi n}{d}\tau x'(t)} \right]. \tag{2.4}$$

This formula may be verified, with its evaluation at the origin being sufficient for that purpose, recalling that $P_N = R_N(0)$. This provides the formula for the total power of the quantization noise, which was initially developed by Shannon (1948b)

$$P_N = \frac{d^2}{2\pi^2} \sum_{n=1}^{\infty} \frac{1}{n^2} = \frac{d^2}{2\pi^2} \times \frac{\pi^2}{6} = \frac{d^2}{12}. \tag{2.5}$$

The power spectral density is then obtained through the Wiener–Khintchin theorem, i.e., the Fourier transform of the autocorrelation function is calculated (Papoulis, 1981)

$$S_N(w) = \frac{d^2}{2\pi^2} \sum_{n=1}^{\infty} \frac{1}{n^3} p_{X'}\left(\frac{\omega d}{2\pi n} \right). \tag{2.6}$$

Equation (2.6) shows that the quantization noise spectrum is related to the probability density function of the signal $x(t)$ derivative. In fact, the power spectral density of the noise is presented as a superposition of curves, with the amplitude decreasing with the cubed index n of the curve. Each curve has the form of the probability density function of the input signal derivative, with a bandwidth that increases with the index of the total sum. The form of the spectrum is also controlled by the quantization step value d.

The error in the approximation of the noise spectrum drops quickly as the modulation index, or phase deviation, increases. This implies a minimum approximation error for a small quantization step. Subsections 2.6.1 and 2.6.2 present examples of the application of the formula developed for the power spectral density of the quantization noise.

2.6.1 Quantization of a Gaussian signal

In order to give an example of how to use (2.6), suppose that the input signal is Gaussian with zero average and variance expressed by $\sigma_X^2 = P_X$

$$p_X(x) = \frac{1}{\sqrt{2\pi P_X}} e^{-x^2/2P_X}, \ P_X = R_X(0). \tag{2.7}$$

For this kind of signal, the probability density function of its derivative can be calculated from the second derivative of the autocorrelation function $R_X(\tau)$

$$p_{X'}(x) = \frac{1}{\sqrt{2\pi P_{X'}}} e^{-x^2/2P_{X'}}, \ P_{X'} = -R_X''(0). \tag{2.8}$$

By substituting (2.8) in (2.6), the power spectral density of the quantization noise can be obtained

$$S_N(w) = \frac{d^2}{2\pi^2} \sum_{n=1}^{\infty} \frac{1}{n^3} \frac{1}{\sqrt{-2\pi R_X''(0)}} e^{(wd)^2/8\pi^2 n^2 R_X''(0)}. \tag{2.9}$$

Figure 2.10 illustrates the application of the formula for the quantization noise spectrum produced by a signal with an input Gaussian distribution for two values of

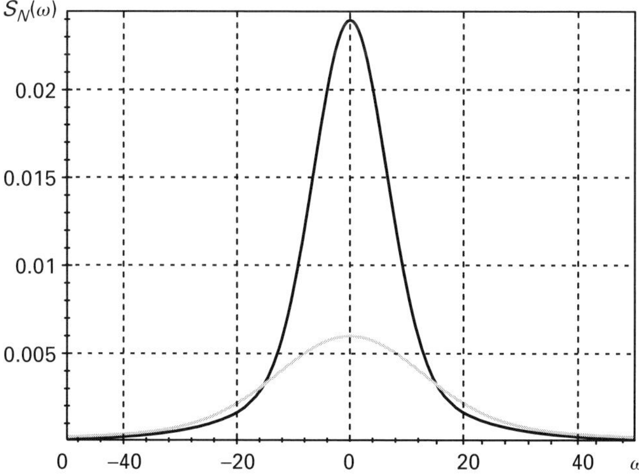

Figure 2.10   Quantization noise spectrum for an input Gaussian signal

the quantization step d. It can be seen that, for a smaller step, the bandwidth of the noise increases, having, on the other hand, a decrease of the amplitude of the power spectral density and a flatter spectrum. The effect of the derivation of the signal is the broadening of the probability density function, which results from the increase of the dynamic range of the signal.

2.6.2   Quantization of the video signal

The analog video signal originated from a CCD, after the sampling, which produces an image composed of a fixed number of pixels, according to the number of lines and the horizontal resolution of the format used, is quantized, i.e., the levels are truncated so that the signal may be stored on a digital medium, such as a hard disk (HD). Each pixel has characteristics of brightness and color. The probability density function for the image is approximately uniform, which gives a roughly uniform power spectral density.

The original analog signal of the image reproduces the colors through the combination of the basic colors of the RGB model. In addition, this signal contains the brightness intensity (luminance). The information about brightness and color is combined during the digitization process, generating three signals, one for each RGB color. The combination of the signals that represent red (R), green (G), and blue produces the desired color.

The color plus brightness signal is called the channel. Thus, the color image of each pixel of the digitalized video is represented by three channels: one for the red, one for the green, and one for the blue. There is a great diversity of types of digital image formats, for which the number of channels used varies according to the format. The gray-scale format uses only one channel with 254 variations of tones of gray, plus black and white. The CMYK format has four channels (cyan, magenta, yellow, and black) and 256 variations of tone for each one of the four colors.

In the case of RGB, the most frequently used system also employs 256 possible variations of tone for each channel, showing the need for eight bits per sample for each channel. The final number of colors obtained, combining the three channels, exceeds 16 million. This number of colors is also approximately the total number of colors perceived by the human eye. That is why systems of this type are called true color (Views, 2006).

2.7   The encoding process

Voice encoders may be classified as waveform encoders, parametric encoders, and hybrid encoders. Waveform encoders attempt to reproduce the signal sample by sample, exploring its statistical, temporal, or spectral characteristics. They are encoders of low delay and small implementation complexity, but require a high broadcasting rate (higher than 16 kbit/s).

Parametric encoders, or vocoders, are based on the voice production model. This model is represented by a set of parameters which have periodical updating. For the definition of these parameters, the signal is segmented at periodical intervals called frames. The parameters are usually updated at every frame. The rate required by the vocoders is low (less than 4.8 kbit/s), but the delay and the complexity are high and the voice sounds synthetic.

Hybrid encoders combine the quality of waveform encoders with the efficiency of parametric encoders. These encoders are more elaborate than vocoders and require encoding rates ranging from 4.8 to 16 kbit/s. The voice quality approximates to that of the waveform encoders, which will be dealt with next.

After the digital encoding, an $X(t)$ signal, with probable levels A and $-A$, presents the following autocorrelation

$$R_X(\tau) = A^2 \left[1 - \frac{|\tau|}{T_b} \right] [u(\tau + T_b) - u(\tau - T_b)],$$

in which T_b is the period of the bit. The autocorrelation of the binary digital signal drops to zero when $\tau = T_b$. Since the signal has zero mean, it must be non-correlated at this stage. The signal power, calculated for $\tau = 0$, is $P_X = A^2$, as expected.

Adaptive PCM (APCM)

In the APCM quantizer, shown in Figure 2.11, the quantization step varies with time to accompany the signal amplitude variations. This adaptation is based on the previous signal samples. The objective of the adaptive technique is the reduction of the dynamic range of the signal, in order to obtain a reduction of the final transmission rate.

Differential PCM (DPCM)

A voice signal, like a video signal, presents significant correlation between successive samples, since the signal amplitude does not vary much from one sample to another. Therefore, this implies signals that are redundant and susceptible to compression. The bigger the signal redundancy, the higher the compression rate obtained.

The objective of the differential technique is the reduction of the signal redundancy. This is obtained by quantizing the difference in amplitude between adjacent samples. As these samples are similar, a smaller number of bits may be used to represent the signal.

The input signal in the quantizer is the difference between the original signal and one of its predictions, resulting in a signal called a prediction error. Figure 2.12 shows a block diagram of a DPCM. The 32 kbit/s ADPCM was standardized by the ITU-T for

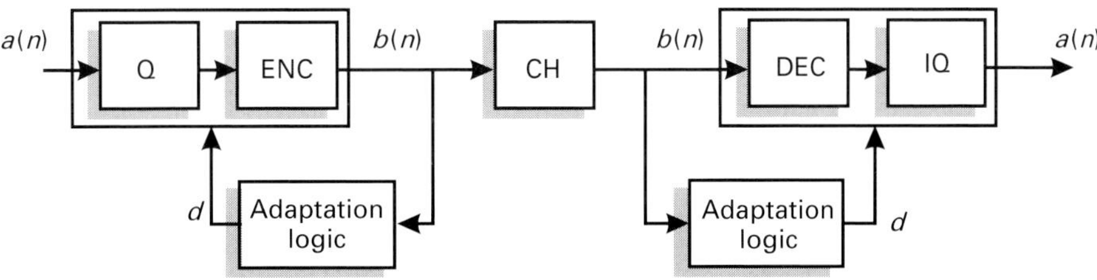

Figure 2.11   Flowchart of an APCM

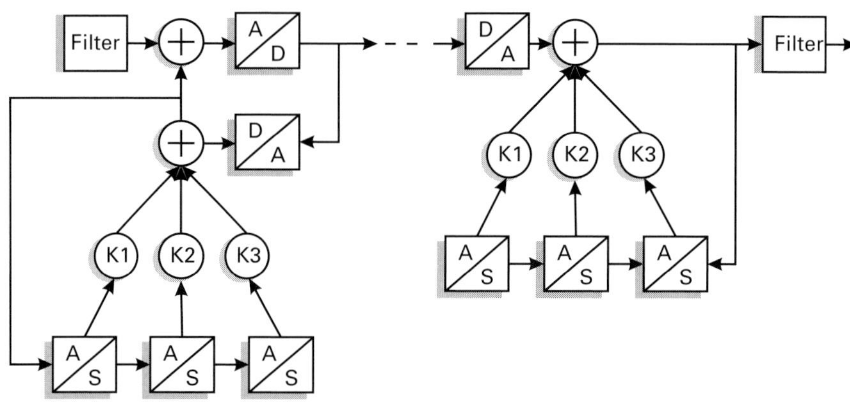

Figure 2.12 Block diagram of a DPCM

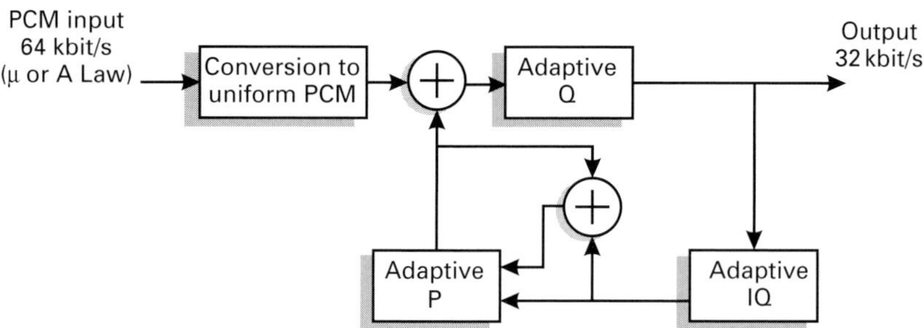

Figure 2.13 ADPCM encoder

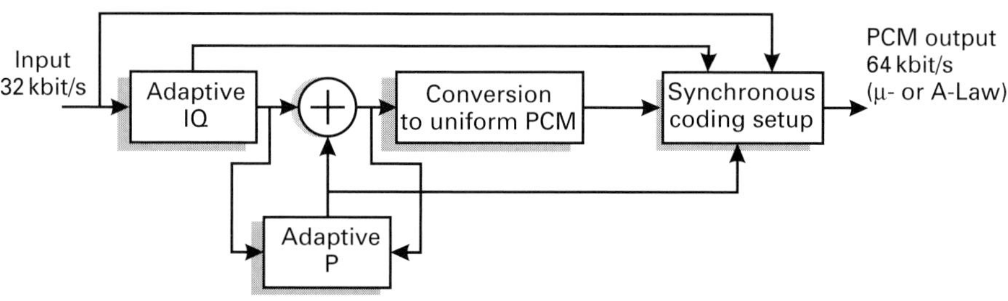

Figure 2.14 ADPCM decoder

applications in the integrated services digital network (ISDN) and wireless telephony. Later, other rates were standardized, varying from 16 to 40 kbit/s.

Adaptive differential PCM (ADPCM) encoders use adaptive quantization and/or prediction. Adaptive prediction consists of the dynamic adjustment of the predictor, according to variations in the voice signal. ADPCM encoders give good quality for rates between 24 and 48 kbit/s. Figures 2.13 and 2.14 show the block diagrams of the ADPCM encoder and decoder, respectively.

2.8 Video encoding standards

The Motion Picture Experts Group (MPEG) was established in January 1988 as a work group to create international standards of audio and video encoders/decoders (codec). The MPEG-1 standard was published in 1989 and approved as the international standard in 1992. Work began on a new standard to replace it in 1990 . This was called MPEG-2 and was approved in 1994. New improvements, which would have lead to the MPEG-3 standard, were later incorporated in the MPEG-2. This caused the acronym MPEG-3 not to be used.

The structure of the MPEG-4 standard was set out in 1993 and the standard was approved in 1998. Work on MPEG-7, which covered audio and video content, began in 1997 and the standard was approved in 2002. Currently, the group is developing the MPEG-21 standard, which is called the multimedia framework.

This group is also working with other standards related to the use of audio and video standards mentioned, such as the Intellectual Property Management Protection (IPMP). MPEG is a member of the International Standards Organization (ISO).

2.8.1 The MPEG-1 standard

MPEG-1 aims at the compression of audio and video images previously digitized. Short video clips made available on the Internet, video compact disks (VCDs), and videos distributed on disks of the CD-ROM type use this standard. The MP3 audio compression standard is also MPEG-1. This is layer 3 of the MPEG-1, thus the acronym MP3. The compression rate is variable, and the same standard, as with the MPEG-1, can be used to compress the original content to a greater or lesser extent, making it possible to reach a compression rate of 200:1 (Jurgen, 1997).

With the maximum rate the images become distorted, because the compression introduces artifacts in the image, which can be seen to a greater or lesser extent, depending on the quality of the compression algorithm and the compression rate used. Thus, most of the videos compressed with the MPEG-1 use a compression rate lower than 50:1. Even at this rate, the horizontal resolution obtained after the compression is low, *circa* 320 lines, and similar to the VHS format. The compression used is the multiframe type (Poynton, 2003b).

2.8.2 The MPEG-2 standard

MPEG-2 was developed in conjunction with the Video Coding Experts group from the International Telecommunications Union (ITU-T). It is also known in the ITU-T community as H.262, the name of the project within this group. DVDs and digital TV broadcasting use this standard, which is more advanced than MPEG-1 and produces better quality images. The compression rate, just like in the MPEG-1, is variable but normally values of around 40:1 are used.

MPEG-2 is more efficient at compressing, without apparent quality loss, than MPEG-1. However, that demands higher computational effort in the process, requiring more powerful hardware than that needed for MPEG-1. The horizontal resolution obtained after the compression is superior to that of the VHS format. The compression used is the multiframe type. Owing to the assembling process of the group of pictures (GOP) used by the multiframe technique, it is more complex to compress MPEG-2 content than it is to retrieve it (Reimers, 2005a).

2.8.3   The MPEG-4 standard

MPEG-4 is the standard created by the MPEG group for the compression of audio and video images previously digitized. The MPEG-4 standard is used on videos transmitted by the Internet, as well as on cell phones that use images. It is also common on several standards of digital TV broadcasting, especially HDTV in its advanced video coding (AVC) version (Sun and Chiang, 2005).

Like the MPEG-1 and MPEG-2 standards, MPEG-4 accepts different profiles, which establish several values for compression rates, according to the application. Nevertheless, unlike MPEG-2, whose quality is equivalent to the DVD quality standard, for the MPEG-4 the variation is much wider, and a great variety of values may be used, which allows the visualization of video images, the capacity of the transmission medium notwithstanding, be it Internet broadband on dial up, for example. It is for this reason, among others, that HDTV broadcasting systems use it as the encoding standard, as an alternative to MPEG-2.

The computational effort employed in the manipulation of MPEG-4 videos is still greater than that required in the MPEG-2 format, which demands more powerful hardware. MPEG-4 is a more advanced standard than MPEG-2. Besides the improvements in the compression processes, which are observed on smaller files compressed without apparent quality loss, it also accepts the use of other media types interacting with the video, such as texts and digital photos, for example, activated by intelligent menus.

These menus, unlike those of movie DVDs, which use the whole screen when activated, with the device returning to the beginning of the disk, may be displayed how the producer wants. This is because the interactivity is not dependent upon the video playing device, but is part of the video itself. The commands are encoded together with the images and not in a separate chapter, dedicated to the menu.

MPEG-4, MPEG-1, and MPEG-2 are acronyms given to sets of several topics which are known as parts. Each part deals with a different aspect of the standard. Thus, in MPEG-4, part 1 describes the synchronization of audio and video; part 2, the compression process of the images; part 3, the compression process of the audio; part 4, procedures to verify the conformity of a given sample with other parts of the pattern; and part 5 indicates the software to demonstrate and illustrate certain parts of the standard. Part 10 of the standard was included when a more optimized version of part 2 (video compression) was developed. This part was named AVC. It also became known as H.264, because this was the name given by the Video Coding Experts group from the ITU-T, who developed it in conjunction with the MPEG group (Simonetta, 2002).

The ISO defined, in 2002, the program QuickTime, by Apple, as the standard for distribution of video on MPEG-4. The horizontal resolution obtained after the compression is variable and can be adjusted for various levels of quality, from slightly inferior to the VHS format to that of the DVD format. The compression used is multiframe type.

2.8.4 The H-264 standard

The AVC, or H-264, or MPEG-4 Part 10, was the standard created in 2003 by the MPEG group in conjunction with the Video Coding Experts group from the ITU-T for digitization of video images. The objective was to develop a pattern which had the quality given by MPEG-2 or MPEG-4, but which could do it using a lower bit rate, without being excessively complex, to favor its implementation on cheaper digital circuits. The flexibility of use of the standard was extended in comparison with the original MPEG-2 and MPEG-4, allowing its use on both high resolution and low resolution systems.

More than 20 new techniques were used in the elaborate compression processes, providing better performance than that of the other standards in several situations. With less than half the bit rate used in the MPEG-2 it is possible to obtain the same quality of image. Much like MPEG-2, AVC establishes several levels of profile, for uses ranging from mobile applications – cell phones for instance, demanding lower computer power from the circuits – to video-conferencing, to applications that demand higher processing power, such as the display of images in high definition, such as in HDTV systems. Many applications have adopted this standard, such as direct broadcasting of TV programs from satellites to residences, digital television terrestrial broadcasting, and image distribution of images on the Internet. Blue-ray disks and HD-DVD also use it for the recording of images (Views, 2006).

3 Fundamentals and standards of video and audio compression

with Jean Felipe Fonseca de Oliveira and José Ewerton Pombo de Farias

3.1 An overview of audio and video compression

MPEG-1 and MPEG-2, developed by the Moving Pictures Expert Group (MPEG) within the International Organization for Standardization, were the first international standards in the field of high-quality digital audio compression. MPEG-1 includes coding of stereophonic audio signals at high sampling rates aiming at transparent quality, whereas MPEG-2 also offers stereophonic audio coding at lower sampling rates. In addition, MPEG-2 introduces multichannel coding, with and without backwards compatibility to MPEG-1, to provide an improved acoustic image for audio-only applications and for enhanced television and video-conferencing systems. MPEG-2 audio coding without backwards compatibility, called MPEG-2 advanced audio coding (AAC), offers the highest compression rates. Typical application areas for MPEG-based digital audio are in the fields of audio production, program distribution and exchange, digital sound broadcasting, digital storage, and various multimedia applications.

Video compression is the process of converting digital video into a format that takes up less storage space or transmission bandwidth. Video compression (or video coding) is an essential technology for applications such as digital television (terrestrial, cable or satellite transmission), optical storage/reproduction, mobile television, video-conferencing, and Internet video streaming. Video compression standards make it possible for products from different manufacturers (e.g. encoders, decoders, and storage media) to interoperate. An encoder converts video into a compressed format and a decoder converts compressed video back into an uncompressed format.

Recommendation H.264: Advanced Video Coding is a document published by the international standards bodies International Telecommunications Union–Telecommunication Standardization Sector (ITU-T) and International Organization for Standardization/International Electrotechnical Commission (ISO/IEC). It defines a format (syntax) for compressed video and a method for decoding this syntax to produce a displayable video sequence. The standard document does not actually specify how to encode (compress) digital video. The compression algorithms are up to the individual manufacturers. In practice, however, it is expected that the encoder is likely to follow the steps of the decoding process. An H.264 video encoder carries out prediction, transformation, and encoding processes to produce a compressed H.264 bit-stream.

An H.264 video decoder carries out the complementary processes of decoding, inverse transformation, and reconstruction to produce a decoded video sequence.

3.2 General concepts and fundamentals

Digital compression techniques play an important role in the field of multimedia communications in which bandwidth still represents a valuable resource. Those techniques are essential to reduce the required transmission rate for a sequence of images without much loss of quality when judged by human viewers.

Modern compression techniques involve the use of complex electronic circuits whose costs can only be kept at reasonable levels through the production of VLSI chips in large quantities. Therefore, standardization is very important.

MPEG-1 enabled the introduction of the first generation of video coder/decoders (CODECs) proposed by MPEG. Its development came as a response to the market demand for an efficient way to store visual information. It was intended for use with digital storage media, such as: compact disk (CD), digital audio tape, magnetic hard disks, and other optic disks. By the time of the development of this standard, CD-ROMs had a playback speed of approximately 1.2 Mbits/s, and the MPEG used this rate as a reference. However, in most of the MPEG-1 applications the video bit rates range from 1 to 1.5 Mbit/s.

The possibility of storing TV signals (625- and 525-line systems) was also pursued, and their compatibility with work stations and personal computers was examined. For this reason MPEG-1 is based on progressively scanned images and does not recognize interlacing. Interlaced sources must be converted to a non-interlaced format before coding. The expected subjective quality of MPEG-1 was comparable to that of the competing analog technology at that time, i.e., video cassette recorder (VCR). Visualization features such as image freezing, fast forward, fast reverse, slow forward, and random access were also required for MPEG-1. It should be noted here that efficient coding and operational flexibility are usually conflicting requirements. The random search feature, for example, requires the insertion of regular images between frames. Those images do not explore the temporal redundancy present in the video signal, and so contribute to a lower compression ratio or a higher bit rate at the output of the coder. The MPEG-1 standard did not include either the coder design or the decoder design. Manufacturers had freedom to adopt algorithms and to optimize them in order to achieve efficiency and functionality.

Most of the current image/video applications use transform coding. This approach is motivated by the fact that adjacent pixels in an image have a certain level of correlation. Also, in a video transmission system, adjacent pixels in successive frames exhibit high correlation. These correlations can be exploited for the prediction of a pixel from its neighbors. A transformation is defined as a mapping of correlated spatial data into transformed coefficients. The transformation must take into account the fact that the information content of a pixel is relatively low, and therefore the visual contribution of a pixel can be anticipated using its neighbors.

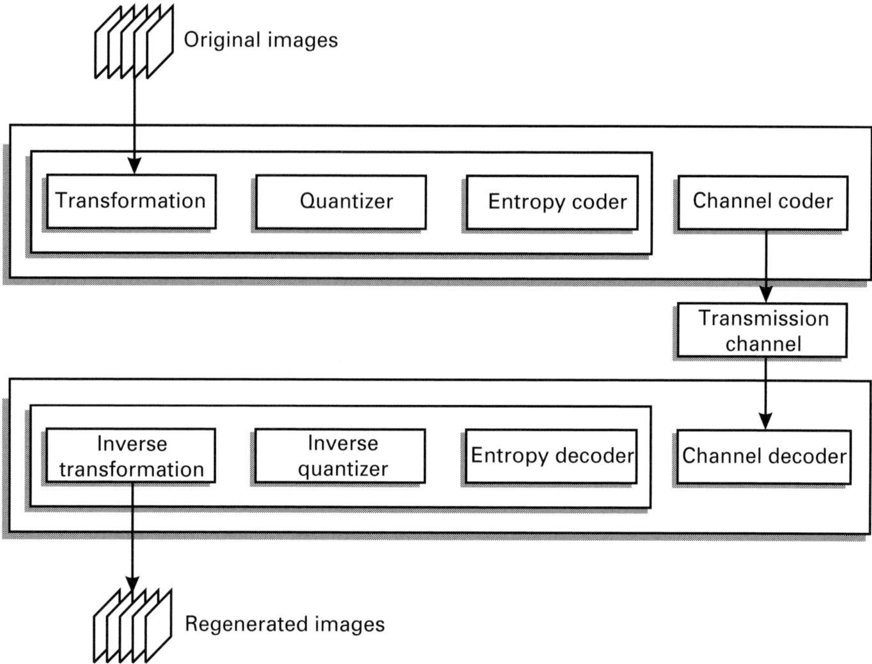

Figure 3.1   Components of a typical image/video transmission system

Figure 3.1 illustrates an image/video transmission system. The function of the source encoder is to achieve compression of the image data through the removal of redundancies. The source encoder reduces the source entropy, i.e., reduces the average number of bits per pixel required to represent the image. On the other hand, the channel encoder adds redundancy to the output of the source encoder in order to increase transmission reliability. The interplay of these two conflicting objectives has been extensively studied in the field of information theory and coding.

3.3   The MPEG-4 standard

MPEG-4 (ISO/IEC, 1999) is an ISO/IEC standard developed by the same research group as the MPEG-1 and MPEG-2 standards (MPEG). MPEG-4 is the result of an international effort involving hundreds of researchers all over the world. The standard was published in October 1998 and received the official designation ISO/IEC 14496. By the first months of the following year, the standard had already been internationally accepted. Many contributions were added to the original standard as extensions.

The MPEG-4 standard provides technological standard elements that enable the integration of production, distribution, and procedures for video content access in the following three major areas (MPEG, 2002):

- digital television;
- interactive graphical applications;
- interactive multimedia.

The main characteristics

The MPEG-4 standard covers a wide range of visual communications applications, supporting a set of visual information coding tools. The main characteristics of the standard that differ from previous ones are (Richardson, 2003b):

- Efficient compression of interlaced and progressive video sequences: the main compression techniques were developed based on ITU-T H.263 (ITU-T, 2000a). The use of new compression techniques has enabled the MPEG-4 standard to significantly exceed the compression efficiency of previous standards.
- Video objects coding: video objects are identifiable areas with irregular boundaries within a video sequence. This is the new coding concept introduced by the MPEG-4 standard. This feature allows the use of determining different parameters for different areas within the same video sequence.
- Effective control of the data transmission: error resilience tools enable the decoder to maintain a good exhibition quality even in transmission channels with a high bit error rate (BER). Scalable coding tools allow a flexible transmission rate that depends on the state of the channel.
- Encoding of animated visual objects: for example, two-dimensional and three-dimensional forms, animated faces, and human bodies.
- Encoding for specific applications: for example, studio-quality video. In this type of application, the visual quality is often more important than the compression efficiency.
- Texture compression.

Applications

There follows, a brief description of the main applications of the MPEG-4 standard (Forum, 2002):

- Real-time transmission of various types of media: MPEG-4 meets the required characteristics that make it the ideal standard for real-time video transmission. Some of these characteristics are:
  - In low rates networks, the applications content can be compressed at rates of up to 24 kbits/s, and in broadband networks, applications can encode the same content at higher rates.
  - The interactive nature of the MPEG-4 standard ensures that the MPEG-4 content can be used in advanced multimedia applications.
  - Due to the scalability of the MPEG-4, the same content can be transmitted to different devices in heterogeneous networks.
- Real-time video transmission to mobile devices: the MPEG-4 standard is capable of transmitting in real-time in all types of communications networks using very low bit

rates. This feature ensures transmission in channels with a high incidence of noise. These qualities are essential to ensure good real-time transmission of multimedia content for mobile devices.

- Interactive television: applications for digital television can take advantage of the MPEG-4 standard. This is because the standard supports high-quality interactive content in traditional television networks and cable television.

- Interactive shopping: support for interactive content opens up a wide range of applications for the MPEG-4 stream. A classic example of an application is an interactive shopping system in which the consumer has access to products through the television.

- Education and training at distance: one of the key features of education and training at a distance is the ability to transmit through different communication network infrastructure and support for interactivity. In a business training scenario, the MPEG-4 content can be transmitted by satellite to the headquarters of the company in a remote location and, from that point, be available through a local area network to the employees of various sectors of the company.

The MPEG-4 documents

This subsection is a survey of the standards for digital video compression. The reader can find details of each standard in the documents indicated herein. Figure 3.2 shows the organization of the MPEG-4 standard. The document ISO/IEC 14496 (ISO/IEC, 1999) which specifies the MPEG-4 standard comprises the following 23 parts (Wikipedia, 2007b):

- Part 1 (ISO/IEC 14496-1): System: this describes the synchronization and multiplexing techniques of video, audio, and data;

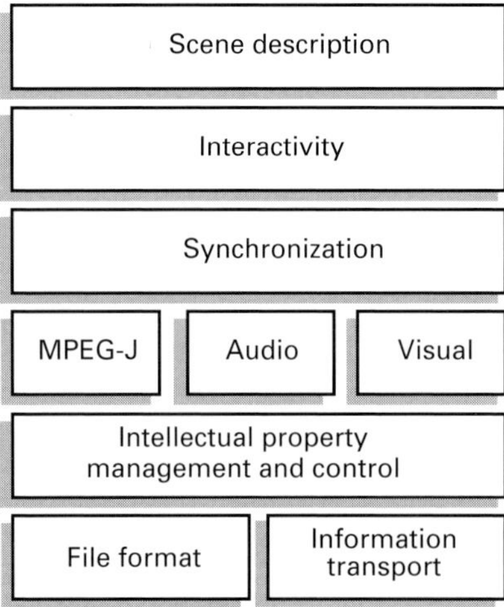

Figure 3.2   MPEG-4 documents

- Part 2 (ISO/IEC 14496-2): Visual: this defines an algorithm for compression of visual information (video, textures, synthetic images);
- Part 3 (ISO/IEC 14496-3): Audio: this defines a set of algorithms for compression for perceptual coding of audio signals;
- Part 4 (ISO/IEC 14496-4): Compliance: this describes procedures for compliance testing;
- Part 5 (ISO/IEC 14496-5): Reference Software: this provides a reference program for the demonstration of the other parts of the standard;
- Part 6 (ISO/IEC 14496-6): Delivery Multimedia Integration Framework (DMIF);
- Part 7 (ISO/IEC 14496-7): Optimized Reference Software: this describes examples of how to optimize implementations regarding Part 5;
- Part 8 (ISO/IEC 14496-8): Transportation through internet protocol (IP) Networks: this specifies methods of transporting MPEG-4 content through IP networks;
- Part 9 (ISO/IEC 14496-9): Reference Hardware Platform: this describes a hardware implementation project to demonstrate how to implement the other parts of the standard;
- Part 10 (ISO/IEC 14496-10): Advanced Video Coding (AVC): this describes a CODEC for video signals that is technically equal to the ITU-T standard H.264;
- Part 11 (ISO/IEC 14496-11): Description of scene and the engine of the application, also called BIFS (Binary Format for Scenes);
- Part 12 (ISO/IEC 14496-12): ISO Base Media File Format: a file format for storage of several types of media;
- Part 13 (ISO/IEC 14496-13): Extensions of management and protection of intellectual property;
- Part 14 (ISO/IEC 14496-14): MPEG-4 File Format: this describes the file format for storing the MPEG-4 contents. This format is based on the specifications of Part 12;
- Part 15 (ISO/IEC 14496-15): AVC File Format: this is defined for storage of video based on Part 10;
- Part 16 (ISO/IEC 14496-16): Animation Framework Extensions (AFX);
- Part 17 (ISO/IEC 14496-17): Subtitle Format;
- Part 18 (ISO/IEC 14496-18): Compression and streaming of fonts (OpenType Fonts);
- Part 19 (ISO/IEC 14496-19): Synthesized Textures Stream;
- Part 20 (ISO/IEC 14496-20): Lightweight Scene Representation (LASeR);
- Part 21 (ISO/IEC 14496-21): MPEG-J Graphical Framework extension (GFX);
- Part 22 (ISO/IEC 14496-22): Open Font Format Specification (OFFS) based on OpenType;
- Part 23 (ISO/IEC 14496-23): Symbolic Music Representation (SMR).

3.3.1   The MPEG-4 Visual standard

Profiles and levels

The definition of video profiles is an important mechanism for interoperability between CODECs developed by various manufacturers. The MPEG-4 Visual standard describes a variety of coding tools. A commercial MPEG CODEC does not support all the tools. A coding tool defines a subset of coding functions to support a particular feature, for example, encoding of basic video, interlaced video, etc. In turn, the manufacturers can choose a profile that contains adequate coding tools for the application to be developed. For a low-performance processor, for example, the profile called Simple Profile can be used. The Simple and Advanced Simple profiles are the one most frequently used by manufacturers of MPEG-4 CODECs nowadays.

The MPEG-4 Visual profiles for natural video sequences are listed in Table 3.1. Table 3.2 lists the profiles for coding of artificial and hybrid video sequences. Artificial video corresponds, for example, to animated sequences produced by computer graphics. Hybrid videos are compositions which may contain natural and artificial video sequences.

The tools for the representation of natural videos in MPEG-4 Visual standard provide a set of techniques that enable efficient storage, transmission, and manipulation of images, textures, and video information in a multimedia environment. These tools ensure the decoding and exhibition of the content units of images and videos, known as video objects. A scene is formed by a set of video objects. An example of a video object in a scene is a person talking (without the background), which can then be used to compose the scene with other video objects together, for example, another person talking. The encoding of rectangular video sequences is handled as a special case of these objects. In order to meet these targets for Part 2, MPEG-4 Visual, the MPEG-4 standard provides solutions for:

- efficient compression of images and video,
- compression of textures for efficient texture mapping in two- and three-dimensional shapes,
- efficient compression of the data flow which feeds time-variant two- and three-dimensional shapes,
- efficient random access for all types of visual objects,
- extended handling functionalities of image and video,
- image and video content based coding,
- video, audio, and textures scalability,
- temporal, spatial, and video quality scalability,
- robustness and recovery in noisy environments.

Video objects

The MPEG-4 Visual standard supports the video representation of objects of natural or artificial origin. These video objects are encoded as separate entities within the bit stream, allowing users to access and manipulate them (cut, paste, zoom, etc.). In MPEG-4, a

Table 3.1. MPEG-4 Visual profiles of natural video sequences

MPEG-4 Visual profile	Main features
Simple	Low-complexity encoding of sequences of rectangular frame
Advanced Simple	Enhanced encoding of sequences of rectangular frame and support to interlaced video
Advanced Real-Time Simple	Encoding of sequences of rectangular frame for real-time applications
Core	Basic encoding of arbitrarily shaped objects
Main	Enhanced encoding of arbitrarily shaped objects
Advanced Coding Efficiency	High-performance encoding of objects of arbitrary shapes
N-bit	Encoding of video objects with sampling different from 8 bits per sample
Simple Scalable	Scalable coding of rectangular video sequences
Fine Granular Scalability	Enhanced scalable coding of rectangular video sequences
Core Scalable	Scalable coding of video objects
Scalable Texture	Scalable coding of still textures
Advanced Scalable Texture	Enhanced scalable coding of still textures using the video objects features
Advanced Core	Combines the features of the Simple, Core and Advanced Scalable Texture profiles
Simple Studio	Video objects high-quality encoding
Core Studio	Video objects high-quality encoding with enhanced compression

Table 3.2. MPEG-4 Visual profiles of artificial video sequences

MPEG-4 Visual profile	Main features
Basic Animated Texture	Still textures two-dimensional object coding
Simple Face Animation	Support for animated human face models
Simple Face and Body Animation	Support for animated human face and body models
Hybrid	Combines the features of the Simple, Core, Basic Animated Texture and Simple Face Animation profiles

video object can also be the traditional rectangular video sequences composed of pixels. However, a video object can also correspond to a set of arbitrary and random shapes, possibly with an associated semantic meaning. The audio-visual representation based on objects is the main difference between the MPEG-4 standard and the previous standards, adding unprecedented features to the process of media encoding. One scene based on objects is constructed using objects that have independent relationships between them in time and space. This technique offers major advantages of representation, among which are:

- different objects can receive different types of coding to match the characteristics of the object itself and its relevance in the scene or video sequence;
- a more harmonious integration of different types of information within the same scene;
- ability to implement interactivity with the objects of the scene – hyperlinking.

The way in which the video objects are identified is not part of the scope of the MPEG-4 standard; this procedure is considered to be a preprocessing stage. The MPEG-4 standard provides ways to represent any composition of objects, whatever the methods used for this purpose. A wide variety of methods are used to obtain arbitrary shaped video objects, including:

- automatic methods;
- assisted segmentation of natural information;
- chroma key techniques;
- artificial generation of information on computers.

The test video set used by the MPEG-4 standard supports rectangular and presegmented video objects. Knowing that the use of MPEG-4 will heavily depend on the availability of robust tools for the analysis and production of video content, it is expected, considering the results already achieved in the field of visual computing, that significant developments will be made in this area. Currently, the bit stream of the verification model of the MPEG-4 Visual standard consists of the following hierarchy of classes:

- Video Session: VS;
- Video Object: VO;
- Video Object Layer: VOL;
- Video Object Plane: VOP.

The Video Session class is the highest entity in the hierarchy presented above and may contain one or more video objects (VO_1, VO_2, ..., VO_N,); in turn, a video object can have one or more layers (VOL_1, VOL_2, ..., VOL_N,), which can be used to improve the spatial or temporal resolution of a video object. The video object plane is the MPEG-4 equivalent of video frames. For instance, see Figure 3.3.

Types of video objects

Five different types of video objects are used to represent natural information (MPEG, 2000):

- Simple Object: a rectangular video object of natural origin with arbitrary height–width relation, designed to be used on low-rate transmissions. It uses simple and low-cost coding tools. It is based on I-VOP (Intra) and P-VOP (Predicted).

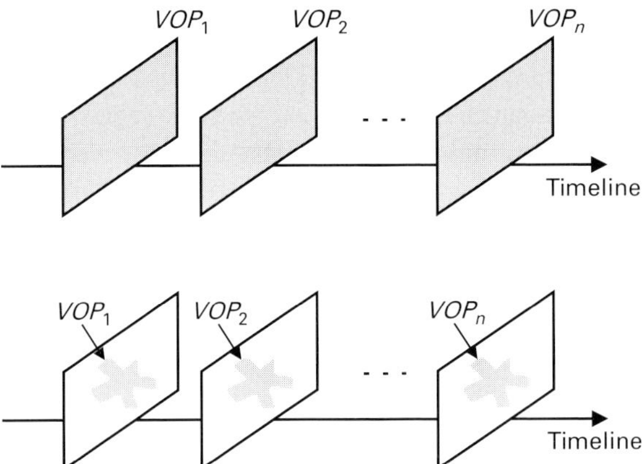

Figure 3.3   Rectangular and arbitrarily shaped VOP

- Simple Scalable: this is a scalable extension of the above type, providing temporal and spatial scalability using Simple Object type as base layer. Layers of enrichment are also rectangular.
- Core Object: this uses a superset of Simple Object tools, enabling better quality by the use of bidirectional prediction (B-VOP). It supports scalability and the use of transparency in the objects.
- Main Object: this is the video object type that produces the highest quality. Compared with the Core Object type, it supports several levels of transparency, and also interlaced content.
- N-bit Object: this has all the characteristics of the Core Object type, with the addition of the possibility of changing the pixel resolution from 4 to 12 bits for the planes of luminance and chrominance.

There is a special type of object for representation of visual information of textures:

- Still Scalable Texture: this provides the use of wavelets for the scalability and incremental download and construction of the object.

The following types of objects use artificial video coding tools; some of them also use textures of natural sequences:

- Animated 2D Mesh: this combines artificial and rectangular video object shapes with video of natural origin. The coding of natural video uses the tools specified for the core object type. This natural video can be mapped into an artificial object allowing deformation by displacement of the points of image. This property allows the production of interesting effects with the sequences of natural video.
- Basic Animation Texture: this allows the functionality of artificially synthesized objects with arbitrary and variable size.
- Simple Face: this has the tools for facial animation. It does not define the appearance of the face; the animation can be applied to any chosen model.

3.3.2   The MPEG-4 video encoding process

The process of video encoding specified by the MPEG-4 standard has the advantage that the efficiency of compression can be improved for some video sequences due to the fact that the tools and techniques for prediction of moving objects are adequate, based on the video characteristics. The main techniques for movement prediction, which can be used to ensure an efficient encoding and a good presentation of video objects, are listed below:

- Blocks of 8×8 or 16×16 pixels for the prediction and compensation of movement.
- $\frac{1}{4}$ pixel precision. This feature implies a negligible increase in computational overhead. A more precise movement description implies a smaller amount of prediction error generating a video of better quality.
- Global Motion Compensation (GMC). This compensates for the global movement of an object using a small number of parameters. The GMC technique is based on the codification of the trajectory of the camera moves and encoding of the textures. Figure 3.4 illustrates this feature.
- Compensation of the global movement for continuous images of large dimension, which describe, generally, background panoramic planes. For each consecutive image in a sequence of video, only eight parameters of the global movement describing the camera moves are encoded and transmitted to subsequent object reconstruction.
- Shape-Adaptive DCT. Shape-Adaptive DCT improves the compression efficiency of arbitrary shape objects. The algorithm SA-DCT is based on predefined sets of orthonormal basis-functions of uni-dimensional DCT.

Figure 3.4 illustrates the basic concept of the MPEG-4 video encoding process. It is assumed that the image can be segmented into the background and the table tennis player. The background plane defines a large area that will be presented throughout the video

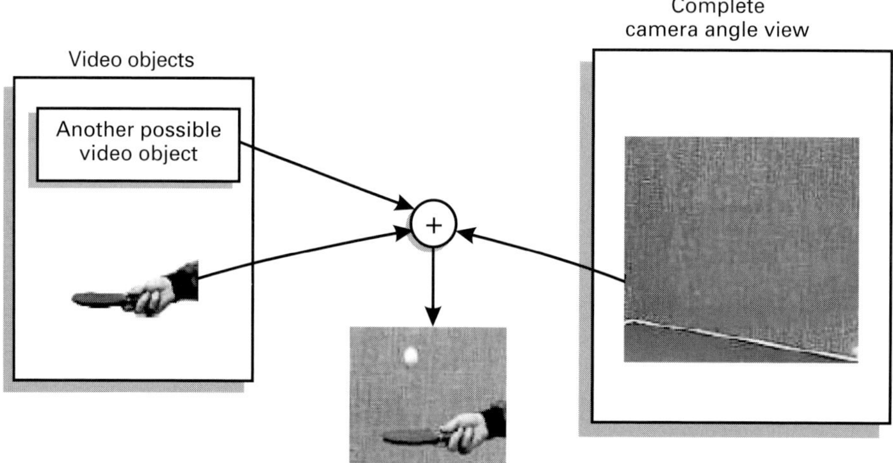

Figure 3.4   MPEG-4 encoding process using GMC

sequence. The background can be greater than the frame resolution and is transmitted once and stored in a specifically selected memory in the receiver. In each consecutive frame the parameters of the camera movements are transmitted. The table tennis player, which is a video object with movement, is transmitted separately as an arbitrarily shaped video object. To reconstruct the original sequence, the receiver combines the background and the arm of the table tennis player. Without discarding these potential features of object based encoding offered by the MPEG-4, standard encoding rectangular video frames is still the most common application. The tools needed to handle these frames, which are seen by the syntax of MPEG-4 standard as rectangular video object planes, are grouped in the Simple Profile. See Figure 3.5. The MPEG-4 standard adopts compression algorithms based on the DCT of motion estimated macroblocks. Most of the tools are similar to those in previous standards, in particular the ITU-T H.263 standard, with additional tools to improve the coding efficiency and transmission.

MPEG-4 encoding tools

- Four motion vectors per macroblock: the motion compensation increases its precision with the decrease of the size of the blocks. The standard size of the blocks for the motion compensation is 16×16 for the luminance component (see Figure 3.6) and 8×8 for the chrominance component, thus, using only one vector of movement per macroblock. With this new tool, the MPEG-4 standard gives the encoder the option to choose the size of the block to be used in the process. Now the luminance blocks can be 8×8 and the blocks of the chrominance component, 4×4, and macroblocks can have up to four motion vectors. This new tool ensures that the resulting motion compensated frame has less energy. There is a significant increase in header size to

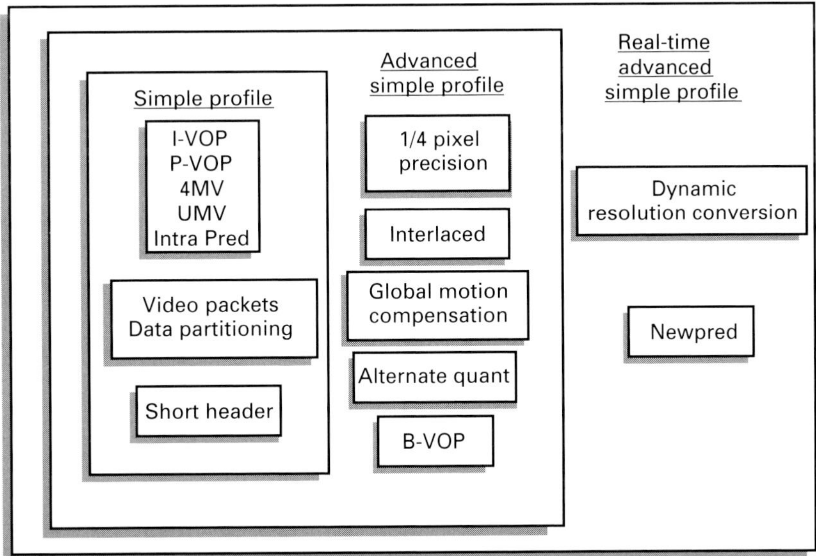

Figure 3.5   MPEG-4 encoding tools

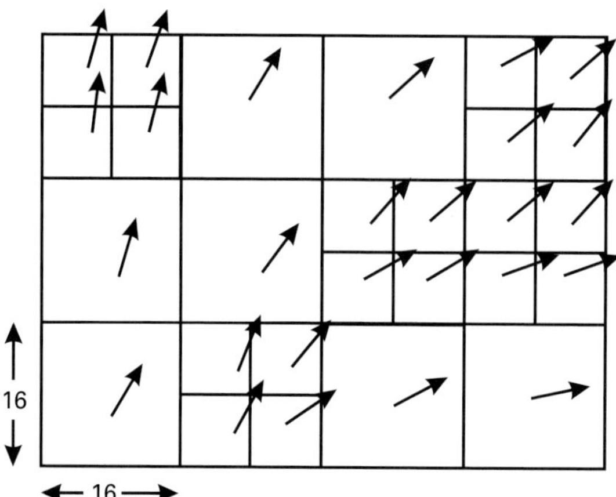

Figure 3.6   Motion vector distribution example on a frame

transmit four vectors for each macroblock. The encoder is responsible for reviewing macroblock by macroblock and detecting the real need to use this technique.

- Unrestricted motion vectors (UMV): this allows vectors pointing out the limits of the reference video object plane. If a sample given by the motion vector is outside the reference video object plane, the value that will be used will be that of the nearest edge. The UMV mode can improve the efficiency of the motion compensation, especially when there are objects moving into and out of the boundaries of the video.
- Data partitioning: this tool enables an encoder to reorganize the information encoded in a video package to reduce the impact of transmission errors.
- Reversible variable length codes (VLCs): an additional set of reversible VLCs can be used to encode the discrete cosine transform (DCT) coefficients. As suggested by the name, these codes can be properly decoded in both directions, enabling the decoder to minimize the area of the image affected by an error.
- P-VOP: this is encoded in the interleaving stage. It uses a previously encoded I-VOP or P-VOP.
- B-VOP: this uses bidirectional prediction to improve the efficiency of the motion compensation. It uses an I-VOP and/or a P-VOP to make the compensation. See Figure 3.7.
- $\frac{1}{4}$ pixel precision motion vectors: reference video object plane samples are interpolated twice. The first time the reference motion vectors have a pixel precision of $\frac{1}{2}$. To obtain a precision of $\frac{1}{4}$ pixel, the samples of the reference video object plane are interpolated again. This procedure increases the complexity and cost of the computational encoding process but confers significant gains in efficiency of coding.
- Alternate quant: this method proposes a new way of the rescheduling (inverse quantization) of the DCT coefficients, except the DC coefficients.
- Newpred: this it is a technique in which the encoder can choose a reference video object plane from any other video object plane already stored in the internal memory

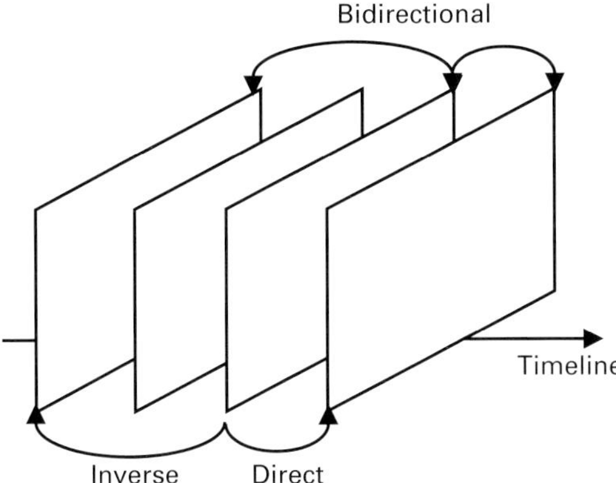

Figure 3.7   Bidirectional prediction scheme

of the encoder for each package of video. This technique is employed to prevent error propagation.

- Dynamic resolution conversion: this technique allows the encoder to encode a video object plane with less spatial resolution. This is extremely useful when there is a need to reduce the output bit rate of the encoder.

3.4   ITU-T H.264/AVC

Since the end of the 1980s, when the video compression technology was in its initial stages, the international video compression standards have been the engine of its commercial success. These standards have played an important role in disseminating this technology, providing interoperability between different products developed by different manufacturers and enabling enough flexibility for optimization and adaptation of the technologies to suit different application requirements.

ITU-T H.264 / MPEG-4 Part 10 Advanced Video Coding, known as H.264/AVC (Joint Video Team of ITU-T and ISO/IEC JTC 1, 2003), is one of the latest proposals for digital video codification. The standard was developed by the Joint Video Team (JVT) consortium formed by specialists from the Video Coding Experts Group (VCEG) of the ITU-T and the ISO/IEC MPEG. As in the previous standards, the H.264/AVC is an alternative that balances compression efficiency and implementation complexity, based on the current technologies.

The main goal of the H.264 standard is to improve the compression efficiency by at least a factor of two compared with MPEG-2, which is the current most widely used standard, without increasing the final cost of the technology. In July 2004, a new part, named Fidelity Range Extensions (FRExt) (Joint Video Team of ITU-T and ISO/IEC JTC 1, 2004), was added to the standard. It specifies new compression efficiency improvements for high-definition video applications, including production and editing of digital video.

3.4.1   History of H.264

The H.264/AVC standard was developed over a period of four years. The development began as a part of the H.26L project of the VCEG in the first months of 1998. In August 1999, the first part of the standard was launched. In 2001, the ISO/IEC MPEG, finally, finished the development of its most recent video codification standard, known as MPEG-4 Part 2, and launched a call for proposals for new contributions to future compression efficiency improvements. Then, the VCEG submitted a project in response to the call and proposed a work team with members of the two committees. Several other proposals were submitted and tested by the MPEG. The MPEG findings based on the results of the tests that confirmed the choice of the VCEG proposal were:

- Applying the DCT to the motion compensated frames is superior to the other techniques with the same objectives, implying that there is no need, at least at this stage, to make structural changes for the next generation of coding standard.
- Some video coding tools that had been excluded in the past, because of their complexity and high implementation cost, could be re-examined for inclusion in future standards. VLSI technology had advanced significantly since the development of the previous standards, which reduced the final implementation costs of these tools, which had been prohibitive.
- The syntax of the new standard should not be backward compatible with the previous standards to ensure maximum freedom in the development of performance improvements at the encoder.
- The ITU-T H.26L proposal reached the best performance in the tests.

ITU-T and ISO/IEC decided to join forces to develop the next generation of video coding standards using as the starting point the H.26L project (Schäfer, 2003). In December 2001, the Joint Video Team, composed of specialists from both entities, was formed, with the goal of finishing the technical development until December 2003. ITU-T planned to adopt the name ITU-T H.264 for the standard, and ISO/IEC planned to adopt the name MPEG-4 Part 10 Advanced Video Coding (AVC) which is part of the set of documents that comprises the whole ISO/IEC 14496 MPEG-4 standard. The H.264 standard has, at least, six nomenclatures: H.264, H.26L, ISO/IEC 14496-10, JVT, MPEG-4 AVC and MPEG-4 Part 10.

The nomenclature used in this book is H.264/AVC, which accommodates both parts. With the wide range of applications considered by the two organizations, the work focus was very wide, e.g. video conferencing, transmission (cable, satellite, terrestrial broadcasting), storage (DVDs, hard drives, video on demand), video streaming, military applications, and digital cinema.

Levels and profiles

The ITU-T H.264 standard specifies a set of three profiles. A coding profile defines the encoder (or decoder) requirements and specifies a set of coding tools in accordance with the purpose of the profile. The H.264/AVC profiles are:

Table 3.3. Levels of the H.264/AVC standard

Level	MB/s max	FS max	Extended main and baseline	High profile	High 10 profile	High profile 4:2:2 and 4:4:4	res@FR
1	1485	99	64 kbit/s	80 kbit/s	192 kbit/s	256 kbit/s	128×96@30.9
1b	1485	99	128 kbit/s	160 kbit/s	384 kbit/s	512 kbit/s	128×96@30.9
1.1	3000	396	192 kbit/s	240 kbit/s	576 kbit/s	768 kbit/s	320×240@10.0
1.2	6000	396	384 kbit/s	480 kbit/s	1152 kbit/s	1536 kbit/s	320×240@20.0
1.3	11880	396	768 kbit/s	960 kbit/s	2304 kbit/s	3072 kbit/s	352×288@30.0
2	11880	396	2 Mbit/s	2.5 Mbit/s	6 Mbit/s	8 Mbit/s	352×288@30.0
2.1	19800	792	4 Mbit/s	5 Mbit/s	12 Mbit/s	16 Mbit/s	352×576@25.0
2.2	20250	1620	4 Mbit/s	5 Mbit/s	12 Mbit/s	16 Mbit/s	720×480@15.0
3	40500	1620	10 Mbit/s	12.5 Mbit/s	30 Mbit/s	40 Mbit/s	720×480@30.0
3.1	108000	3600	14 Mbit/s	17.5 Mbit/s	42 Mbit/s	56 Mbit/s	1280×720@30.0
3.2	216000	5120	20 Mbit/s	25 Mbit/s	60 Mbit/s	80 Mbit/s	1280×1024@42.2
4	245760	8192	20 Mbit/s	25 Mbit/s	60 Mbit/s	80 Mbit/s	2048×1024@30.0
4.1	245760	8192	50 Mbit/s	62.5 Mbit/s	150 Mbit/s	200 Mbit/s	2048×1024@30.0
4.2	522240	8704	50 Mbit/s	62.5 Mbit/s	150 Mbit/s	200 Mbit/s	2048×1088@60.0
5	589824	22080	135 Mbit/s	168.75 Mbit/s	405 Mbit/s	540 Mbit/s	3680×1536@26.7
5.1	983040	36864	240 Mbit/s	300 Mbit/s	720 Mbit/s	960 Mbit/s	4096×2304@26.7

- Baseline profile: this supports Inter and Intra coding and entropy coding using context adaptive variable length coding (CAVLC). It is applied in video telephony, video conferencing, and wireless communications.
- Main profile: this supports interlaced video, B slice encoding, weighted prediction Intra coding, and context-adaptive binary arithmetic coding (CABAC). This profile is used in applications of broadcasting television and video storage.
- Extended profile: this does not support interlaced video and CABAC, but supports permutation methods of encoded bit-streams and better error resilience. This profile is used in media streaming applications.

The encoder and decoder performance limits are defined by a set of levels. Currently, the H.264/AVC standard defines a set of 16 encoding levels, shown at Table 3.3. The level also defines the size of the input, in storage memory, of the decoder. In the table in

- MB/s max is the maximum number of macroblocks;
- FS max is the maximum frame size; and
- res@FR is the resolution and frame rate.

Fidelity range extensions (FRext)

Even taking a wide variety of possible applications, the H.264/AVC standard was basically focused on quality of video for entertainment, based on 8 bits per sample and the chrominance subsampling format 4:2:0. Due to the development time limitations, the initial standard did not include support for the majority of the professional video editing and production environments, nor had the development focused on high video resolution sizes. To solve these problems, the original work was extended to add new features to the functionalities of the original standard.

This new effort started in May 2003 and finished in July 2004. The official announcement of the new functionalities of the H.264/AVC standard came in September 2004. Originally, this extension was named Professional Extensions, but this was changed to Fidelity Range Extensions (Joint Video Team of ITU-T and ISO/IEC JTC 1, 2004) to better express the purpose of the added tools. The main new functionalities are:

- Support of adaptive sizing of the spatially transformed blocks.
- Support of a new perceptual adaptive model for the scalable quantization matrices.
- A set of four new profiles known as High Profiles:
  - High Profile (HiP): this profile covers applications that require high definition without the need of a higher chroma format or better video samples precision. It supports up to 8 bits per sample and the 4:2:0 chrominance subsampling format.
  - High 10 Profile (Hi10P): this supports up to 10 bits per sample and the 4:2:0 chrominance subsampling format.
  - High 4:2:2 Profile (Hi422P): this supports up to 10 bits per sample and the 4:2:2 chrominance subsampling format.
  - High 4:4:4 Profile (Hi444P): this supports up to 10 bits per sample and the 4:4:4 chrominance subsampling format. This profile is optimized to RGB domain coding to avoid conversion errors.

The response of the market to this new standard was immediate. The High Profile standard has been used in several applications, for example:

- HD-DVD;
- Blue-ray disk;
- The DVB future video coding standard.

The network adaptation layer (NAL)

The ITU-T H.264/AVC standard developed, besides its specifications of video coding, the NAL with the aim of adapting the encoder output bit stream. The NAL provides information about the type of the transport layer of the network on which the video will be transmitted, or the storage device that will be used (Zhang *et al.*, 2004). The full architecture of the NAL together with the Video Coding Layer (VCL) can be seen in Figure 3.8. The main functionalities of the NAL are:

- Support of any type of wireless or wired real-time Internet service using RTP/IP;
- File management, for example, ISO MP4;
- H.32X for conversational services;
- MPEG-2 System encapsulation for radio broadcasting services.

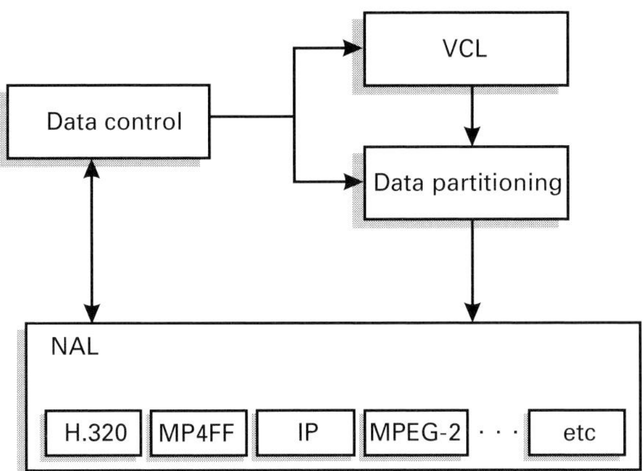

Figure 3.8   The NAL

The bit streams are stored in NAL units with a fixed size. The format of NAL units is the same for both packed data transport and continuous-flow storage systems, differing only in the prefix code that marks the beginning of the bit stream in continuous-flow systems.

The deblocking loop filter

A process of filtering is applied to each decoded macroblock in order to reduce the effect of blocking on the picture, which is characterized by the appearance of the boundaries of the macroblocks of the picture. The blocking mainly happens due to sensitive variations in the DC level of neighboring macroblocks. The deblocking filter is applied after the inverse transform and before the reconstruction and storage of the estimated macroblock for future predictions. In decoding, the filter is applied in the previous step of the reconstruction and display of the macroblock (Richardson, 2003c). The deblocking filter has two main advantages:

- The edges of the blocks are softened, improving the appearance of the decoded frame mainly for sequences with high rates of compression.
- The filtered macroblock, at the encoder, is subsequently used in the process of motion compensation, resulting in a lower level residual error in the prediction process.

The filtering is applied from the vertical edges to the horizontal edges of 4×4 blocks of a macroblock in the following order:

- Vertical edges of luminance coefficients block (a, b, c and d in Figure 3.9);
- Horizontal edges of luminance coefficients block (e, f, g and h in Figure 3.9);
- Vertical edges of chrominance coefficients block (i and j in Figure 3.9);
- Horizontal edges of chrominance coefficients block (k and l in Figure 3.9).

Each operation of filtering affects up to three pixels on each side of the border between two blocks.

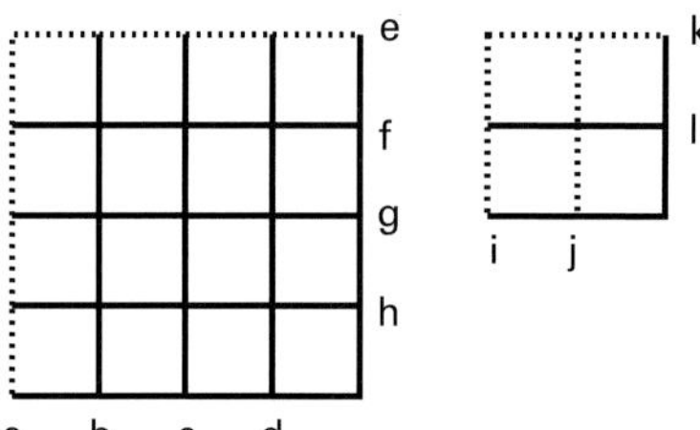

Figure 3.9   Boundary filtering order

Boundary strength

The choice of the deblocking filter operation mode depends on the boundary strength and the gradient of the samples located on the borders of neighboring blocks. The boundary strength (BS) is chosen in accordance with the rules of Table 3.4.

Integer transform and quantization

Similarly to the previous standards, the H.264/AVC also uses transform techniques for the encoding process of the residual frames (resulting from the motion compensation process). H.264 specifies that the transform is applied in 4×4 blocks, and using the DCT. The method used is a separable integer transform of the 4×4 blocks, with, basically, the same properties of the DCT.

The inverse transform is defined only by integer operations, which ensures recovery with no loss of the transformed information. An additional 2×2 transform is applied to the four discrete cosine coefficients of each chroma component. If a macroblock is encoded in the Intra-16×16, a similar 4×4 transform is applied to the discrete cosine coefficients of the luminance signal. See the organization of the blocks in the process of transformation at Figure 3.10. The H.264 standard uses scalar quantization for the quantization of the transformed coefficients (Zhang *et al.*, 2004).

One of the 52 quantizers is chosen for each macroblock according to the quantization parameter. The quantizers are organized so that there is an increase of 12.5% in the quantization step when the quantization parameter is increased by a unit. The transformed and quantized coefficients of a block are usually scanned with the zig-zag option and transmitted after the encoding entropy stage. An alternative scalability is proposed for blocks that are part of an interlaced mode encoded macroblock. The DCT chrominance coefficients are scanned from top to bottom and left to right in accordance with the sequence in Figure 3.10. All of the transformations in the standard H.264 can be implemented using only addition and shifting operations of the bits.

Entropy encoding

The H.264/AVC standard specifies two entropy-encoding algorithms known as CABAC and CAVLC. The CAVLC algorithm switches between a set of VLC tables according to the syntax elements already transmitted. The VLC tables are specially developed to adaptively determine the quantity of bits per word transmitted, and the entropy encoding

Table 3.4. Rules of choice of the parameter boundary strength (BS)

p or q is encoded in the intra mode **and** the edge is a macroblock edge	$BS = 4$ (stronger filtering)
p or q is encoded in the intra mode **and** the edge is **not** a macroblock edge	$BS = 3$
Neither p nor q is encoded in the intra mode; neither p nor q contain coded coefficients	$BS = 2$
Neither p nor q is encoded in the intra mode; neither p nor q contains coded coefficients; p or q has different reference frames **or** a different number of reference frames **or** different motion vector values	$BS = 1$
Neither p nor q is encoded in the intra mode; neither p nor q contains coded coefficients; p or q has the same reference frames and identical motion vectors	$BS = 0$ (no filtering)

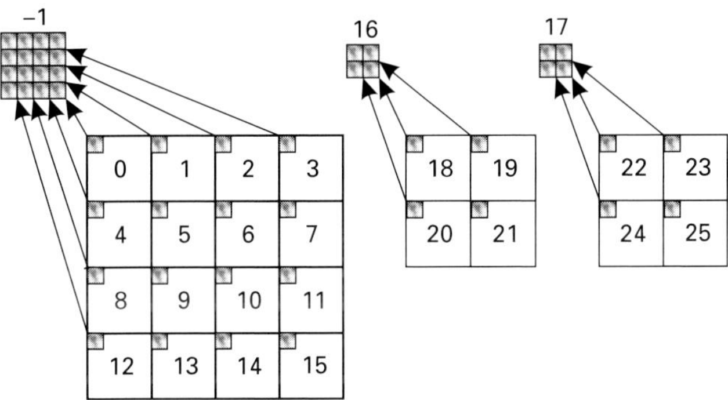

Figure 3.10 Block organization inside a macroblock

is improved compared with schemes with static VLC tables. Entropy encoding efficiency can be further improved by applying the CABAC algorithm. The use of arithmetic coding allows a non-integer number of bits to be associated to each symbol of the alphabet, which is beneficial for symbol probabilities higher than 0.5. The use of adaptive codes allows adaptation for non-stationary statistics.

Another important property of CABAC is context modeling. The statistics of already encoded syntax elements are used to estimate the conditional probabilities. The conditional probabilities are used to switch between several models of estimated probabilities. In the H.264/AVC standard, the core of arithmetic coding algorithms and their associated estimated probabilities are specified as low-complexity methods without multiplication operations, using only shifting and look-up tables. Compared with CAVLC, the CABAC algorithm provides a gain of output bit rate saving from 10% to 15% when encoding a television signal with the same quality (Schäfer *et al.*, 2003).

Table 3.5. Comparison of the average output bit rate saving for the main encoders

Coder	MPEG-4 ASP	H.263	MPEG-2
H.264/AVC	38.62%	48.80%	64.46%
H.264/AVC		16.65%	42.95%
H.264/AVC			30.61%

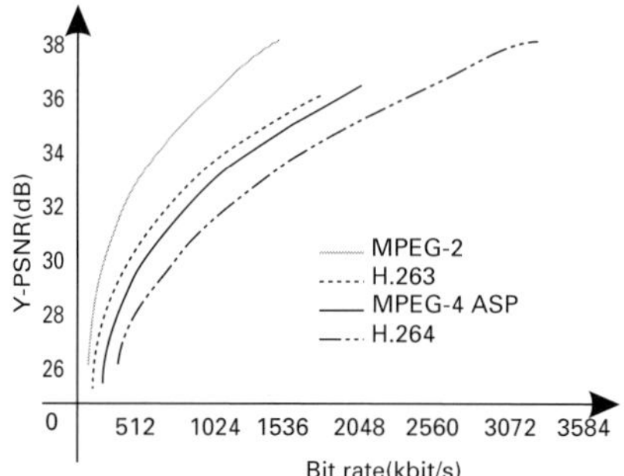

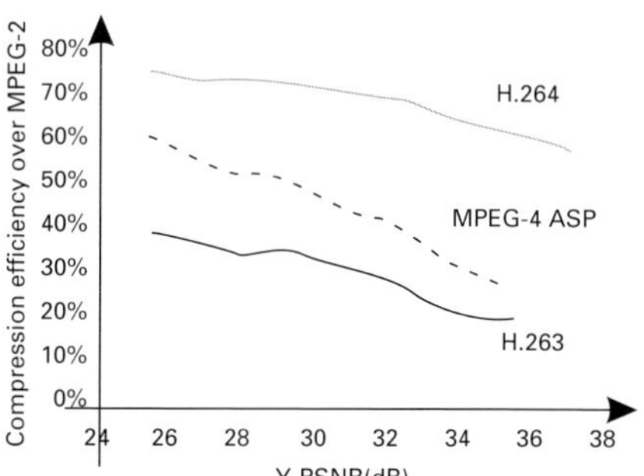

Figure 3.11   Peak signal-to-noise ratio and compression efficiency

3.4.2   Comparison of H.264/AVC standard with other standards

To demonstrate the encoding efficiency of H.264, comparative tests were conducted with other important and successful standards of video encoding. The encoders involved in the tests were:

- MPEG-2 MP@ML;
- H.263++ HLP;
- MPEG-4 Visual ASP, configured with motion estimation precision of $\frac{1}{4}$;
- H.264/AVC Main Profile, configured to use five reference frames.

Test sequences were chosen from a set of common standard QCIF (176×144) and CIF (352×288) sequences. Only the first frame of each sequence was coded as an I frame, B frames were placed between two successive P frames. The B frames were not stored in the frame memory, thus, preventing them from being used as reference frames. An exhaustive search procedure was used in order to estimate the motion for an encoder specified in Sullivan and Wiegand (1998) and Wiegand and Andrews (1998).

The graphs in Figure 3.11 show the bit rate saving relative to the worst performance video encoder, the MPEG-2. The result is plotted depending on the peak signal-to noise ratio (PSNR) of the luminance component for H.263, MPEG-4 ASP and H.264. Table 3.5 describes the average bit rate saving achieved by each encoder. These results are relative to the average of the results of all tested sequences. The highly flexible model of motion estimation and the efficient scheme of context adaptive arithmetic encoding are the primary factors in the best results of the standard H.264/AVC.

4   Channel coding for digital television

with Valdemar Cardoso da Rocha Jr.

4.1   Introduction

Reliable information transmission and storage has been a constant challenge for engineers and researchers in telecommunications due to the ever increasing demands for higher processing speeds and higher reliability. Error-correcting codes (Shannon, 1948a, Peterson and Weldon, 1972) are nowadays found in most electronic digital processing equipment, from the audio compact disk (CD) (Wicker and Bhargava, 1994) to deep-space probes (Hagenauer, 1992), and more recently in digital television.

The purpose of this chapter is to focus on specific coding schemes employed for digital television, covering a minimum of background material on error-correcting codes. The term reliability in telecommunications refers to the level of signal immunity to noise and interference, and is not concerned with secrecy or authenticity of the data. Cyclic codes (Lin and Palais, 1986) include important subclasses of codes, e.g., the Bose–Chaudhury–Hocquenghem (BCH) codes, the Reed–Solomon codes, and practical low-density parity-check (LDPC) codes, and for this reason are the subject of the next section. In Section 4.5 another important class of codes, the convolutional codes, will be addressed.

4.2   Cyclic codes

Cyclic codes are among the most important codes for practical applications in engineering (Lin and Palais, 1986). Cyclic codes have been used as part of many communication protocols, in music CDs, in magnetic recording (Immink, 1991), etc. The preference for cyclic codes is a consequence of their mathematical structure based on discrete mathematics, which allows a considerable simplification in the implementation of encoders and decoders for such codes. A formal mathematical treatment of q-ary cyclic codes relies on polynomial rings (Peterson and Weldon, 1972), modulo $x^n - 1$, with coefficients in the Galois field $GF(q)$, in which n denotes the code block length (Berlekamp, 1968).

DEFINITION 4.1 *A code is defined as a cyclic code when its code words are invariant to a cyclic permutation, i.e. a cyclic permutation applied to any code word gives as a result a code word in this code.*

Thus, for example, if $\mathbf{v} = (v_0, v_1, v_2, \ldots, v_{n-1})$ is a code word and is right-shifted cyclically by i positions, then $\mathbf{v}^i = (v_{n-i}, v_{n-i+1}, \ldots, v_0, v_1, \ldots, v_{n-i-1})$ is also a code word, in which the indices are reduced modulo n. An n-tuple, e.g. $\mathbf{v}$, can be represented by a polynomial of degree at most $n - 1$ as follows

$$v(x) = v_0 + v_1 x + v_2 x^2 + \cdots + v_{n-1} x^{n-1}.$$

The notation (n, k, d) is used to denote a code of block length n, having k information digits and minimum distance d (Peterson and Weldon, 1972). The code words of a cyclic code are multiples of a well-defined *generator polynomial* $g(x)$, of degree $n - k$. Conversely, every polynomial divisible by $g(x)$, with coefficients belonging to the code alphabet and degree at most $n - 1$, is a code word. The generator polynomial is a factor of $x^n - 1$.

Encoder with $(n - k)$-stage shift-register

This encoding procedure is based on the property of cyclic codes by which all code words in a cyclic code are multiples of the code generator polynomial $g(x)$. The k information digits are represented by a polynomial $I(x)$ of degree at most $k - 1$, referred to as the information polynomial. Multiplying $I(x)$ by x^{n-k} the polynomial $I(x)x^{n-k}$ results, which has degree at most $n - 1$ and does not contain any terms of degree lower than $n - k$. By dividing $I(x)x^{n-k}$ by $g(x)$ it follows that

$$I(x)x^{n-k} = Q(x)g(x) + R(x),$$

where $Q(x)$ and $R(x)$ are the quotient polynomial and the remainder polynomial, respectively. Since $R(x)$ has degree lower than that of $g(x)$, it follows that $R(x)$ has degree at most $n - k - 1$. If $R(x)$ is subtracted from $I(x)x^{n-k}$, the result is a multiple of $g(x)$, i.e. it is a code word. It also follows that the polynomial $R(x)$ represents the parity-check digits and has no overlapping non-zero terms with the polynomial $I(x)x^{n-k}$. These operations of multiplication and division of polynomials are illustrated in Figure 4.1.

The circuit shown in Figure 4.1 employs a binary shift-register with $n - k$ stages and premultiplies the information polynomial by x^{n-k}. The switches $g_1, g_2, \ldots, g_{n-k-1}$ are closed if the corresponding coefficients in $g(x)$ are equal to 1. Otherwise, they are left open. Initially, the shift-register contents are zeroes. With switch S_1 closed and switch S_2 (the output switch) in position 1, the information digits are simultaneously fed to the circuit and to the output. After k information digits are transmitted, the parity-check

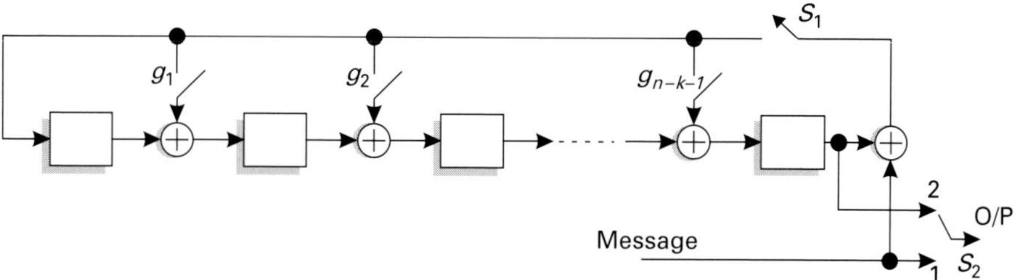

Figure 4.1   Binary encoder for a (n, k, d) cyclic code employing a shift-register with $n - k$ stages

digits (coefficients of the remainder polynomial) are in the shift-register. Then, switch S_1 is opened and switch S_2 is moved to position 2. During the next $n - k$ time slots (clock pulses) the parity-check digits are transmitted. This procedure is repeated for subsequent information blocks of k digits.

There is another encoding procedure for cyclic codes, which is based on the parity-check polynomial $h(x) = (x^n - 1)/g(x)$, which employs a binary shift-register with k stages (Lin and Palais, 1986). Some important classes of cyclic codes are treated in Subsections 4.2.1–4.2.3.

4.2.1 Bose–Chaudhury–Hocquenghem (BCH) codes

BCH codes were discovered independently by Hocquenghem in 1959 (Hocquenghem, 1959) and Bose and Chaudhury in 1960 (Bose and Ray-Chaudhury, 1960). BCH codes are cyclic codes and represent one of the most important classes of block codes. For any positive integers, m and t, there exists a q-ary BCH code with the following parameters

$$n = q^m - 1, \ n - k \leq mt, \ d \geq 2t + 1.$$

In practice, most BCH codes are binary, i.e. $q = 2$. BCH codes are specified in terms of roots of the generator polynomial as follows.

DEFINITION 4.2 *A t-error correcting q-ary primitive BCH code, of block length* $n = q^m - 1$, *has the terms* α^{m_0}, $\alpha^{m_0+1}, \ldots, \alpha^{m_0+2t-1}$ *as roots of its generator polynomial, for any positive integer* m_0, *in which* α *is a primitive element of GF(q^m).*

It follows that the generator polynomial $g(x)$ of a BCH code can be written as the least common multiple (LCM) of minimal polynomials $\phi_i(x)$ (Lin and Palais, 1986), i.e. as

$$g(x) = \text{LCM}\{\phi_{m_0}(x), \phi_{m_0+1}(x), \ldots, \phi_{m_0+2t-1}(x)\},$$

in which $\phi_i(x)$ denotes the minimal polynomial of α^{m_0+i}, $0 \leq i \leq 2t - 1$. Non-primitive BCH codes are those BCH codes for which α is not a primitive element of $GF(q^m)$ and, consequently the block length is given by the order of α in $GF(q^m)$. BCH codes with $m_0 = 1$ are called *narrow-sense* BCH codes. An alternative characterization of BCH codes was introduced by Blahut (1983), in which a frequency domain representation is employed. Binary BCH codes were proposed for the DVBS-2, the European standard for satellite television transmission, standard as an outer code in a serial concatenation coding scheme with an LDPC code as the inner code.

Example 4.1 *Construct the generator polynomial of a* $t = 3$ *error-correcting binary BCH code of block length* $n = 2^4 - 1 = 15$, *having* $\alpha, \alpha^2, \alpha^3, \alpha^4, \alpha^5$ *and* α^6 *as roots of its generator polynomial, in which* α *is a primitive element of GF(2^4).*

Using Table 4.1 to simplify results, it follows that

$$g(x) = \phi_1(x)\phi_3(x)\phi_5(x)$$

Table 4.1. Galois field $GF(2^4)$ generated by $p(X) = X^4 + X + 1$

Exponential form	Polynomial form	Binary form
0	0	0000
1	1	0001
α	α	0010
α^2	α^2	0100
α^3	α^3	1000
α^4	$\alpha + 1$	0011
α^5	$\alpha^2 + \alpha$	0110
α^6	$\alpha^3 + \alpha^2$	1100
α^7	$\alpha^3 + \alpha + 1$	1011
α^8	$\alpha^2 + 1$	0101
α^9	$\alpha^3 + \alpha$	1010
α^{10}	$\alpha^2 + \alpha + 1$	0111
α^{11}	$\alpha^3 + \alpha^2 + \alpha$	1110
α^{12}	$\alpha^3 + \alpha^2 + \alpha + 1$	1111
α^{13}	$\alpha^3 + \alpha^2 + 1$	1101
α^{14}	$\alpha^3 + 1$	1001

in which

$$\phi_1(x) = (x - \alpha)(x - \alpha^2)(x - \alpha^4)(x - \alpha^8) = x^4 + x + 1,$$

$$\phi_3(x) = (x - \alpha^3)(x - \alpha^6)(x - \alpha^{12})(x - \alpha^9) = x^4 + x^3 + x^2 + x + 1,$$

$$\phi_5(x) = (x - \alpha^5)(x - \alpha^{10}) = x^2 + x + 1,$$

i.e.

$$g(x) = (x^4 + x + 1)(x^4 + x^3 + x^2 + x + 1)(x^2 + x + 1)$$
$$= x^{10} + x^8 + x^5 + x^4 + x^2 + x + 1.$$

4.2.2   Reed–Solomon (RS) codes

Non-binary BCH codes for which $m = m_0 = 1$, defined over a q-ary alphabet by the parameters

$$n = q - 1, \; n - k = 2t, \; d = 2t + 1,$$

are called RS codes (Reed and Solomon, 1960). The generator polynomial for an RS code has the form

$$g(x) = (x - \alpha)(x - \alpha^2)(x - \alpha^3) \cdots (x - \alpha^{2t}),$$

in which α is an element of order n in $GF(q)$. RS codes with $q = 2^r$ are used in practice with the q-ary symbols represented as r-byte binary tuples. Since $d = n - k + 1$, RS

Table 4.2. Galois field $GF(2^3)$ generated by $p(X) = X^3 + X + 1$

Exponential form	Polynomial form	Binary form
0	0	000
1	1	001
α	α	010
α^2	α^2	100
α^3	$\alpha + 1$	011
α^4	$\alpha^2 + \alpha$	110
α^5	$\alpha^2 + \alpha + 1$	111
α^6	$\alpha^2 + 1$	101

codes are also maximum distance separable codes (MacWilliams and Sloane, 1977). Some efficient binary codes, capable of correcting both random and burst errors, result from the binary mapping of an RS code over $GF(2^r)$. Such codes can correct any combination of at most t erroneous binary r-tuples in a block containing rn binary digits. In a serial concatenation coding scheme (Forney, 1966), usually an RS code is used as an outer code.

Example 4.2 *Construct the generator polynomial of a $(7, 5, 3)$ RS code that operates over the Galois field $GF(2^3)$, generated by the primitive polynomial (Lin and Palais, 1986) $p(x) = x^3 + x + 1$. Since this RS code has $n - k = 2t = 2$, it is able to correct one error in any of the seven positions of a code word. Using Table 4.2 to simplify results, the corresponding generator polynomial is*

$$g(x) = (x - \alpha)(x - \alpha^2) = x^2 - (\alpha + \alpha^2)x + \alpha^3 = x^2 + \alpha^6 x + \alpha^3.$$

4.2.3 Low-density parity-check (LDPC) codes

LDPC codes are linear block codes discovered by Gallager in 1960 (Gallager, 1963), but were largely ignored for about 30 years, except for a paper by Tanner in 1981 (Tanner, 1981) which was also ignored until recently. LDPC codes have a decoding algorithm the complexity of which grows linearly with the code block length (Gallager, 1963). At the time of their discovery there was no computational means to implement them in practice or to perform computer simulations. The investigation of codes on graphs and iterative decoding led to the rediscovery of LDPC codes, and was accompanied by further theoretical advances. It was shown that long LDPC codes with iterative decoding achieve a performance, in terms of error rate, very close to the Shannon capacity (Mackay and Neal, 1997) LDPC codes have as their direct competitors the turbo codes (Berrou *et al.*, 1993), however, LDPC codes have the following advantages with respect to turbo codes:

- a long interleaver is not required for LDPC codes to achieve low error rates;
- for comparable decoder complexity, LDPC codes achieve lower block error rates and their error floor occurs at lower bit error rates.

Classical LDPC codes are defined by their parity-check matrix H (Lin and Palais, 1986). Let ρ and γ denote positive integers, in which ρ is small in comparison with the code block length and γ is small in comparison with the number of rows in H.

DEFINITION 4.3 *A binary LDPC code is defined as the set of code words that satisfy a parity-check matrix H, in which H has ρ 1s per row and γ 1s per column. The number of 1s in common between any two columns in H, denoted by λ, is at most 1, i.e. $\lambda \leq 1$.*

After their rediscovery, a number of good LDPC codes were constructed by computer search, which meant that such codes lacked in structure and consequently had complex encoding. Systematic algebraic construction of LDPC codes based on finite geometries was introduced in (Kou *et al.*, 2000). As discussed later in this chapter, the Chinese digital television broadcast standard (Song *et al.*, 2007) employs a structured LDPC code construction clearly derived from finite geometries.

4.3   Decoding cyclic codes

The procedures for decoding linear block codes are also applicable to cyclic codes. However, the algebraic structure of cyclic codes allows for some significant simplifications on decoder implementation. The syndrome calculation consists just of computing the remainder of the division of the received n-tuple in polynomial form by the generator polynomial. The syndrome polynomial is denoted by $s(x)$. If $s(x) = 0$, the received n-tuple is accepted as a valid code word, otherwise, i.e. if $s(x) \neq 0$, the decoder declares the occurrence of errors. It follows that a circuit to implement error-detection with cyclic codes is indeed very simple. Error location in a received n-tuple, however, is in general a more difficult task to solve in practice. The most important algebraic decoding techniques are those based on the Berlekamp–Massey (BM) algorithm (Massey, 1969) and on the Euclidean algorithm (Clark and Cain, 1981).

4.3.1   Algebraic decoding

In terms of implementation complexity, encoding and decoding are in general very asymmetric operations, with decoding being far more complex than encoding. For practical reasons, very often a suboptimum coding scheme is chosen by taking into consideration that performance is satisfactory and the decoder implementation is amenable. Strictly algebraic decoding algorithms (Berlekamp, 1968, Massey, 1969) that are efficient to handle hard-decision errors have proved very hard to adapt to perform soft-decision decoding.

The Berlekamp–Massey (BM) algorithm

The BM algorithm was discovered by Elwyn Berlekamp in 1968 (Berlekamp, 1968) in the context of decoding BCH codes. In 1969 James Massey (Massey, 1969) showed that the iterative algorithm introduced by Berlekamp actually provided a general solution to the problem of synthesizing the shortest linear feedback shift-register capable of generating a prescribed finite sequence of digits.

The BM algorithm plays a key role in the decoding of algebraic codes, including RS codes and BCH codes. In the case of binary BCH codes, there is no need to calculate the error values, since it is sufficient to determine the positions of the errors to perform error correction in $GF(2)$. However, in the case of non-binary BCH codes and RS codes, both error location and error values have to be determined to perform error correction.

Consider a (n, k, d) algebraic code, with generator polynomial having roots $\alpha, \alpha^2, \ldots, \alpha^{2t}$. Let $c(x)$ denote a code word polynomial, let $e(x)$ denote the error polynomial and let $r(x)$, in which

$$r(x) = c(x) + e(x),$$

with addition over $GF(q)$, denote the received n-tuple in polynomial form. The algebraic decoding procedure consists of the following steps.

(1) Compute the first $2t$ coefficients of the syndrome polynomial $s(x) = s_0 + s_1 x + \cdots + s_{2t-1}x^{2t-1} + \cdots$, in which

$$s_0 = r(\alpha),$$

$$s_1 = r(\alpha^2),$$

$$\vdots \quad \vdots$$

$$s_{2t-1} = r(\alpha^{2t}).$$

(2) Feed the BM algorithm with the sequence $s_0, s_1, \ldots, s_{2t-1}$ and compute the error-locator polynomial $\sigma(x)$, of degree τ, $\tau \leq t$, in which

$$\sigma(x) = 1 + \sigma_1 x + \sigma_2 x^2 + \cdots + \sigma_\tau x^\tau.$$

(3) Find the roots of $\sigma(x)$, denoted by $\beta_1, \beta_2, \ldots, \beta_\tau$, the inverses of which indicate the error locations.

(4) Determine the error magnitudes in case of non-binary codes.

Berlekamp (1968) introduced a procedure for computing error magnitudes for non-binary cyclic codes and defined the polynomial

$$Z(x) = 1 + (s_0 + \sigma_1)x + (s_1 + \sigma_1 s_0 + \sigma_2)x^2 + \cdots$$

$$+ (s_{\tau-1} + \sigma_1 s_{\tau-2} + \sigma_2 s_{\tau-3} + \cdots + \sigma_\tau)x^\tau. \tag{4.1}$$

Error magnitudes at positions β_i, $1 \leq i \leq \tau$, are calculated as

$$e_i = \frac{Z(\beta_i^{-1})}{\prod_{j=1, j \neq i}^{\tau}(1 + \beta_j \beta_i^{-1})}. \tag{4.2}$$

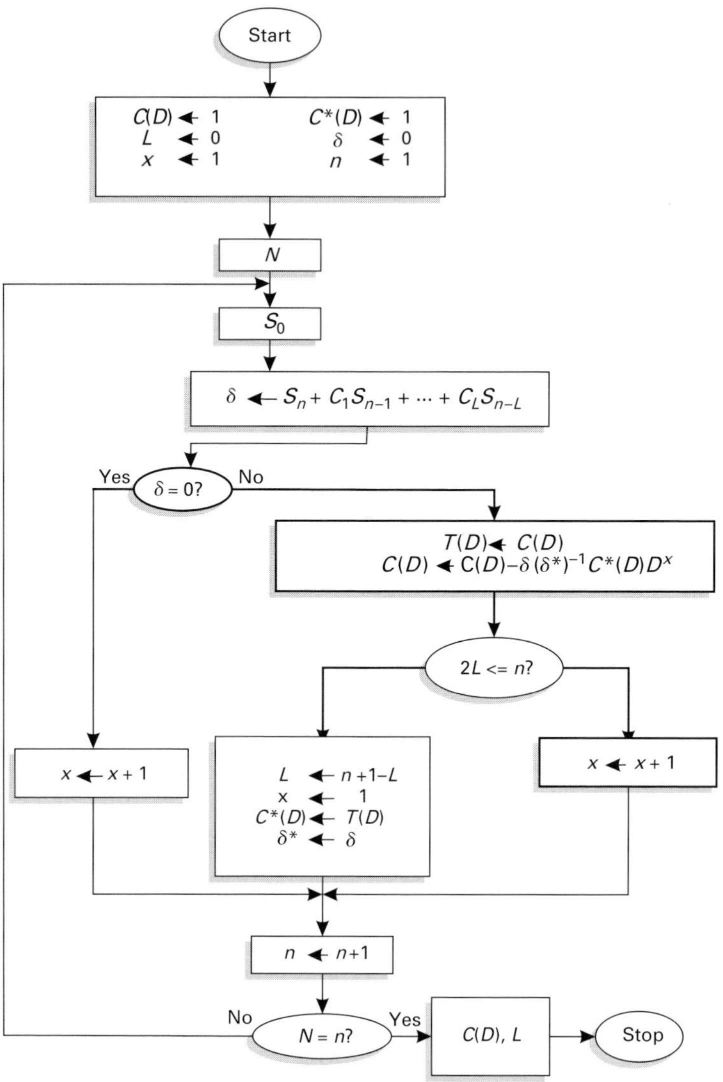

Figure 4.2   Flowchart of the BM algorithm

The block diagram in Figure 4.2 (Massey, 1998) illustrates the steps for running the BM algorithm. This procedure leads to an estimate of the error pattern of minimum weight, which solves the system of syndrome equations. This will be the true error pattern that occurred on the channel if the number of errors τ in this pattern is $\tau \le t$.

Example 4.3   *Consider decoding the* $(15, 7, 5)$ *binary BCH code, i.e. with* $t = 2$*, and generator polynomial* $g(x) = \phi_1(x)\phi_3(x) = (x^4 + x + 1)(x^4 + x^3 + x^2 + x + 1)$ *(please refer to Example 4.1). Let the received polynomial be* $r(x) = x^2 + x^8$*. The syndrome*

Table 4.3. Evolution of the BM algorithm for the input sequence $s_0, s_1, s_2, s_3 = 1, 1, \alpha^5, 1$

n	s_n	δ	$T(D)$	$C(D) = \sigma(X)$	L	$C^*(D)$	δ^*	x
0	1	1		1	0	1	1	1
1	1	0	1	$1 + D$	1	1	1	1
2	α^5	α^{10}	1	$1 + D$	1	1	1	2
3	1	0	$1 + D$	$1 + D + \alpha^{10}D^2$	2	$1 + D$	α^{10}	1

coefficients are calculated as:

$$s_0 = r(\alpha) = 1,$$

$$s_1 = r(\alpha^2) = 1,$$

$$s_2 = r(\alpha^3) = \alpha^5,$$

$$s_3 = r(\alpha^4) = 1.$$

Table 4.3 is used to compute the error-location polynomial with the BM algorithm for the input sequence $s_0, s_1, s_2, s_3 = 1, 1, \alpha^5, 1$. The polynomial $C(D) = 1 + D + \alpha^{10}D^2$ is the error-locator polynomial, i.e. $\sigma(x)$ is equal to $C(D)$ with x replacing D, or

$$\sigma(x) = 1 + x + \alpha^{10}x^2,$$

the roots of which are found by an exhaustive search to be $\beta_1 = \alpha^7$ and $\beta_2 = \alpha^{13}$. The error positions are the exponents of α in the representation of β_1^{-1} and β_2^{-1}, i.e. $\beta_1^{-1} = \alpha^{-7} = \alpha^8$ and $\beta_2^{-1} = \alpha^{-13} = \alpha^2$. There are thus two errors, occurring at x^2 and x^8, respectively. After performing error correction, the decoded polynomial is the all-zero polynomial, i.e. $c(x) = 0$.

Example 4.4 *Consider decoding the $(7, 3, 5)$ Reed–Müller (RM) code over $GF(8)$, with generator polynomial $g(x) = (x - \alpha)(x - \alpha^2)(x - \alpha^3)(x - \alpha^4)$. Let the received polynomial be $r(x) = x^2 + \alpha^3 x^5$. The syndrome coefficients are calculated as (please refer to Example 4.2):*

$$s_0 = r(\alpha) = \alpha^4,$$

$$s_1 = r(\alpha^2) = \alpha^3,$$

$$s_2 = r(\alpha^3) = \alpha^3,$$

$$s_3 = r(\alpha^4) = \alpha^4.$$

Table 4.4 is used to compute the error-location polynomial with the BM algorithm for the input sequence $s_0, s_1, s_2, s_3 = \alpha^4, \alpha^3, \alpha^3, \alpha^4$. The polynomial $C(D) = 1 + \alpha^3 D + D^2$ is the error-locator polynomial, i.e. $\sigma(x)$ is equal to $C(D)$ with x replacing D, or

$$\sigma(x) = 1 + \alpha^3 x + x^2,$$

Table 4.4. Evolution of the BM algorithm for the input sequence $s_0, s_1, s_2, s_3 = \alpha^4, \alpha^3, \alpha^3, \alpha^4$

n	s_n	δ	$T(D)$	$C(D) = \sigma(X)$	L	$C^*(D)$	δ^*	x
0	α^4	α^4		1	0	1	1	1
1	α^3	1	1	$1 + \alpha^4 D$	1	1	α^4	1
2	α^3	α^5	1	$1\alpha^6 D$	1	1	α^4	2
3	α^4	α^2	$1 + \alpha^6 D$	$1 + \alpha^6 D + \alpha D^2$	2	$1 + \alpha^6 D$	α^5	1
				$1 + \alpha^3 D + D^2$				

the roots of which are found by an exhaustive search to be $\beta_1 = \alpha^2$ and $\beta_2 = \alpha^5$. The error positions are the exponents of α in the representation of β_1^{-1} and β_2^{-1}, i.e. $\beta_1^{-1} = \alpha^{-2} = \alpha^5$ and $\beta_2^{-1} = \alpha^{-5} = \alpha^2$. There are thus two errors, occurring at x^2 and x^5, respectively. The error magnitudes are found with the help of (4.1) and (4.2) as follows.

$$Z(X) = 1 + (s_0 + \sigma_1)x + (s_1 + \sigma_1 s_0 + \sigma_2)x^2$$
$$= 1 + (\alpha^4 + \alpha^3)x + (\alpha^3 + \alpha^3 \alpha^4 + 1)x^2$$
$$= 1 + \alpha^6 x + \alpha^3 x^2.$$

Then

$$e_1 = \frac{Z(\beta_1^{-1})}{(1 + \beta_2 \beta_1^{-1})} = \frac{1 + \alpha^6 \alpha^5 + \alpha^3 \alpha^{10}}{(1 + \alpha^5 \alpha - 2)} = \frac{\alpha}{\alpha} = 1,$$

$$e_2 = \frac{Z(\beta_2^{-1})}{(1 + \beta_1 \beta_2^{-1})} = \frac{1 + \alpha^6 \alpha^2 + \alpha^3 \alpha^4}{(1 + \alpha^2 \alpha^{-5})} = \frac{\alpha}{\alpha^5} = \alpha^3.$$

After performing error correction, the decoded polynomial is the all-zero polynomial, i.e. $c(x) = 0$.

4.4   Soft-decision decoding

A soft-decision decoder is characterized by the fact that it retains received symbol reliability information supplied by the demodulator. This reliability information is used by the decoder to improve its decisions when deciding which code word was transmitted. Very often the use of soft-decision decoding leads to a gain of at least 2.0 dB with respect to hard-decision decoding. A hard-decision decoder simplifies its operation and subsequent implementation requirements by discarding the received symbol reliability information.

4.4.1   The decoding of LDPC codes

The efficient decoding of LDPC codes relies on the sum–product algorithm (SPA) (Moreira and Farrell, 2006). The SPA is a symbol-by-symbol soft-in soft-out iterative decoding algorithm. Iterations are employed on the received symbols to improve the reliability of the decoded symbols, based on the code parity-check matrix H. A stop condition is defined for the iterations. When the stop condition is reached the available symbol reliability values are used to make hard decisions and to output a decoded binary n-tuple. Let $\mathbf{v} = (v_1, v_2, \ldots, v_i, \ldots, v_n)$ denote a code word and let $\mathbf{y}$ denote a received n-tuple the coordinates of which are soft values. The SPA is implemented by computing the marginal probabilities $P(v_i|\mathbf{y})$, for $1 \leq i \leq n$. A detailed description of SPA decoding of LDPC codes is available in MacKay and Neal (1999).

4.5 Convolutional codes

Convolutional codes were discovered by Peter Elias in 1954 (Elias, 1954) and since then many researchers have dedicated time to understand the properties and the structure of such codes (Lin and Palais, 1986, Johannesson and Zigangirov, 1999, Heegard and Wicker, 1999). Convolutional codes offer an alternative for error control which is substantially different from that offered by block codes.

Wozencraft and Reiffen (1961) described the first practical decoding algorithm for convolutional codes. In 1967, Viterbi (1985) discovered another way of decoding convolutional codes which he thought to be asymptotically optimum. In 1973, Forney showed that the Viterbi algorithm was a maximum likelihood decoding algorithm for convolutional codes (Forney, 1973). More recently, Berrou *et al.* (1993) introduced a turbo decoding algorithm for a code construction employing convolutional codes. This is by far the most remarkable result in coding theory since Shannon's papers. A turbo decoder allows performance very close to the Shannon limit (channel capacity) (Berrou and Glavieux, 1996), in the presence of additive white Gaussian noise. Turbo decoding was also shown to work equally well with block codes (Pyndiah *et al.*, 1994). Convolutional codes have championed the code race in space research in combination with RS codes. In the following, the basic theory of convolutional codes is revisited.

4.5.1 Basic concepts

A (n, k, m) binary convolutional encoder (BCE) is a linear device with memory that accepts blocks of k binary digits or message bits as input, and outputs blocks of n binary digits which depend on the current input bits and on a number m of previous blocks of message bits. Usually n and k, $n > k$, are small positive integers. Usually, the larger the value of the memory m, the smaller the error probability that results after decoding. The quantity k/n is called the code asymptotic rate. A binary convolutional code (BCC) is the set of all code words that can be produced at the output of the corresponding BCE. Figure 4.3 illustrates a BCE.

A BCE may have a finite impulse response (FIR), also referred as a non-recursive BCE, or an infinite impulse response (IIR). FIR encoders are also called non-recursive encoders and IIR encoders are also called recursive encoders. A BCC is called systematic if every block of k input message bits is present among the n output bits of the corresponding code word.

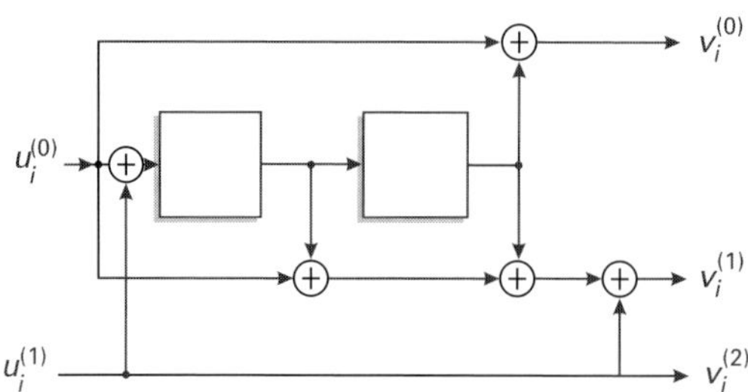

Figure 4.3   BCE of rate 2/3

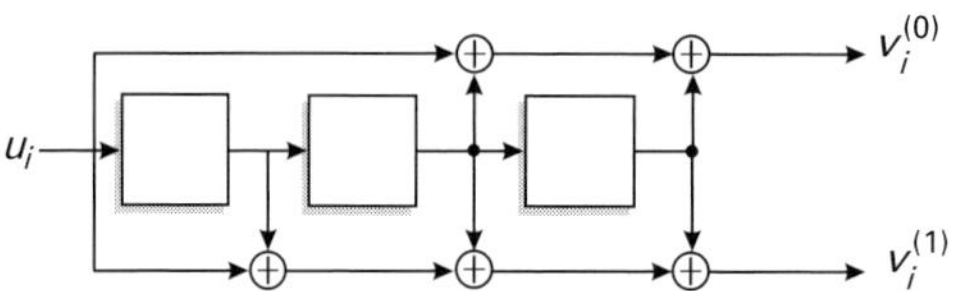

Figure 4.4   Non-recursive (FIR) non-systematic encoder of rate 1/2

4.5.2   Non-recursive convolutional codes

A convolutional encoder is called an FIR encoder if and only if its output can be expressed as a linear combination of current inputs and a finite number of past inputs. The linear combination is expressed in terms of input symbols and coefficients of the encoder generator sequences.

Consider the FIR encoder illustrated in Figure 4.4. The corresponding convolutional code has parameters $(2, 1, 3)$ and an asymptotic rate of $1/2$. The encoder input consists of message blocks containing one symbol and the encoder output consists of blocks of two symbols. The message sequence $\boldsymbol{u} = (u_0, u_1, u_2, \ldots)$ is fed to the encoder input in a bit by bit manner. Since the encoder is a linear system, it follows that its two output sequences, namely

$$\boldsymbol{v}^{(0)} = (v_0^{(0)}, v_1^{(0)}, v_2^{(0)}, \ldots)$$

and

$$\boldsymbol{v}^{(1)} = (v_0^{(1)}, v_1^{(1)}, v_2^{(1)}, \ldots),$$

can be written as the convolution of the input sequence $\boldsymbol{u}$ with the corresponding encoder impulse response. For $(n, 1, m)$ convolutional codes, the impulse response for a given encoder output is the output produced by feeding the encoder input with the sequence $\boldsymbol{u} = (100 \ldots)$. Since the encoder has a memory m it follows that the respective impulse responses are

$$\boldsymbol{g}^{(0)} = (g_0^{(0)}, g_1^{(0)}, g_2^{(0)}, \ldots, g_m^{(0)})$$

and

$$\boldsymbol{g}^{(1)} = (g_0^{(1)}, g_1^{(1)}, g_2^{(1)}, \ldots, g_m^{(1)}).$$

For the specific case of the encoder in Figure 4.4 it follows that

$$\boldsymbol{g}^{(0)} = (1011),$$

$$\boldsymbol{g}^{(1)} = (1111).$$

The impulse responses are also referred as the code generator sequences.

In terms of convolutions the encoding equations can be written as

$$\begin{aligned} \boldsymbol{v}^{(0)} &= \boldsymbol{u} * \boldsymbol{g}^{(0)}, \\ \boldsymbol{v}^{(1)} &= \boldsymbol{u} * \boldsymbol{g}^{(1)}, \end{aligned}$$

in which $*$ denotes the discrete convolution and all operations are reduced modulo 2. The convolution operation $\boldsymbol{v}^{(i)} = \boldsymbol{u} * \boldsymbol{g}^{(i)}$ implies that, for all $l \geq 0$,

$$v_l^{(j)} = \sum_{i=0}^{m} u_{l-i} g_i^{(j)} = u_l g_0^{(j)} + u_{l-1} g_1^{(j)} + \cdots + u_{l-m} g_m^{(j)}, \quad j = 0, 1,$$

in which $u_{l-i} = 0$ for all $l < i$. Thus, for the encoder depicted in Figure 4.4 it follows that

$$\begin{aligned} v_l^{(0)} &= u_l + u_{l-2} + u_{l-3}, \\ v_l^{(1)} &= u_l + u_{l-1} + u_{l-2} + u_{l-3}, \end{aligned}$$

which are obtained in a straightforward manner by inspection of the encoder circuit. After encoding, the output sequences $\boldsymbol{v}^{(0)}$ and $\boldsymbol{v}^{(1)}$ are multiplexed to form a single sequence which constitutes a code word, for transmission through a channel. The code word is expressed as

$$\boldsymbol{v} = (v_0^{(0)} v_0^{(1)}, v_1^{(0)} v_1^{(1)}, v_2^{(0)} v_2^{(1)}, \ldots).$$

In a linear system, time domain operations involving convolutions can be equivalently replaced by polynomial operations (Lin and Palais, 1986). Since a convolutional encoder performs linear operations on the input bits, each one of the encoder output sequences is representable by a linear operation in which convolutions are replaced by polynomial multiplications. For example, for a $(2, 1, m)$ convolutional code the encoding equations are expressed in terms of a delay operator D as follows:

$$\boldsymbol{v}^{(0)}(D) = \boldsymbol{u}(D) \boldsymbol{g}^{(0)}(D),$$

$$\boldsymbol{v}^{(1)}(D) = \boldsymbol{u}(D) \boldsymbol{g}^{(1)}(D),$$

in which the information sequence is denoted as

$$\boldsymbol{u}(D) = u_0 + u_1 D + u_2 D^2 + \cdots,$$

and

$$v^{(0)}(D) = v_0^{(0)} + v_1^{(0)}D + v_2^{(0)}D^2 + \cdots,$$

$$v^{(1)}(D) = v_0^{(1)} + v_1^{(1)}D + v_2^{(1)}D^2 + \cdots,$$

denote the encoder output sequences or encoded sequences, and the generator polynomials are denoted as

$$g^{(0)}(D) = g_0^{(0)} + g_1^{(0)}D + g_2^{(0)}D^2 + \cdots + g_m^{(0)}D^m,$$

$$g^{(1)}(D) = g_0^{(1)} + g_1^{(1)}D + g_2^{(1)}D^2 + \cdots + g_m^{(1)}D^m.$$

The facts that a convolutional encoder performs linear operations on its inputs and that $u^{(i)}(D)$, $0 \leq i \leq k - 1$, denotes the ith input sequence and $v^{(j)}(D)$, $0 \leq j \leq n - 1$, denotes the jth output sequence, allow the corresponding generator polynomial to be interpreted as the transfer function relating the ith input and the jth output.

By considering a convolutional encoder with k inputs and n outputs, there will be $k \cdot n$ transfer functions. These $k \cdot n$ transfer functions can be represented simultaneously by a transfer function matrix, also known as the generator polynomial matrix, denoted as $\mathbf{G}(D)$, i.e.

$$\mathbf{G}(D) = \begin{bmatrix} g_0^{(0)}(D) & g_0^{(1)}(D) & \cdots & g_0^{(n-1)}(D) \\ g_1^{(0)}(D) & g_1^{(1)}(D) & \cdots & g_1^{(n-1)}(D) \\ \vdots & \vdots & \ddots & \vdots \\ g_{k-1}^{(0)}(D) & g_{k-1}^{(1)}(D) & \cdots & g_{k-1}^{(n-1)}(D) \end{bmatrix}, \tag{4.3}$$

in which $g_i^{(j)}(D)$ denotes the generator polynomial relating the encoder ith input with its jth output. The encoding equations for a (n, k, m) convolutional code can be expressed in terms of the transfer function matrix as

$$\mathbf{V}(D) = \mathbf{U}(D)\mathbf{G}(D), \tag{4.4}$$

in which $\mathbf{U}(D) = [u^{(0)}(D), u^{(1)}(D), \ldots, u^{(k-1)}(D)]$ denotes a k-tuple in the input sequence and in which $\mathbf{V}(D) = [v^{(0)}(D), v^{(1)}(D), \ldots, v^{(n-1)}(D)]$ denotes the corresponding output sequence.

Example 4.5

(a) *The generator polynomial matrix G(D) for the FIR encoder in Figure 4.3 is*

$$G(D) = \begin{bmatrix} 1 + D^2 & 1 + D + D^2 & 0 \\ D^2 & D + D^2 & 1 \end{bmatrix}.$$

and

(*b*) *The generator polynomial matrix $G(D)$ for the FIR encoder in Figure 4.4 is*

$$\boldsymbol{G}(D) = \begin{bmatrix} 1 + D^2 + D^3 & 1 + D + D^2 + D^3 \end{bmatrix}.$$

Consider a (n, k, m) convolutional code with the polynomial generator matrix given in (4.3).

DEFINITION 4.4 *The constraint length for the ith encoder input as specified by its polynomial generator matrix is defined as*

$$v_i = \max_{0 \le j \le n-1} \{\deg \mathbf{g}_i^{(j)}(D)\}, \quad 0 \le i \le k - 1.$$

DEFINITION 4.5 *The total constraint length is defined as the sum of the constraint lengths for all ith inputs*

$$v = \sum_{i=0}^{k-1} v_i.$$

DEFINITION 4.6 *The memory m of the polynomial generator matrix is defined as the maximum constraint length, i.e.*

$$m = \max_{(0 \le i \le k-1)} \{v_i\}.$$

In other words, for a (n, k, m) convolutional code with $k > 1$, the ith shift-register contains v_i past information symbols. The coordinates in the vector $(v_1, v_2, \ldots, v_k)$ represent the respective lengths of each one of the k encoder shift-registers, i.e. the ith shift-register has v_i memory elements.

The binary m-tuple

$$(u_{l-1}^{(0)} u_{l-2}^{(0)} \cdots u_{l-v_1}^{(0)} u_{l-1}^{(1)} u_{l-2}^{(1)} \cdots u_{l-v_2}^{(1)} \cdots u_{l-1}^{(k-1)} u_{l-2}^{(k-1)} \cdots u_{l-v_k}^{(k-1)})$$

represents the *state* of the encoder at time instant l when the inputs to the encoder are

$$u_l^{(0)} u_l^{(1)} \cdots u_l^{(k-1)}.$$

It follows that a binary convolutional encoder with memory v has a total of 2^v states.

Example 4.6 *For a $(n, 1, m)$ convolutional code it follows that $v = v_1$ and that the encoder state at time instant l is*

$$(u_{l-1} u_{l-2} \cdots u_{l-m}).$$

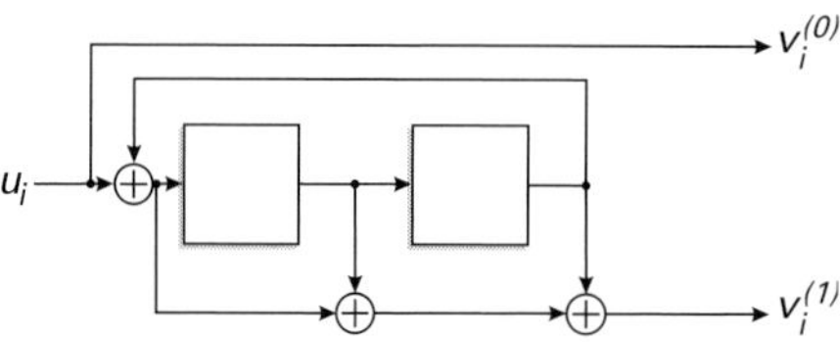

Figure 4.5   Encoder for a binary RSC code of rate 1/2

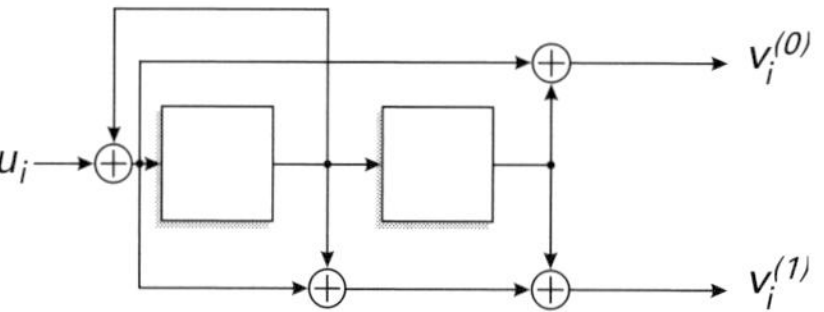

Figure 4.6   Encoder for a binary recursive non-systematic convolutional code of rate 1/2

4.5.3   Recursive systematic convolutional (RSC) codes

RSC codes (Heegard and Wicker, 1999, Berrou and Glavieux, 1996), also referred as convolutional codes with an IIR, may result from FIR convolutional codes as shown next. Figure 4.5 shows an encoder for a binary RSC code of rate $\frac{1}{2}$. Figure 4.6 shows an encoder for a binary recursive non-systematic convolution code of rate $\frac{1}{2}$. In a systematic (n, k, m) convolutional code every n-digit output sub-block contains k digits, which are identical to the k digits in the input information sequence that produced that sub-block. The polynomial generator matrix of a systematic convolutional code has the form

$$\mathbf{G}(D) = [\mathbf{I} \quad \mathbf{P}(D)],$$

in which $\mathbf{I}$ denotes the identity $k \times k$ matrix and $\mathbf{P}(D)$ denotes a $k \times (n - k)$ matrix. In order to determine the systematic form associated with a given polynomial generator matrix of a convolutional code it is necessary first to define the equivalent polynomial generator matrix.

DEFINITION 4.7 *Two polynomial generator matrices are equivalent if they generate the same convolutional code.*

Example 4.7   *Consider the polynomial generator matrix associated with the encoder in Figure 4.10, i.e.*

$$\mathbf{G}(D) = \left[1 + D^2 \quad 1 + D + D^2 \right].$$

The output produced by this encoder is given by

$$V(D) = U(D)G(D), \tag{4.5}$$

$$= U(D)\begin{bmatrix} 1+D^2 & 1+D+D^2 \end{bmatrix},$$

$$= U(D)(1+D^2)\begin{bmatrix} 1 & \dfrac{1+D+D^2}{1+D^2} \end{bmatrix}, \tag{4.6}$$

$$= U'(D)\begin{bmatrix} 1 & \dfrac{1+D+D^2}{1+D^2} \end{bmatrix}, \tag{4.7}$$

$$= U'(D)G_1(D), \tag{4.8}$$

in which

$$G_1(D) = \begin{bmatrix} 1 & \dfrac{1+D+D^2}{1+D^2} \end{bmatrix}$$

$$U'(D) = U(D)T(D)$$

$$T(D) = (1+D^2).$$

The set of output sequences $V(D)$ may be obtained by either (4.5) or (4.8). Alternatively, two polynomial generator matrices, $G(D) = T(D)G_1(D)$ and $G_1(D)$, are equivalent if the matrix $T(D)$ has an inverse. Matrix $G_1(D)$ in the previous example is in systematic form. However, the entries in this matrix are not polynomials but rather are rational functions. The parity sequence at the output in (4.6) results from multiplying the input sequence by the polynomial $(1+D+D^2)$ and dividing by the polynomial $(1+D^2)$. Polynomial multiplication and polynomial division are easily implementable with shift-registers (Peterson and Weldon, 1972).

An RSC encoder has rational functions in the variable D with binary coefficients as elements of its polynomial generator matrix. Equivalently, an RSC encoder has polynomial ratios with binary coefficients as entries in its polynomial generator matrix. As an example, the transfer function

$$\frac{f_0 + f_1 D + \cdots + f_m D^m}{1 + q_1 D + \cdots + q_m D^m}$$

can be implemented with the structure shown in Figure 4.7 (Johannesson and Zigangirov, 1999). The output in Figure 4.7 is a linear function of its input and of the contents in the shift-register. At time instant j the output v_j is given by

$$v_j = \sum_{i=0}^{m} f_i w_{j-i}.$$

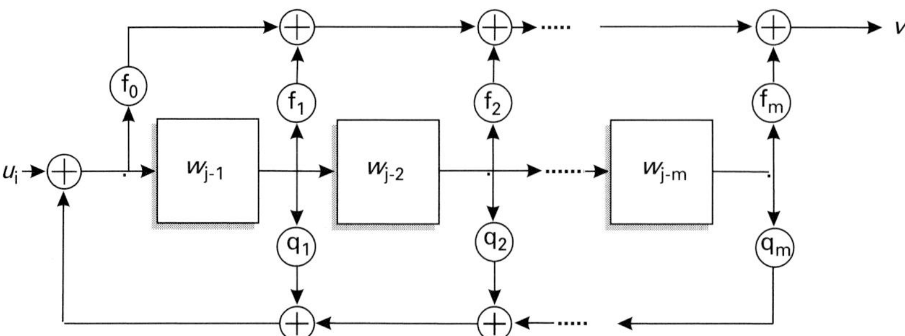

Figure 4.7   This structure shows an implementation of a rational transfer function

By making use of the polynomial representation it follows that

$$v(D) = \sum_{j=-\infty}^{\infty} v_j D^j = \sum_{j=-\infty}^{\infty} \sum_{i=0}^{m} f_i w_{j-i} D^j \tag{4.9}$$

$$= \sum_{k=-\infty}^{\infty} \left(\sum_{i=0}^{m} f_i D^i \right) w_k D^k = f(D)w(D), \tag{4.10}$$

in which k was used to replace $j - i$, with

$$f(D) = f_0 + f_1 D + \cdots + f_m D^m,$$

and

$$w(D) = \sum_{k=-\infty}^{\infty} w_k D^k.$$

From Figure 4.7 it follows that

$$w_j = u_j + \sum_{i=1}^{m} q_i w_{j-i}.$$

Assuming that $q_0 = 1$, it follows that

$$u_j = \sum_{i=0}^{m} q_i w_{j-i},$$

and in analogy with the steps taken in (4.10) it follows that

$$u(D) = q(D)w(D), \tag{4.11}$$

in which

$$u(D) = \sum_{j=-\infty}^{\infty} u_j D^j$$

and

$$q(D) = 1 + q_1 D + \cdots + q_m D^m.$$

By combining (4.10) and (4.11) it follows that

$$v(D) = u(D)\frac{f(D)}{q(D)} = u(D)\frac{f_0 + f_1 D + \cdots + f_m D^m}{1 + q_1 D + \cdots + q_m D^m}. \tag{4.12}$$

Let

$$g(D) = \frac{f(D)}{q(D)}.$$

It then follows from (4.12) that

$$v(D) = u(D)g(D).$$

The function $g(D)$ is called a rational transfer function. In general, a matrix $G(D)$ in which the entries are rational functions is called a matrix of rational transfer functions.

Example 4.8 *Matrices of rational transfer functions for the encoders in Figure 4.5 and in Figure 4.6 are, respectively,*

$$\mathbf{G}(D) = \begin{bmatrix} 1 & \dfrac{1 + D + D^2}{1 + D^2} \end{bmatrix}$$

and

$$\mathbf{G}(D) = \begin{bmatrix} \dfrac{1 + D^2}{1 + D} & \dfrac{1 + D + D^2}{1 + D} \end{bmatrix}.$$

4.5.4 Representation of convolutional codes

State-transition diagram

The state-transition diagram or simply the *state diagram* for a convolutional code is a graph consisting of nodes and branches. The nodes represent the states of the encoder and the branches, denoted by lines with arrows, represent the transitions among the states. Each branch in the state diagram has a label of the form $\mathbf{X}/\mathbf{Y}$, in which $\mathbf{X} = X_1 X_2 \ldots X_k$ denotes the input k-tuple and $\mathbf{Y} = Y_1 Y_2 \ldots Y_n$ denotes the corresponding encoder output n-tuple. Given the current encoder state, the input information sequence determines the path followed through the state diagram and the corresponding output sequence. Each new k-tuple block of information digits fed to the encoder input causes a state transition in the state diagram. It follows that there are 2^k branches leaving each state, corresponding to each distinct k-tuple block of information digits.

A convolutional encoder can be represented by a finite state machine. The current encoder shift-register contents and feedback connections determine the mapping relating the next k-tuple block of information digits and the corresponding n-tuple block of output digits. The current encoder state and the current encoder output are uniquely determined by the previous encoder state and the current encoder input. The encoder makes a state transition when it is fed with a k-tuple block of information digits.

Example 4.9 *For a $(n, 1, m)$ binary convolutional code there are only two branches leaving each state, each branch labeled by an input/output pair.*

Example 4.10 *Consider the encoder illustrated in Figure 4.4. The corresponding state diagram is shown in Figure 4.8, in which the branch labels represent input/output pairs. The state diagram has eight states labeled as 000, 001, 010, 011, 100, 101, 110 and 111, in which each state represents the contents of the encoder shift-register. There are two branches leaving each state, corresponding to the two possible values for the input message digit. There are also two branches entering each state. Being a typical finite state machine, transitions between states are constrained by the feedback connections. Given the current state $X_1 X_2 X_3$, the next state can be either $0 X_1 X_2$ (corresponding to a 0 at the input) or $1 X_1 X_2$ (corresponding to a 1 at the input).*

Trellis diagram

A trellis diagram is derived from a state diagram by expressing all possible state transitions and respective input/output sequences as a function of time. In general, for a (n, k, m) convolutional code and an input information sequence of length kL, there are 2^k branches leaving and 2^k branches entering each state. Furthermore, there are 2^{kL} distinct paths through the trellis corresponding to 2^{kL} code words.

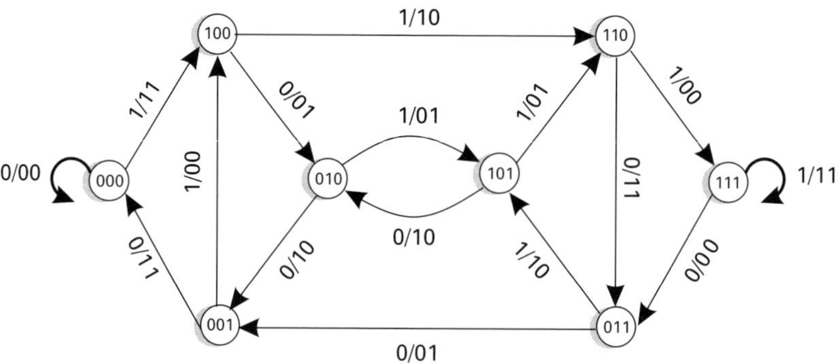

Figure 4.8   State diagram for the encoder illustrated in Figure 4.4

Example 4.11 *Consider the trellis diagram in Figure 4.9, corresponding to the encoder illustrated in Figure 4.10, with polynomial generator matrix*

$$\mathbf{G}(D) = \begin{bmatrix} 1 + D^2 & 1 + D + D^2 \end{bmatrix},\tag{4.13}$$

and an information sequence of length $L = 5$. The trellis diagram has $L + m + 1$ time units labeled from 0 to $L + m$ in Figure 4.10. The states are denoted as $00, 01, 10, 11$. The branch labels represent encoder outputs. The dotted lines represent input symbols that are 1s and continuous lines represent 0 inputs respectively. The bold face numbers at the bottom row represent time intervals or depth of the trellis.

Consider that 00 is the encoder initial state and that 00 is also the final state, the the first m time units correspond to the decoder starting at the 00 state and the last m time units correspond to the return of the encoder to the 00 state. This implies that not all states can be reached in the first m or in the last m time units. However, all states situated m time units after encoding started (00 state) and m time units before encoding is ended (00 state) can be reached, in accordance with the state diagram transition rules. At each state, there are two branches entering and two branches leaving that state. A branch denoted by a dotted line corresponds to a 1 as input symbol and a branch denoted by a continuous line corresponds to a 0 as input symbol. Each branch is labeled with an n-digit output $\mathbf{v}_i$ and corresponding to each one of the 2^L code words there is a unique path of length $N = n(L+m)$ digits through the trellis. The information sequence $\mathbf{u} = (11101)$

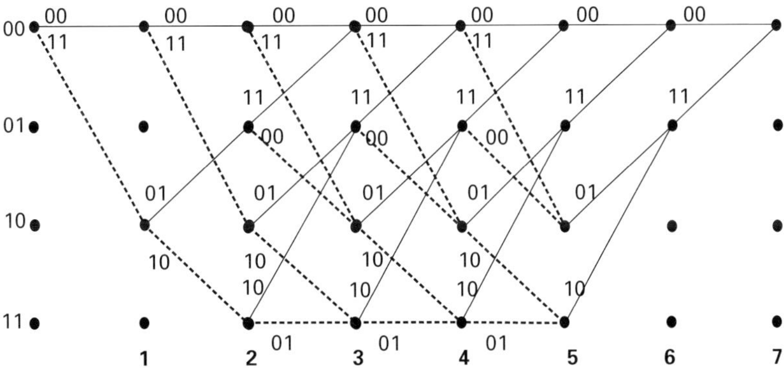

Figure 4.9 Trellis diagram for the encoder shown in Figure 4.10

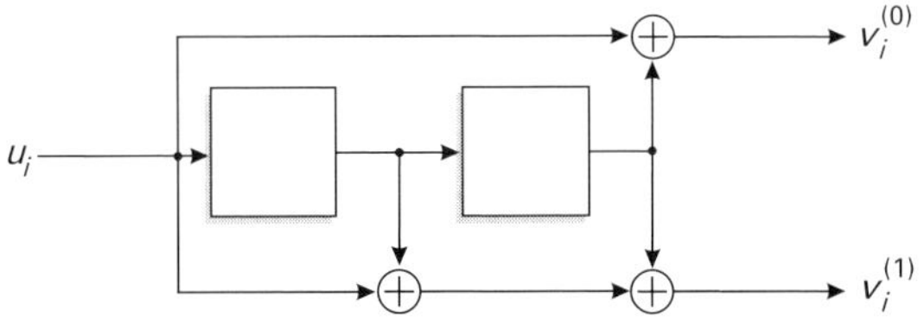

Figure 4.10 Non-systematic FIR encoder, used to construct the trellis in Figure 4.9

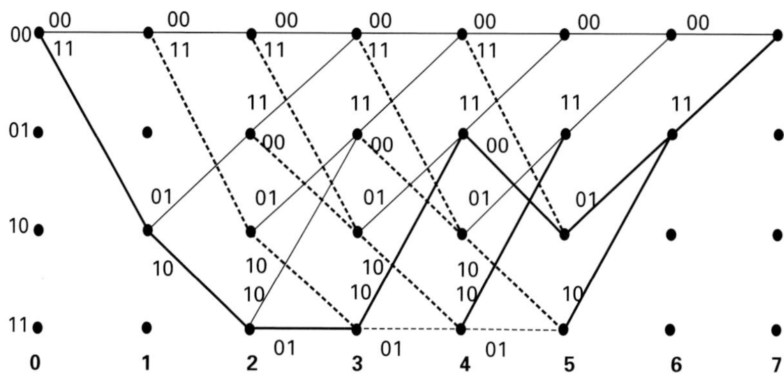

Figure 4.11   The sequence of labels (11 10 01 10 00 01 11) is indicated by thick lines

is used as an example in Figure 4.11 associated with the code word 11 10 01 10 00 01 11,
produced by the input information sequence (1 1 1 0 1); *it is shown emphasized by thick
lines.*

4.5.5   Decoding convolutional codes

At present, in most practical applications of convolutional codes the Viterbi algorithm
(VA) (Viterbi, 1985) is employed for decoding. This preference comes from the fact
that the bit error rate at the output of a Viterbi decoder is considered adequate in
such applications. Lower bit error rates, however, may call for a sequential decoder.
As mentioned earlier, the VA performs maximum likelihood decoding and benefits from
the trellis structure of convolutional codes. Given a sequence of convolutionally encoded
symbols contaminated by noise, at the decoder the VA estimates the most likely sequence
path through the corresponding code trellis. Due to the scope of this chapter and the lack
of space to treat the decoding of convolutional codes at its deserved depth, no further
details of the VA will be given. However, the interested reader now has available a wealth
of publications covering the VA, for example, Lin and Palais (1986), Wicker (1995) and
Johannesson and Zigangirov (1999).

4.6   Concatenated codes

The technique called concatenation (Forney, 1966) was introduced by Forney and
finds application in various digital television standards, as described in the remaining
part of this chapter. Forney's concatenation technique is nowadays referred as *serial
concatenation*. A serial concatenation coding scheme, as depicted in Figure 4.12,
consists, on the transmitter side, of a cascade of an outer encoder for a block code capable
of correcting byte errors, usually an RS code, with an interleaver (Ramsey, 1970, Forney,

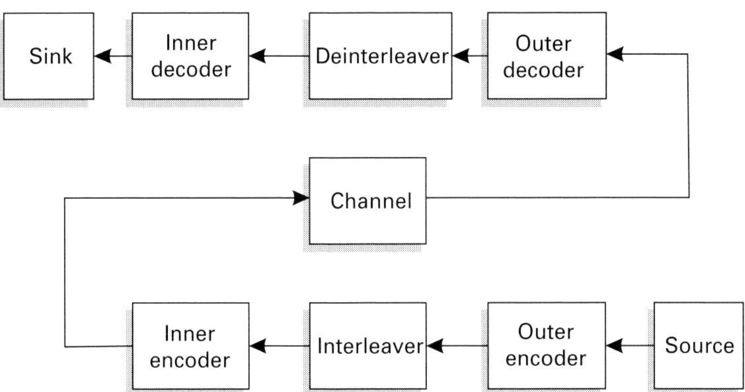

Figure 4.12   Serial concatenation coding scheme

1971) which is followed by an inner encoder for a convolutional code. The convolutional code is employed to combat uniformly distributed errors; however, a decoding failure may result in bursts of errors being delivered at the inner decoder output. On the receiver side, after being processed by the inner decoder, the bits pass through a deinterleaver, the function of which is to spread the bursts of errors on a byte basis in order to improve the burst error-correction capability of the outer decoder.

More recently (Song *et al.*, 2007), serial concatenation employing two block codes has been described, but with a different perspective. The inner code is an LDPC code, capable of correcting most of the errors but suffering from an *error-floor* (Lin and Palais, 1986) limitation, and the outer code is a BCH code, capable of correcting practically all the errors that escaped correction by the LDPC code.

The coding scheme known as turbo coding (Berrou *et al.*, 1993) employs a *parallel concatenation* but that is not analyzed in this book.

4.7   Error correction in digital television standards

The following digital television broadcast systems are described in detail in Chapters 6, 7, 8, 9, 10. The main interest at this point is to call the reader's attention to the specific error-correcting codes employed by such systems (Drury *et al.*, 2001, Francis and Green, 2007).

4.7.1   Digital Video Broadcast Terrestrial (DVB-T) system

For error protection in the DVB-T system a concatenated forward error correction (FEC) scheme is used, as illustrated in Chapter 7. The randomized transport stream (TS) packets are encoded by an RS encoder for a $(255, 239)$ RS code, the symbols of which are 8-bit bytes, i.e. the RS code is defined over $GF(2^8)$. This means that $255 - 239 = 16$ check symbols are appended to the information bytes. This code allows correction of up to $16/2 = 8$ symbols of corrupted data in a block of 255 symbols representing a code word that is a total of up to 64 bits within 8 corrupted bytes. As the TS packets are only 188 bytes long, the first 51 bytes are set to zero, but are not transmitted. Therefore, the actual RS code is a shortened (204, 188) code. The encoded bytes are then applied to a convolutional interleaver (Forney, 1971). Inner error protection is provided by a convolutional code. A non-systematic convolutional code with code rate $\frac{1}{2}$ is used, with generator polynomials $G_1 = (1, 1, 1, 1, 0, 0, 1)$, or 171 in octal, and $G_2 = (1, 0, 1, 1, 0, 1, 1)$, or 133 in octal, with output denoted by $\mathbf{XY} = (X_1 Y_1, X_2 Y_2, X_3 Y_3 \ldots)$. In order to make transmission more efficient, according to the noise level in the channel the number of bits transmitted may be varied by discarding selected bits from the encoder output. The technique for discarding encoded bits is called puncturing (Cain $et\ al.$, 1979).

Example 4.12 *The sequence of encoded bits $X_1 Y_1 X_2 Y_2 X_3 Y_3 X_4 Y_4 \ldots$ appears at the encoder output of a rate $\frac{1}{2}$ convolutional code. A rate $\frac{2}{3}$ output is produced from the rate $\frac{1}{2}$ convolutional code by puncturing symbols X_{2i}, $i = 1, 2, \ldots$ at the encoder output, thus producing the punctured sequence $X_1 Y_1 Y_2 X_3\ Y_3 Y_4 \ldots$ A pair of consecutive sub-blocks corresponds to two information symbols that, after puncturing, will contain three symbols thus producing a $\frac{2}{3}$ code rate.*

A $\frac{3}{4}$ code rate is produced from the rate $\frac{1}{2}$ encoder output by puncturing positions X_{2+3i} and Y_{3+3i} for $i = 0, 1, 2, \ldots$, to produce the punctured coded sequence $X_1 Y_1 Y_2 X_3 X_4 Y_4 Y_5 X_6 \ldots$ which means that for every three input bits, only four output bits are kept rather than the six bits that are actually generated.

Table 4.5 shows the recommended code rates produced by puncturing. The punctured decoder output is alternately read as I or Q inputs, for example $\frac{2}{3}$, for the rate in Table 4.5 the punctured output sequence is $X_1 Y_1 Y_2 X_3\ Y_3 Y_4 \ldots$ and the values of I and Q are read as follows.

$$
\begin{array}{cccccc}
I & Q & I & Q & I & Q & \ldots \\
X_1 & Y_1 & Y_2 & X_3 & Y_3 & Y_4 & \ldots
\end{array}
$$

The outer encoder is followed by an interleaver, as explained in Chapter 7, the output of which feeds the inner encoder input.

The inner decoding function is performed by a Viterbi decoder. The Viterbi decoder has a constraint length of 7 and uses polynomials 171 (octal) and 133 (octal). To handle the puncture rates, as specified by the standard, puncturing is performed externally to the Viterbi decoder and, in place of the missing symbols, null symbols are inserted along with an erase input to indicate the position of the null symbols. The format of the input data to the Viterbi decoder can be either hard or soft values.

The soft coding format tends to give a better bit error rate performance, because it gives a confidence value for each bit ranging from a maximum confidence zero to a maximum confidence one.

The computation of the soft data is done using the log likelihood ratio (LLR). The I and Q data from the demodulator are used to create the confidence value. Decoding is

Table 4.5. Parameters of punctured convolutional codes

No.	Code rate	I and Q	d_{free}
1	$\frac{1}{2}$	$I = X_1; Q = Y_1$	10
2	$\frac{2}{3}$	$I = X_1 Y_2 Y_3; Q = Y_1 X_3 Y_4$	6
3	$\frac{3}{4}$	$I = X_1 Y_2; Q = Y_1 X_3$	5
4	$\frac{5}{6}$	$I = X_1 Y_2 Y_4; Q = Y_1 X_3 Y_5$	4
5	$\frac{7}{8}$	$I = X_1 Y_2 Y_4 Y_6; Q = Y_1 Y_3 X_5 X_7$	3

done by calculating the minimum distance from the received symbol to all the possible transmitted symbols and then calculating the probability that a particular bit is either a 0 or a 1. The data from the Viterbi decoder are deinterleaved before the decoded data block is fed to the RS decoder. The RS decoder for the $(204, 188)$ code indicates the number of errors and how many bits have changed from ones to zeroes and vice versa. The transport packets are then recovered.

4.7.2 Digital Video Broadcast Satellite (DVB-S) system

The DVB-S system (ETSI, 1997a) uses an error protection scheme similar to the DVB-T system. The outer code is a shortened $(204, 188)$ RS code, allowing the correction of up to a maximum of 8 bytes for each 188-byte packet. Inner error protection is provided by convolutional encoding as seen earlier. The DVB-S system has a number of puncture schemes based on a rate $\frac{1}{2}$ code and these are $\frac{2}{3}$, $\frac{3}{4}$, $\frac{5}{6}$, and $\frac{7}{8}$.

The I and Q signals (see Chapter 7) are converted to soft decisions for the Viterbi decoder, also called the inner decoder. Even though the Viterbi decoder can perform hard-decision decoding, a better bit error rate performance is achieved when soft decisions are employed. The decoded output of the Viterbi decoder is deinterleaved and followed by the RS decoder, or outer decoder of the serial concatenation coding scheme. The data are recovered from the random pattern introduced in the transmitter to create the original video/data packets.

4.7.3 Digital Video Broadcast Satellite (DVB-S2) system

DVB-S2 (ETSI, 2006a) constitutes the second generation of the DVB-S scheme for satellite communications. DVBS-2 is backward compatible with DVB-S, has a system capacity increased by 40% and enhanced error protection, in which the latter is achieved by the use of recent results in error-correcting coding techniques.

The FEC coding for DVBS-2 uses a serial concatenation of a BCH code with an LDPC code, and a bit interleaver. The FEC block length is either 64 800 bits or 16 200 bits. The BCH code is used to reduce error floors at low bit error rates. The recommended code rates are $\frac{1}{3}$, $\frac{2}{5}$, $\frac{1}{2}$, $\frac{3}{5}$, $\frac{2}{3}$, $\frac{3}{4}$, $\frac{4}{5}$, $\frac{5}{6}$, $\frac{8}{9}$, and $\frac{9}{10}$. This type of coding scheme is capable of ensuring error protection close to the Shannon limit (MacKay and Neal, 1997). The error correction is improved so much that DVBS-2 claims the signal-to-noise ratio (SNR) is very close the theoretical best limit or Shannon limit of this transmission channel.

4.7.4   Digital Video Broadcast Cable (DVB-C) system

The coding part of the DVB-C (ETSI, 1998) system (see Chapter 7) relies on an RS encoder and interleaver at the transmitter, but no convolutional code is employed, i.e. it coincides with the DVB-S system up to the interleaver output. On the receiver side, after demodulation, the I and Q signals go through a matched filter and through to the FEC decoding sections consisting of the deinterleaver and the RS decoder.

4.7.5   Digital Video Broadcast Hand-held (DVB-H) system

The DVB-H system (ETSI, 2004) provides technical specifications for extending broadcast services to hand-held receivers. The DVB-H standard is based on the DVB-T terrestrial specification. A distinguishing feature of the DVB-H system is Multiprotocol-encapsulation (MPE), described in Chapter 7, which when combined with FEC provides the required level of protection against errors. The MPE–FEC combination uses an FEC frame consisting of 255 columns and up to 1024 rows. The frame contains two sections: one section is called the *application data table*, using 191 columns, and the other section is called the *RS data table*, using the remaining 64 columns for parity bytes. The application table is where the input bytes (IP data) are stored and any unused positions are padded with zeros. The input digits are written to the application table on a column-by-column basis. The (255, 191) RS code is applied on a row-by-row basis. The DVB-H specification allows for puncturing parity bytes, in which case not all parity bytes need to be transmitted.

The FEC frame utilizes the (255, 191) RS code with 8-bit byte symbols, which allows the correction of up to 32 random erroneous bytes in a received word of 255 bytes. When reliable erasure information is used, such as provided by the cyclic redundancy check (CRC32) of the MPE and/or MPE FEC sections, the RS code allows the correction of up to 64 random erroneous bytes. Used in conjunction with CRC32, the bytes fed to the RS decoder are marked as reliable or unreliable. Thus, the positions of the most probable erroneous bytes are known. This information is sent to the RS decoder via the erasures input, to indicate to the RS decoder the location of the errors.

4.7.6   Advanced Television Systems Committee (ATSC)

An RS code is used to provide burst noise protection by appending 20 bytes at the end of the 187 data packets for the (207, 187) RS code. The data are interleaved before being applied to a rate $\frac{2}{3}$ trellis encoder. For trellis coding, each incoming byte is split up into a stream of four 2-bit messages. For every 2-bit message entering the encoder, 3 bits are output based on the past history of previous incoming 2-bit messages and on the convolutional encoder employed. These 3-bit output sub-blocks represent the eight level symbols for the 8-VSB (vestigial sideband) modulation employed.

The trellis decoder uses the received 3-bit sub-blocks employed in 8-VSB modulation to reconstruct the original data stream from one 2-bit message to the next, and uses the signal's past behavior to correct eventual errors. The remaining blocks are the deinterleaver and the RS decoder, which work in a similar manner to previously-mentioned systems and employ codes that match their transmitter counterparts.

4.7.7 Integrated Services Digital Broadcasting (ISDB)

ISDB is the acronym used for the set of Japanese standards that covers terrestrial (ISDB-T) (ARIB, 2004b), satellite (ISDB-S) (ARIB, 2004a) and cable (ISDB-C) communication. Multiple blocks are remultiplexed into a single TS. The TS is first processed through an encoder for the $(204, 188)$ RS code. A key feature of ISDB is hierarchical transmission. The TS packets are divided into sets of packets according to program information, into a maximum of three parallel processing sections, known as hierarchical separation. Each section performs energy dispersal, byte interleaving, and convolutional encoding. The convolutional encoder could have different coding rates and a different modulation scheme.

ISDB-S also defines the 204-byte code length used in RS encoding. Reception uses similar concatenated FEC blocks to other schemes, such as DVB, but with hierarchical modes needing to be determined and the correct FEC parameters used appropriately.

4.7.8 International System for Digital Television (ISDTV)

ISDTV is the name of the digital television system adopted in Brazil. While differing in some important aspects from the Japanese standard, for coding and modulation ISDTV, also known as ISDB-Tb, relies heavily on the ISDB standard and employs the same technology as used by ISDB-T for coding and modulation of digital television signals. ISDTV is covered in detail in Chapter 9.

4.7.9 Chinese Digital Television Terrestrial Broadcasting (DTMB) system

The DTMB (Song *et al.*, 2007) FEC scheme employs concatenated coding with a BCH code as the outer code and LDPC codes as inner codes. The BCH code is a $(762, 752)$ code derived from the $(1023, 1013)$ BCH code by deleting the first 261 information positions. The encoding procedures for shortened BCH codes is found, for example, in (Peterson and Weldon 1972). Equivalently, an encoder for the $(1\,023, 1\,013)$ BCH code can be used, being fed with 261 leading information digits equal to zero.

The LDPC codes adopted in the DTMB have a block length of 7493 bits. After puncturing five parity-check bits, it becomes a 7488-bit block. Three distinct FEC coding rates, namely, 0.4, 0.6, and 0.8, are used and are obtained after making the appropriate changes in the original LDPC code. A structured LDPC code is employed the generator matrix G of which is as follows.

Table 4.6. The three LDPC codes specified by DTMS

(r, s)	(n, k)	Code rate
$(24, 35)$	$(7\,488, 3\,008)$	0.4
$(36, 23)$	$(7\,488, 4\,512)$	0.6
$(48, 11)$	$(7\,488, 6\,016)$	0.8

$$
G = \begin{bmatrix}
G_{1,1} & G_{1,2} & \cdots & G_{1,s} & I_{127} & O_{127} & \cdots & O_{127} \\
G_{2,1} & G_{2,2} & \cdots & G_{2,s} & O_{127} & I_{127} & \cdots & O_{127} \\
\vdots & \vdots & G_{i,j} & \vdots & \vdots & \vdots & \vdots & \vdots \\
G_{r,1} & G_{r,2} & \cdots & G_{r,s} & O_{127} & O_{127} & \cdots & I_{127}
\end{bmatrix},
$$

in which I_{127}, O_{127}, and $G_{i,j}$ are 127×127 matrices denoting, respectively, a unit matrix, an all-zero matrix, and a circulant matrix, for $1 \leq i \leq r$, $1 \leq j \leq s$. A particular choice of the (r, s) pair defines one LDPC code as indicated in Table 4.6.

4.7.10   Data Over Cable Service Interface Specification (DOCSIS)

The DOCSIS suite of specifications (ITU-T, 2000a) was developed at Cable Television Laboratories in Louisville, KY, USA, and has been adopted by both the ANSI-accredited standards body, the Society of Cable Telecommunications Engineers (SCTE), and the International Telecommunications Union (ITU). The main cable digital standards are J.83 A/C and J.83 B.

- **J.83 Annex A/C**: The J.83 A/C cable digital standard is used mainly in Europe. Systematic shortened RS encoding is applied to each randomized MPEG-2 transport packet. An encoder for the (259, 239) RS code is used, with extra bytes discarded after coding to result in a shortened (204, 188) RS code. The RS coding is capable of correcting up to eight erroneous bytes per code word. Convolutional interleaving based upon the Forney approach, with depth $I = 12$, is applied to the error-protected packets.
- **J.83 Annex B**: The J.83 B cable digital standard is used mainly in the USA. The FEC layer uses a concatenated coding technique with four processing layers. An encoder is used for the (128, 122) RS code, which is capable of correcting up to three symbol errors per code word, with a symbol being seven bits. A (127, 122) RS code over $GF(2^8)$ is used with an additional overall parity symbol making a (128, 122) code. The RS encoder is followed by a variable depth convolutional interleaver to provide error protection against burst noise induced errors. The interleaver operates in two separate modes or levels. When operating in level 1, a single interleaving depth $(I = 128, J = 1)$ is supported. Level 2 allows variable interleaving depth for 64- and 256-QAM. A randomizer follows the interleaver to provide a uniform distribution of the symbols in the constellation. Randomization is not carried out on an FEC frame trailer. An FEC

frame comprises 60 RS frames followed by a 42-bit sync trailer for 64-QAM and 88 RS frames followed by a 40-bit sync trailer for 256-QAM. The inner code used in this FEC encoder is a trellis coded modulator (TCM). For 64-QAM, the TCM generates five QAM symbols for every 28-bit sequence it receives. This set of 28-bits forms a trellis group which is divided into two subgroups, A and B. The TCM employs differential precoding followed by 4/5 punctured binary convolutional code on the least significant bits of the A and B subgroups. The overall code rate is 28/30 for 64-QAM. Similarly, for 256-QAM, the TCM generates five QAM symbols for every 38-bit sequence it receives. The overall code rate is 38/40 for 256-QAM.

4.7.11 Digital Multimedia Broadcasting (DMB)

Terrestrial Digital Multimedia Broadcast (T-DMB) (ETSI, 2005, 2006b) was derived from the European Digital Audio Broadcast (DAB) digital radio standard, with modifications to support multimedia reception on portable or mobile devices. RS encoding and decoding, and convolutional interleaving, constitute the error protection in T-DMB. Similarly to DVB systems, the (204, 188) RS code is used and the interleaver uses a Forney algorithm (Forney, 1971) with interleaving depth $I = 12$.

5   Digital and analog transmission systems

5.1   Introduction

Digital television signals are mainly transmitted over the air, which is also called terrestrial transmission, through cable systems, by geostationary satellites, using a microwave carrier, and over the Internet. For each transmission channel, a specific modulation scheme must be selected as the most appropriate, depending on the type of noise, power limitation, fading characteristics, transmission rate or cost.

Carrier waves are employed to allow efficient radiation of radio signals, because it is important to match the wavelength and the principal dimension of the transmitting and receiving antennas. The information is represented by an electrical signal, called a modulating signal, used to modify one or more parameters of the carrier. Appendix B presents a review of signal analysis and the use of the frequency spectrum by several services.

Modulation is the variation of one or more characteristics of the carrier waveform as a function of the modulating signal. The sinusoidal waveform is traditionally used as the carrier, and the modulation can be performed in three distinct ways:

- Amplitude modulation (AM), if amplitude is the carrier parameter that is varied. This is the oldest modulation scheme; its fatherhood was disputed by Lee DeForest and Howard Armstrong.
- Quadrature modulation (QUAM), if both the amplitude and the phase of the carrier are varied simultaneously. Single sideband (SSB) was the first QUAM scheme.
- Angle modulation, if either the phase (PM) or the frequency (FM) is the carrier parameter that changes. The first FM modulator was designed by Howard Armstrong in 1933.

5.2   Amplitude modulation (AM)

AM is a modulation scheme in which the instantaneous amplitude of the carrier wave varies in accordance with the modulating signal (Alencar and da Rocha, 2005). AM, also known as double-sideband amplitude modulation (AM-DSB), is widely used. The scheme has advantages such as economy, simplicity of transmitter and receiver design, and easy maintenance. The AM carrier is a sinusoid represented as $c(t) = A\cos(\omega_c t + \phi)$,

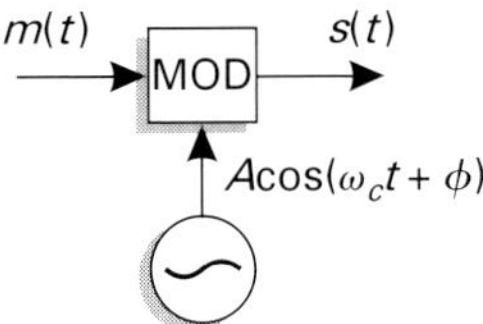

Figure 5.1   AM modulator

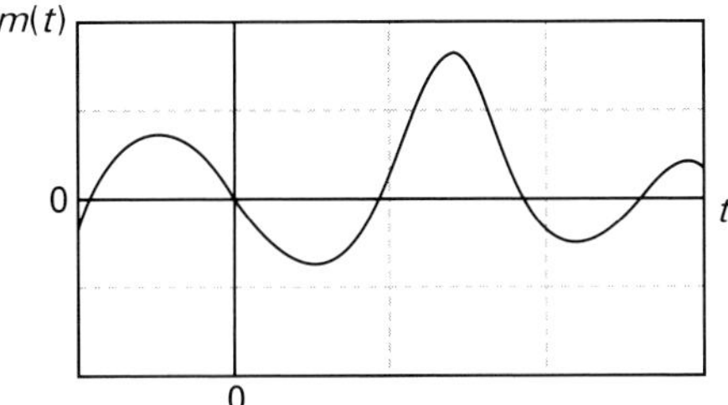

Figure 5.2   Random modulating signal $m(t)$

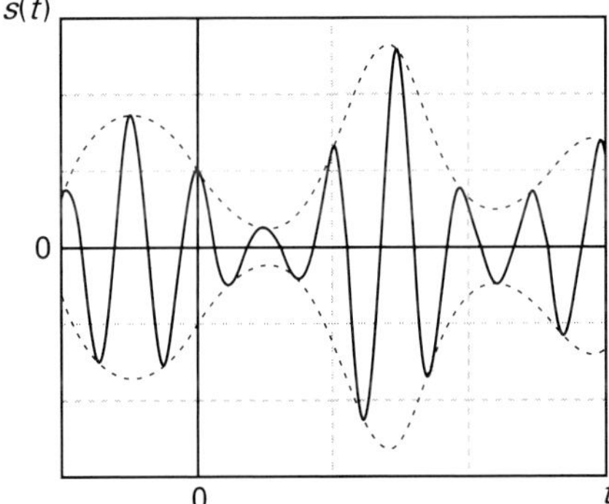

Figure 5.3   Modulated carrier for AM

in which A denotes the carrier amplitude, ω_c denotes the angular frequency (measured in radians per second, i.e. rad/s) and ϕ denotes the random carrier phase. Figure 5.1 illustrates an AM modulator.

The carrier operates at a frequency that is much higher than the frequency of the information, or modulating, signal, denoted $m(t)$, which is random in nature (Gagliardi, 1978). The random modulating signal is shown in Figure 5.2 and the modulated carrier is illustrated in Figure 5.3.

Commercial radio AM and the transmission of the video signal in some analog television standards require the carrier amplitude to vary proportionally with the instantaneous signal amplitude $m(t)$

$$a(t) = A + Bm(t) = A[1 + \Delta_{AM}m(t)], \qquad (5.1)$$

in which $\Delta_{AM} = B/A$ is the AM modulation index. The instantaneous amplitude produces the modulated waveform as follows:

$$s(t) = a(t)\cos(\omega_c t + \phi), \tag{5.2}$$

which can be written as

$$s(t) = A[1 + \Delta_{AM}m(t)]\cos(\omega_c t + \phi). \tag{5.3}$$

The modulation index indicates how strong the modulating signal is with respect to the carrier. The modulation index should not exceed 100% in order to avoid distortion in a demodulated signal whenever an envelope detector is employed. Its value is regulated by the relevant country's regulatory body.

5.2.1   Random signals and AM

It is important to point out that the mathematical treatment used in this section involves the concept of the random signal, which is developed in Appendix C. The generality of the results, derived using an analysis based on random signals, compensates for the extra effort required to understand the theory of stochastic processes. The theoretical development that follows is more elegant than the standard treatment based on deterministic signals, and approximates practical signals, which are of random nature.

Consider the modulated carrier given by (5.2), in which $a(t) = A[1 + \Delta_{AM}m(t)]$. Let the carrier phase ϕ be a random variable with a uniform distribution in the interval $[0, 2\pi]$. The signal $m(t)$ is assumed to be a stationary random process with zero mean and statistically independent of ϕ.

The modulated carrier is a random process, because it has a random phase and is modulated by a stochastic process. The stationarity of the modulated carrier, which means that its averages do not change with time, can be verified. The autocorrelation of a stationary random process, which is discussed in Appendix C, is given by

$$R_S(\tau) = E[s(t)s(t + \tau)]. \tag{5.4}$$

Substituting (5.2) into (5.4), it follows that

$$R_S(\tau) = E\left[(a(t)\cos(\omega_c t + \phi))(a(t + \tau)\cos(\omega_c(t + \tau) + \phi))\right]. \tag{5.5}$$

After replacing the sum of cosines with a product of cosines, it follows that

$$R_S(\tau) = \tfrac{1}{2}E\left[a(t)a(t + \tau)\left(\cos(\omega_c \tau) + \cos(2\omega_c t + \omega_c \tau + 2\phi)\right)\right]. \tag{5.6}$$

Using properties of the mean value, considering that $a(t)$ and ϕ are independent random variables and that the mean value of the carrier is zero, it follows that

$$R_S(\tau) = \tfrac{1}{2}R_A(\tau)\cos(\omega_c \tau), \tag{5.7}$$

in which the autocorrelation of the modulating signal $R_A(\tau)$ is defined as

$$R_A(\tau) = E[a(t)a(t + \tau)]. \tag{5.8}$$

Substitution of (5.1) for $a(t)$ in (5.8), gives

$$R_A(\tau) = E\left[A(1 + \Delta_{AM} m(t))A(1 + \Delta_{AM} m(t + \tau))\right]. \tag{5.9}$$

Using the properties of the expected value and recalling that $m(t)$ is stationary and zero mean, i.e. that $E[m(t)] = E[m(t + \tau)] = 0$, it follows that

$$R_A(\tau) = A^2[1 + \Delta_{AM}^2 R_M(\tau)], \tag{5.10}$$

in which $R_M = E[m(t)m(t + \tau)]$ represents the autocorrelation of the message signal. Finally, the autocorrelation of the amplitude modulated carrier is given by

$$R_S(\tau) = \frac{A^2}{2}\left[1 + \Delta_{AM}^2 R_M(\tau)\right]\cos\omega_c\tau. \tag{5.11}$$

The power of an AM carrier is given by the value of its autocorrelation for $\tau = 0$,

$$P_S = R_S(0) = \frac{A^2}{2}(1 + \Delta_{AM}^2 P_M), \tag{5.12}$$

in which $P_M = R_M(0)$ represents the power in the message signal $m(t)$. The power of the unmodulated carrier is given by $A^2/2$, and represents a significant portion of the total transmitted power.

The power spectral density of the AM modulated carrier can be obtained as the Fourier transform of the autocorrelation function $R_S(\tau)$. This result is known as the Wiener–Khintchin theorem, see Appendix C, i.e.

$$S_S(\omega) = \frac{1}{2\pi}\left[\frac{1}{2}S_A(\omega) * (\pi\delta(\omega + \omega_c) + \pi\delta(\omega - \omega_c))\right]. \tag{5.13}$$

It can be obtained with a spectrum analyzer, as shown in Figure 5.4.

Applying the impulse filtering property, it follows that

$$S_S(\omega) = \tfrac{1}{4}\left[S_A(\omega + \omega_c) + S_A(\omega - \omega_c)\right], \tag{5.14}$$

in which $S_A(\omega)$ is the Fourier transform of $R_A(\tau)$.

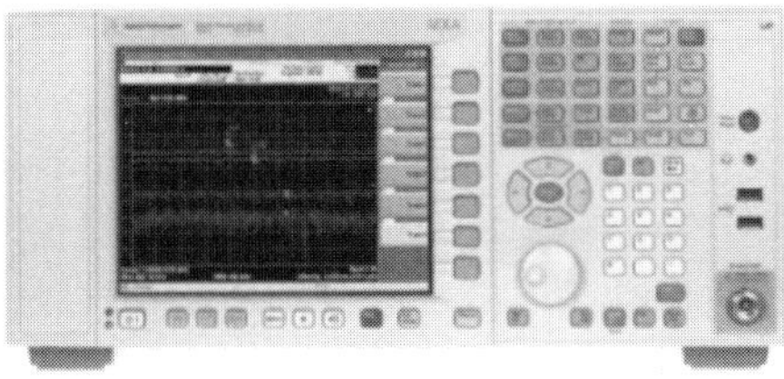

Figure 5.4   Spectrum analyzer (©Agilent Technologies, Inc. 2008. Reproduced with permission, courtesy of Agilent Technologies, Inc.)

The power spectral density of the modulating signal can be derived by writing the expression for $R_A(\tau)$ and then calculating its Fourier transform. Thus

$$S_A(\omega) = A^2[2\pi\delta(\omega) + \Delta_{AM}{}^2 S_M(\omega)], \tag{5.15}$$

in which $S_M(\omega)$ is the power spectral density of the message signal.

Finally, the power spectral density of the modulated AM carrier is given by

$$S_S(\omega) = \frac{A^2}{4}\left[2\pi(\delta(\omega + \omega_c) + \delta(\omega - \omega_c)) + \Delta_{AM}{}^2(S_M(\omega + \omega_c) + S_M(\omega - \omega_c))\right]. \tag{5.16}$$

The power spectral densities of the message signal and of the modulating signal are shown in Figure 5.5(a) and (b), respectively. The power spectral density of the modulated carrier is illustrated in Figure 5.6.

The bandwidth required for the transmission of an AM signal is twice the bandwidth of the message signal. In AM radio broadcasts the maximum frequency of the message signal is limited to 5 kHz, and consequently the AM bandwidth for commercial radio broadcast is 10 kHz. For the NTSC and PAL-M analog television broadcast systems, which use amplitude modulation vestigial sideband (AM-VSB) for transmitting the luminance signal, the bandwidth is 6 MHz.

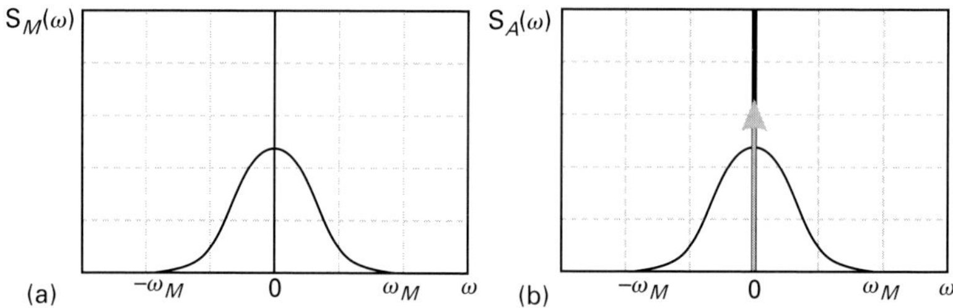

Figure 5.5　　Power spectral density the (a) message, and (b) modulating signals

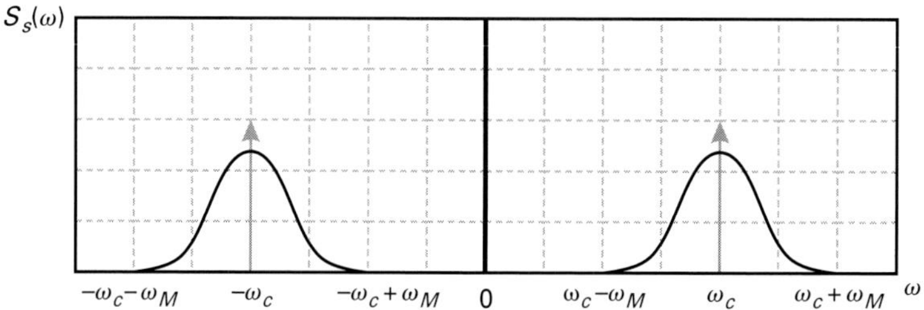

Figure 5.6　　Power spectral density of an AM signal

5.2.2 Digital AM signal

The digital AM signal, also called the amplitude shift keying (ASK) signal, can be produced by multiplying the digital modulating signal by the carrier. The ASK signal is shown in Figure 5.7.

A digital signal can be depicted using a constellation diagram, which consists of representing the modulated signal in axes which are in phase (I axis) and in quadrature (with a phase shift of $\pi/2$, or Q axis) with respect to the carrier phase. The digital signal, which amplitude modulates the carrier, can be written as

$$m(t) = \sum_{k=-\infty}^{k=\infty} m_k p(t - kT_b), \tag{5.17}$$

in which m_k represents the kth randomly generated symbol, from a discrete alphabet, $p(t)$ is the pulse shape of the transmitted digital signal, and T_b is the bit interval.

The modulated signal is given by

$$s(t) = \sum_{k=-\infty}^{\infty} m_k p(t - kT_b) \cos(\omega_c t + \phi). \tag{5.18}$$

As an example, Figure 5.8 shows the constellation diagram of a 4-ASK signal, the symbols of which are $m_k \in \{-3A, -A, A, 3A\}$. All the signal points are on the cosine axis (in phase with the carrier). There is no quadrature component in this case.

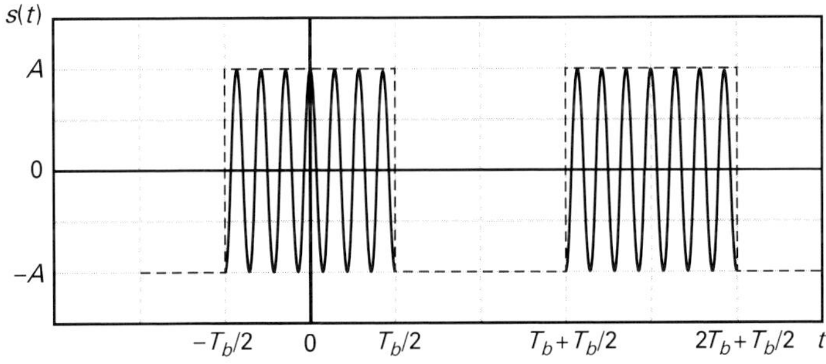

Figure 5.7   Example of a binary ASK signal

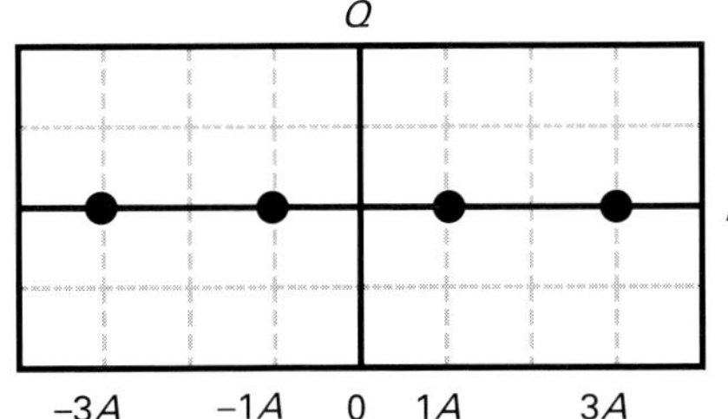

Figure 5.8   Constellation for a 4-ASK signal

For a digital modulating signal, the transmitted power is computed considering the average power per symbol. For an M-ASK signal, considering equiprobable symbols, the modulated carrier power is given by

$$P_S = \tfrac{1}{2} \sum_{k=1}^{M} m_k^2 p(m_k). \tag{5.19}$$

The probability of error for the coherent binary 2-ASK is (Haykin, 1988)

$$P_e = \frac{1}{2}\operatorname{erfc}\left(\sqrt{\frac{E_b}{N_0}}\right). \tag{5.20}$$

In (5.20), E_b is the binary pulse energy, N_0 represents the noise power spectral density and $\operatorname{erfc}(\cdot)$ is the complementary error function,

$$\operatorname{erfc}(x) = \frac{2}{\sqrt{\pi}} \int_x^{\infty} e^{-t^2}\, dt. \tag{5.21}$$

For a rectangular pulse, $E_b = A^2 T_b$, in which A is the pulse amplitude and T_b is the pulse duration.

For the M-ary scheme, the error probability is (Benedetto and Biglieri, 1999)

$$P_e = \frac{M-1}{M}\operatorname{erfc}\left(\sqrt{\frac{E_b}{N_0}\frac{3\log_2 M}{M^2-1}}\right). \tag{5.22}$$

Figure 5.9 shows the bit error probability (BEP) for an M-ASK, with $M = 2, 4, 8, 16$. The error probability increases with the number of symbols.

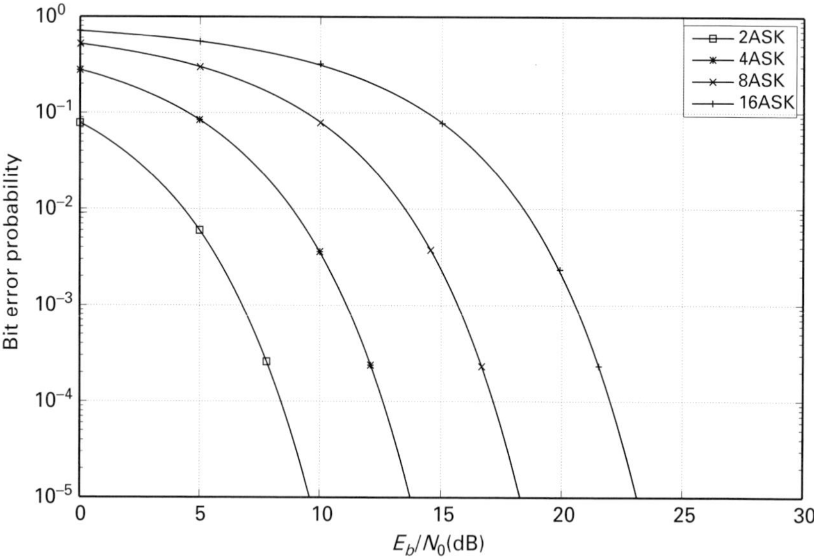

Figure 5.9   BEP for an M-ASK

5.2.3 Suppressed carrier amplitude modulation (AM-SC)

For the standard AM system most of the power is spent in transmitting the carrier. This is a waste of power if one reasons that the carrier by itself conveys no information.

In the AM-SC system the carrier is not sent as part of the AM signal and all of the transmitter power is available for transmitting information over the two sidebands. In terms of efficient use of transmitter power the AM-SC system is more efficient than an AM system. However, a receiver for an AM-SC signal is significantly more complex than an AM receiver.

The AM-SC technique essentially translates the frequency spectrum of the modulating signal by multiplying it by a sinusoid, the frequency of which has a value equal to the desired frequency translation. In other words, the original message, or modulating, signal $m(t)$ becomes $m(t)\cos(\omega_c t + \phi)$, after multiplication by the carrier.

In order to compute the power spectral density of the AM-SC signal, it is first necessary to calculate its autocorrelation function. Let $s(t) = m(t)\cos(\omega_c t + \phi)$, in which $m(t)$ is a stationary random process with zero mean, ϕ is a random variable uniformly distributed in the interval $[0, 2\pi]$ and statistically independent of $m(t)$:

$$R_S(\tau) = E[s(t)s(t + \tau)]. \tag{5.23}$$

Substituting the formula for the modulated signal, one obtains

$$R_S(\tau) = E\left[m(t)m(t + \tau)\cos(\omega_c t + \phi)\cos(\omega_c(t + \tau) + \phi)\right]. \tag{5.24}$$

Following the same procedure used for the commercial AM scheme, the autocorrelation can be written as

$$R_S(\tau) = \tfrac{1}{2}R_M(\tau)\cos\omega_c(\tau). \tag{5.25}$$

The power of the AM-SC signal can be obtained from the autocorrelation function computed for $\tau = 0$:

$$P_S = R_S(0) = \tfrac{1}{2}P_M. \tag{5.26}$$

The power spectral density is obtained by means of the Fourier transform of the autocorrelation function:

$$S_S(\omega) = \tfrac{1}{4}[S_M(\omega + \omega_c) + S_M(\omega - \omega_c)]. \tag{5.27}$$

5.2.4 AM-VSB modulation

The AM-VSB scheme is used to transmit the luminance video signal of the NTSC and PAL-M analog television standards. The process of generating an SSB signal, in order to make efficient use of the available bandwidth, has some practical problems when performed by means of a sideband filter, because in most applications a very sharp filter cut-off characteristic is required. In order to overcome this filter design difficulty, a compromise solution was established between AM-DSB and SSB, called AM-VSB (Taub and Schilling, 1971).

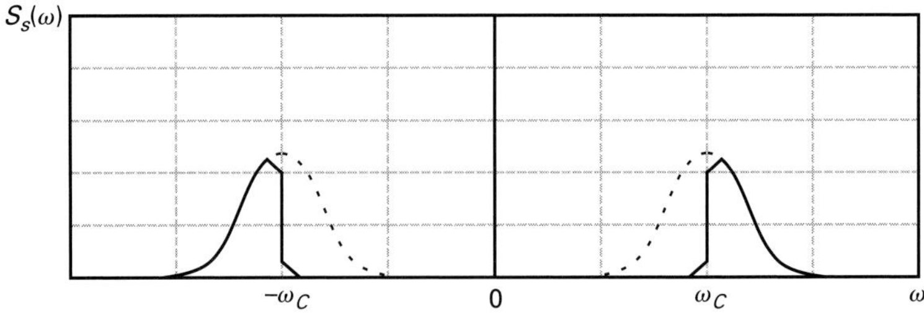

Figure 5.10   Spectra of a VSB modulated signal

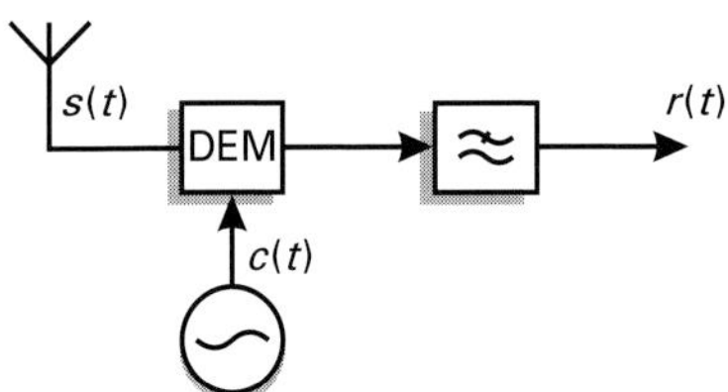

Figure 5.11   Block diagram of an AM demodulator

As shown in Figure 5.10, the AM-VSB signal can be obtained by the partial suppression of the upper (lower) sideband of an AM-SC. The AM-VSB scheme is used for transmitting television video signals. The main reasons for the choice of an AM-VSB system are: bandwidth reduction, practical implementation advantages, and low cost.

5.2.5   Amplitude demodulation

The modulated carrier can be synchronously demodulated to recover the message signal, as shown in Figure 5.11. The incoming signal

$$s(t) = A\left[1 + \Delta_{AM} m(t)\right] \cos(\omega_c t + \phi) \tag{5.28}$$

is multiplied (mixed) by a locally generated sinusoidal signal

$$c(t) = \cos(\omega_c t + \phi) \tag{5.29}$$

and low-pass filtered, to give

$$r(t) = \frac{A}{2}\left[1 + \Delta_{AM} m(t)\right]. \tag{5.30}$$

The DC level is then blocked to give the original signal $m(t)$ multiplied by a constant. The demodulated signal is then sent to the amplification stage.

When the phase of the local oscillator is not the same as the received signal phase, the signal is attenuated, which means that fading occurs. The attenuation is proportional to the cosine of the phase difference. If the frequency of the local oscillator is different from the frequency of the received carrier, the demodulated signal is frequency translated at the output.

5.2.6 Performance of AM

In order to analyze the performance of modulation systems it is usual to employ the so-called quadrature representation for the noise. The quadrature representation for the noise $n(t)$, as a function of its in-phase $n_I(t)$ and quadrature $n_Q(t)$ components, is given by

$$n(t) = n_I(t)\cos(\omega_c t + \phi) + n_Q(t)\sin(\omega_c t + \phi). \tag{5.31}$$

It can be shown that the in-phase and quadrature components have the same mean value and variance as the noise $n(t)$, i.e. $\sigma_I^2 = \sigma_Q^2 = \sigma^2$, and σ^2 represents the variance of the noise $n(t)$. Assuming that an AM-SC was transmitted and corrupted by additive noise in the channel, the received signal is given by $r(t)$ as follows:

$$r(t) = \alpha m(t)\cos(\omega_c t + \phi) + n(t), \tag{5.32}$$

in which α represents the path attenuation of the signal.

The signal-to-noise ratio (SNR_I) at the demodulator input is given by

$$\text{SNR}_I = \frac{\alpha^2/2}{\sigma^2}, \tag{5.33}$$

in which $\sigma^2 = 2\omega_M N_0/2\pi$, for a modulating signal with a highest frequency equal to ω_M. As a consequence of synchronous demodulation, the demodulated signal after filtering is given by

$$\hat{m}(t) = \alpha m(t) + n_I(t), \tag{5.34}$$

thus producing the following output signal-to-noise ratio:

$$\text{SNR}_O = \frac{\alpha^2}{\sigma^2}. \tag{5.35}$$

Therefore, the demodulation gain η, for an AM-SC system is equal to

$$\eta = \frac{\text{SNR}_O}{\text{SNR}_I} = 2. \tag{5.36}$$

AM-VSB modulation has a demodulating gain approximately equal to that of AM-SC modulation.

Following a similar procedure, it can be shown that the demodulation gain of ordinary AM, employing coherent demodulation, is given by

$$\eta = \frac{2\Delta_{AM}^2 P_M}{1 + \Delta_{AM}^2 P_M} \tag{5.37}$$

It can be seen that the demodulation gain of the commercial AM is at least 3 dB lower than the corresponding AM-SC demodulation gain, eventually attaining the value of 6 dB in practice (Lathi, 1989).

5.3   Quadrature amplitude modulation (QUAM)

The QUAM scheme has become a popular way to transmit information, because it occupies the constellation diagram in a very efficient manner. The technique is to use sine and cosine orthogonality properties to allow the transmission of two different signals in the same carrier, which occupies a bandwidth that is equivalent to the AM signal. The information is transmitted by both the carrier amplitude and phase.

The quadrature modulated signal $s(t)$ can be written as

$$s(t) = b(t)\cos(\omega_c t + \phi) + d(t)\sin(\omega_c t + \phi), \tag{5.38}$$

in which the random modulating signals $b(t)$ and $d(t)$ can be correlated or uncorrelated.

It is also possible to write the modulated signal as a function of the amplitude and phase resultants, as follows:

$$s(t) = \sqrt{b^2(t) + d^2(t)}\, \cos\left[\omega_c t - \tan^{-1}\left(\frac{d(t)}{b(t)}\right) + \phi\right], \tag{5.39}$$

in which the modulating signal, or amplitude resultant, can be expressed as $a(t) = \sqrt{b^2(t) + d^2(t)}$ and the phase resultant is $\theta(t) = -\tan^{-1}(d(t)/b(t))$.

The autocorrelation function for the quadrature modulated carrier can be computed from the definition

$$R_S(\tau) = E[s(t) \cdot s(t + \tau)]. \tag{5.40}$$

Substituting (5.38) into (5.40), and expanding the product, gives

$$\begin{aligned}
R_S(\tau) = E\,[&b(t)b(t + \tau)\cos(\omega_c t + \phi)\cos(\omega_c(t + \tau) + \phi) \\
+ &d(t)d(t + \tau)\sin(\omega_c t + \phi)\sin(\omega_c(t + \tau) + \phi) \\
+ &b(t)d(t + \tau)\cos(\omega_c t + \phi)\sin(\omega_c(t + \tau) + \phi) \\
+ &b(t + \tau)d(t)\cos(\omega_c(t + \tau) + \phi)\sin(\omega_c t + \phi)].
\end{aligned} \tag{5.41}$$

Using trigonometric properties and collecting terms which represent known autocorrelation functions, it follows that

$$R_S(\tau) = \left[\frac{R_B(\tau) + R_D(\tau)}{2}\right]\cos\omega_c\tau + \left[\frac{R_{DB}(\tau) - R_{BD}(\tau)}{2}\right]\sin\omega_c\tau. \tag{5.42}$$

It is observed that the QUAM signal suffers both amplitude and phase modulation. The QUAM autocorrelation function can be simplified if one considers zero mean uncorrelated modulating signals. The cross-correlations are null, $R_{BD}(\tau) = E[b(t)d(t + \tau)] = 0$, and $R_{DB}(\tau) = E[b(t + \tau)d(t)] = 0$. The resulting autocorrelation is then given by

$$R_S(\tau) = \frac{R_B(\tau)}{2}\cos\omega_c\tau + \frac{R_D(\tau)}{2}\cos\omega_c\tau. \tag{5.43}$$

The carrier power, for the general case, is given by the formula

$$P_S = R_S(0) = \frac{P_B + P_D}{2}. \tag{5.44}$$

The power spectral density is obtained by applying the Fourier transform to the autocorrelation function, which gives

$$S_S(\omega) = \frac{1}{4} \left[S_B(\omega + \omega_c) + S_B(\omega - \omega_c) + S_D(\omega + \omega_c) + S_D(\omega - \omega_c) \right]$$

$$= \frac{j}{4} \left[S_{BD}(\omega - \omega_c) + S_{BD}(\omega + \omega_c) + S_{DB}(\omega + \omega_c) - S_{DB}(\omega - \omega_c) \right], \quad (5.45)$$

in which $S_B(\omega)$ and $S_D(\omega)$ represent the respective power spectral densities for $b(t)$ and $d(t)$, $S_{BD}(\omega)$ is the cross-spectral density between $b(t)$ and $d(t)$, and $S_{DB}(\omega)$ is the cross-spectral density between $d(t)$ and $b(t)$.

For uncorrelated signals, which is a common assumption for transmission of digital information, the previous formula can be simplified to

$$S_S(\omega) = \tfrac{1}{4} \left[S_B(\omega + \omega_c) + S_B(\omega - \omega_c) + S_D(\omega + \omega_c) + S_D(\omega - \omega_c) \right]. \quad (5.46)$$

The QUAM scheme is shown in Figure 5.12. The oscillator generates two carriers, with a phase difference of $\pi/2$ rad. Color information for both the NTSC and PAL-M standards is transmitted using QUAM. In this case, instead of transmitting the red, green, and blue (RGB) signals, two color difference signals are produced, which modulate the in-phase and quadrature carriers.

5.3.1 Single sideband amplitude modulation (AM-SSB)

AM-SSB is, in reality, a type of QUAM which uses either the lower or upper AM sideband for transmission. The AM-SSB signal can be obtained by filtering out the undesired sideband of an AM-SC signal.

The SSB signal saves bandwidth and power, compared to other systems, but needs frequency and phase synchronization to be recovered (Carlson, 1975). In order to understand the process of generating the SSB signal one uses the Hilbert transform, which is presented in Appendix B.

One way to obtain the SSB signal is by filtering one of the AM-SC sidebands. Another method uses the properties of the Hilbert transform. The SSB signal associated with $m(t)$ is

$$s(t) = m(t) \cos(\omega_c t + \phi) + \hat{m}(t) \sin(\omega_c t + \phi), \quad (5.47)$$

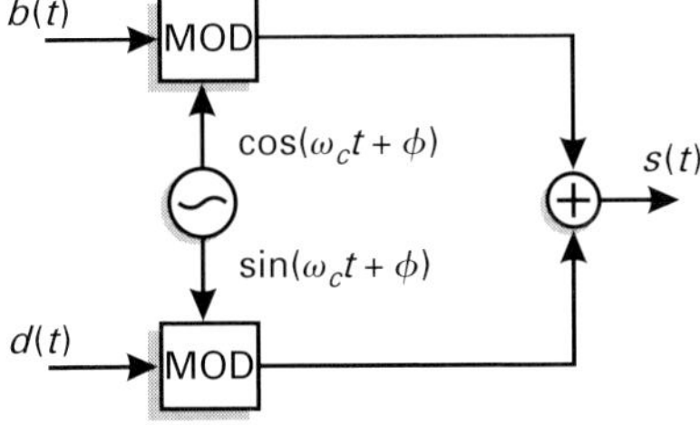

Figure 5.12   Block diagram for the quadrature modulator

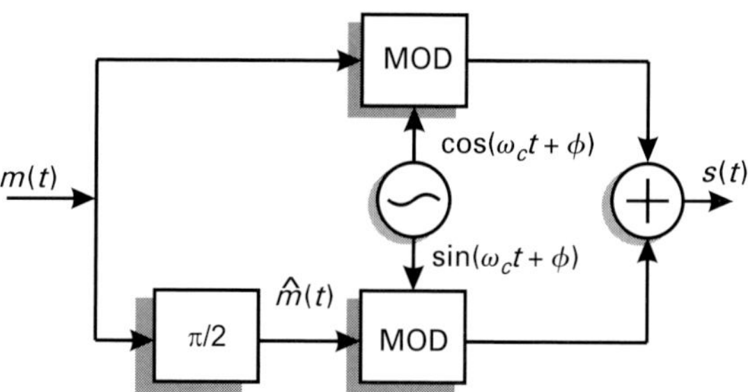

Figure 5.13   Block diagram for the SSB modulator

in which $\hat{m}(t)$ is obtained by phase shifting all frequency components of $m(t)$ by $\pi/2$. The block diagram of the modulator is shown in Figure 5.13.

Autocorrelation and power spectral density of the SSB signal

For a modulating signal $m(t)$ which is a zero mean, stationary, stochastic process, the usual procedure to obtain the power spectral density is to compute its autocorrelation function. The modulated SSB signal is given by

$$s(t) = m(t) \cos(\omega_c t + \phi) + \hat{m}(t) \sin(\omega_c t + \phi). \tag{5.48}$$

The autocorrelation function is calculated, as usual, by the formula

$$R_S(\tau) = E[s(t)s(t + \tau)]. \tag{5.49}$$

Substituting $s(t)$, given by (5.48), into the preceding equation, gives

$$\begin{aligned}
R_S(\tau) = {} & E\left[m(t)m(t + \tau)\cos(\omega_c t + \phi)\cos(\omega_c(t + \tau) + \phi)\right] \\
& + E\left[m(t)\hat{m}(t + \tau)\cos(\omega_c t + \phi)\sin(\omega_c(t + \tau) + \phi)\right] \\
& + E\left[\hat{m}(t)m(t + \tau)\sin(\omega_c t + \phi)\cos(\omega_c(t + \tau) + \phi)\right] \\
& + E\left[\hat{m}(t)\hat{m}(t + \tau)\sin(\omega_c t + \phi)\sin(\omega_c(t + \tau) + \phi)\right].
\end{aligned} \tag{5.50}$$

After the corresponding simplifications, one obtains

$$\begin{aligned}
R_S(\tau) = {} & \frac{1}{2}R_{MM}(\tau)\cos\omega_c\tau + \frac{1}{2}R_{M\hat{M}}(\tau)\sin\omega_c\tau \\
& - \frac{1}{2}R_{\hat{M}M}(\tau)\sin\omega_c\tau + \frac{1}{2}R_{\hat{M}\hat{M}}(\tau)\cos\omega_c\tau.
\end{aligned} \tag{5.51}$$

It is not difficult to demonstrate for the cross-correlation functions that $R_{MM}(\tau) = R_{\hat{M}\hat{M}}(\tau)$ and $R_{M\hat{M}}(\tau) = -R_{\hat{M}M}(\tau)$. Therefore, the cross-power spectra can be computed with the Wiener–Khintchin theorem, using the following equation for the Hilbert filter

$$H(\omega) = \begin{cases} -j & \omega \geq 0 \\ +j & \omega < 0 \end{cases}. \tag{5.52}$$

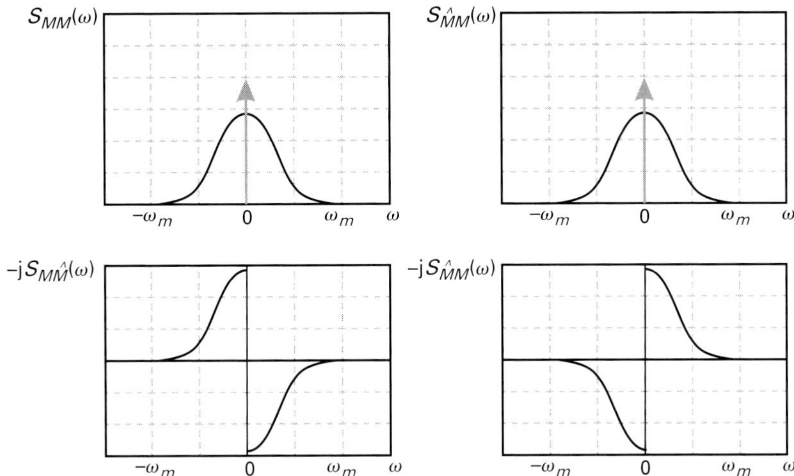

Figure 5.14　Procedure to obtain the cross-power spectra

This leads to

$$S_{\hat{M}\hat{M}}(\omega) = S_{MM}(\omega),\tag{5.53}$$

$$S_{M\hat{M}}(\omega) = j[u(-\omega) - u(\omega)] \cdot S_{MM}(\omega),\tag{5.54}$$

$$S_{\hat{M}M}(\omega) = j[u(\omega) - u(-\omega)] \cdot S_{MM}(\omega).\tag{5.55}$$

Figure 5.14 illustrates the procedure to obtain the cross-power spectral density from the original power spectral density.

After computing the Fourier transform and substituting the previous cross-power spectra, one obtains the following power spectral density for the SSB signal:

$$S(\omega) = S_M(\omega - \omega_c)u(-\omega + \omega_c) + S_M(\omega + \omega_c)u(\omega + \omega_c),\tag{5.56}$$

which represents the lower sideband of the SSB system, obtained from the original spectrum $S_M(\omega)$.

5.3.2　Quadrature amplitude demodulation

The received QUAM signal is synchronously demodulated to recover the message signal, as shown in Figure 5.15. The incoming signal

$$s(t) = b(t)\cos(\omega_c t + \phi) + d(t)\sin(\omega_c t + \phi)\tag{5.57}$$

is multiplied by two locally generated sinusoidal signals $\cos(\omega_c t + \phi)$ and $\sin(\omega_c t + \phi)$ and low-pass filtered. This results in the recovery of the original $b(t)$ and $d(t)$ signals, which are then amplified.

If the local oscillator phase is different from the received carrier phase, fading can affect the signal. The signal is attenuated by a term proportional to the cosine of the

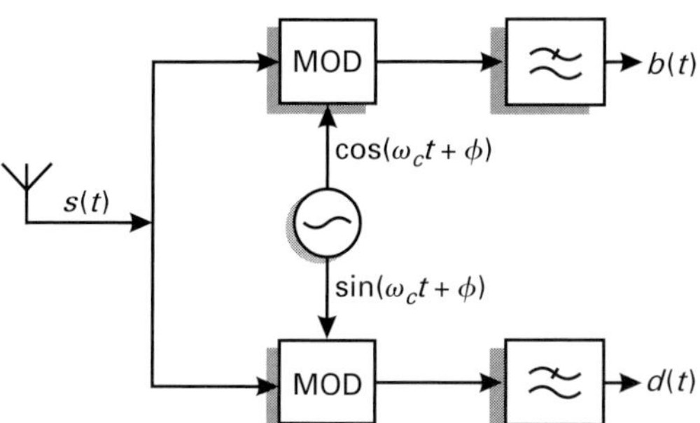

Figure 5.15   Block diagram of a QUAM demodulator

phase difference. As in synchronous AM demodulation, a frequency drift of the local oscillator can shift the demodulated signal frequency and disturb the reception.

5.3.3   Performance evaluation of SSB

The performance of the SSB signal does not differ from the AM-SC performance. The SSB modulated signal, presented previously, can be written as

$$s(t) = m(t)\cos(\omega_c t + \phi) + \hat{m}(t)\sin(\omega_c t + \phi), \tag{5.58}$$

where $m(t)$ represents the message signal and $\hat{m}(t)$ its Hilbert transform.

The spectral analysis of the SSB signal indicates that its bandwidth is the same as that of the baseband signal (ω_M). Therefore, the SSB signal power is certainly identical to the modulating signal power. The received signal is expressed as

$$r(t) = [m(t) + n_I(t)]\cos(\omega_c t + \phi) + [\hat{m}(t) + n_Q(t)]\sin(\omega_c t + \phi). \tag{5.59}$$

Because the noise occupies a bandwidth ω_M and demodulation is synchronous, the demodulation gain η for the SSB signal is the same as the one obtained for the AM-SC, i.e.

$$\eta = \frac{\text{SNR}_O}{\text{SNR}_I} = 2. \tag{5.60}$$

5.3.4   Digital quadrature modulation (QAM)

Quadrature amplitude modulation with a digital signal, called QAM, is a versatile modulation scheme. The computation of the autocorrelation function and the power spectrum density follows the previously established procedures.

For a QAM scheme the modulating signals can be represented as

$$b(t) = \sum_{n=-\infty}^{\infty} b_n p(t - nT_b) \tag{5.61}$$

and

$$d(t) = \sum_{n=-\infty}^{\infty} d_n p(t - nT_b),\tag{5.62}$$

in which T_b is the symbol period, and b_n and d_n represent the input symbol sequences.

The equation for the modulated carrier is similar to that for QUAM

$$s(t) = b(t)\cos(\omega_c t + \phi) + d(t)\sin(\omega_c t + \phi).\tag{5.63}$$

The constellation diagram for a 4-QAM signal is shown in Figure 5.16, in which $b_n = \{A, -A\}$ and $d_n = \{A, -A\}$. It can be shown, using the same methods as previously, that the modulated signal power for this special case is given by $P_S = A^2/2$. This signal is also known as $\pi/4$-QPSK and is largely used in schemes for digital television and mobile cellular communication systems. If the constellation is rotated by $\pi/4$ it produces another QAM modulation scheme, which presents more ripple than the previous one, because one of the carriers is shut off whenever the other carrier is transmitted.

Figure 5.17 shows the constellation diagram for a 16-QAM signal, whose points have coordinates $b_n = \{-3A, -A, A, 3A\}$, for the abscissa, and $d_n = \{-3A, -A, A, 3A\}$, for the ordinate. It can be observed that the constellation goes on occupying all the space

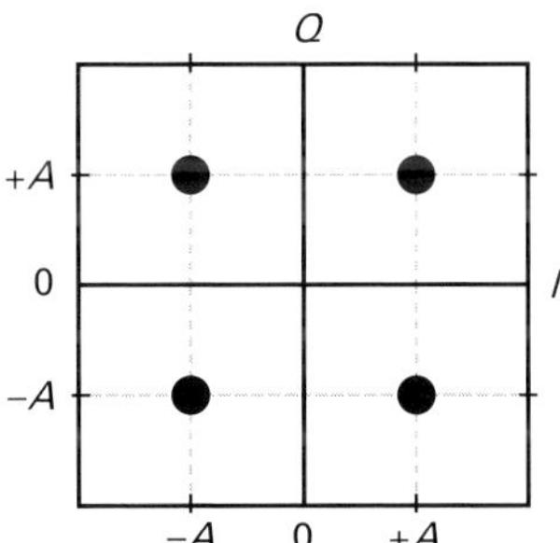

Figure 5.16   Constellation diagram for the 4-QAM signal

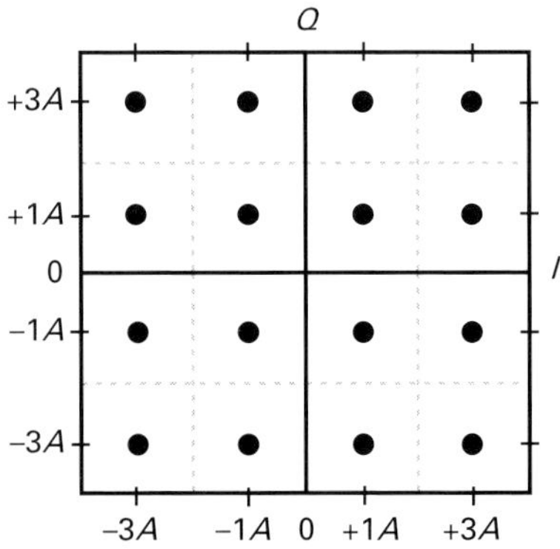

Figure 5.17   Constellation diagram for the 16-QAM signal

as long as the number of symbols increases. This represents an efficient way to allocate the symbols in order to save power and maximize the distance between them.

The efficient occupation of the signal space renders the QAM technique more efficient than ASK and PSK, regarding BEP versus transmission rate. On the other hand, the scheme is vulnerable to non-linear distortion, which occurs in power amplifiers used in television systems. Predistortion can be used to overcome this problem. The scheme is also affected by fading. It is evident that other problems can occur and degrade the transmitted signal. It is important for the communication engineer to identify them, in order to establish the best strategy to combat them.

The random phase variation effect, or jitter, occurs when the local synchronization system is not able to perfectly follow the received signal phase variations. This introduces randomness in the detection process, which increases the bit symbol error probability.

The symbol error probability is usually controlled by the smaller distance between the constellation symbols. The formulas for the error probability produce curves that decrease exponentially, or quasi-exponentially, with the signal-to-noise ratio.

The probability of error for the 4-QAM signal is (Haykin, 1988)

$$P_e \approx \mathrm{erfc}\left(\sqrt{\frac{E_b}{N_0}}\right), \tag{5.64}$$

in which E_b is the pulse energy and N_0 represents the noise power spectral density. For the M-QAM scheme, the bit error probability is given by (Haykin, 1988)

$$P_e = 1 - \left[1 - \left(1 - \frac{1}{\sqrt{M}}\right)\mathrm{erfc}\left(\sqrt{\frac{E_b}{N_0}}\right)\right]^2. \tag{5.65}$$

Figure 5.18 shows the BEP for the QAM modulation scheme.

5.4   Angle modulated systems

Angle modulation is a technique which encompasses FM and PM schemes. John R. Carson, was the first to compute the bandwidth of an angle modulated scheme in a paper published in 1922 (Carson, 1922). At the time, however, he could see no advantage in FM over AM, and the subject was forgotten for a while, until Edwin H. Armstrong invented the first equipment to frequency modulate a carrier (Armstrong, 1936). In an unpublished memorandum of August 28, 1939, Carson established that the bandwidth was equal to twice the sum of the peak frequency deviation and the highest frequency of the modulating signal (Carlson, 1975).

The objective of this section is to present a general mathematical model to analyze stochastic angle modulated signals. Several transmission systems use either FM or PM, including analog and digital mobile cellular communication systems, satellite transmission systems, television systems, and wireless telephones.

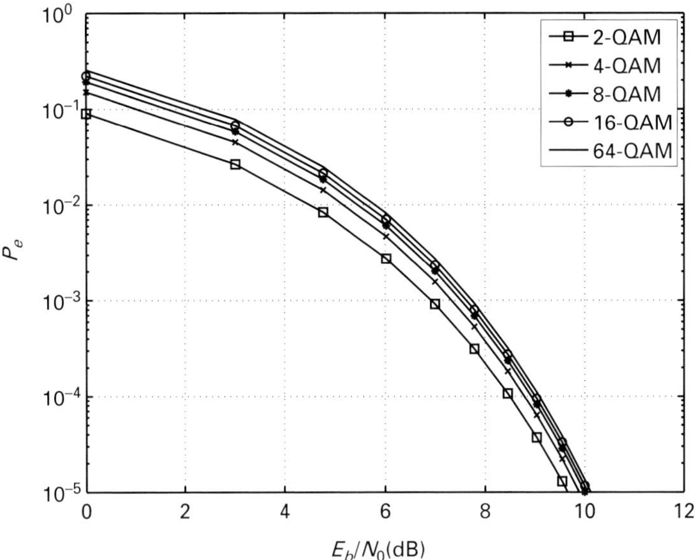

Figure 5.18   BEP for an M-QAM modulation scheme

5.4.1   Angle modulation with random signals

This subsection presents a general and elegant method to compute the power spectral density of angle modulated signals. The modulating signal is considered a stationary random process $m(t)$ that has autocorrelation $R_M(\tau)$.

Woodward's theorem on FM

The first computation of the power spectral density for the FM scheme was provided by P. M. Woodward, who proved that the spectrum of a high-index FM waveform has a shape that approximates that of the probability distribution of its instantaneous frequency (Woodward, 1952). In the following the spectrum of an angle modulated signal is estimated, based on the first-order probability density function of the modulating signal. A proof is presented that includes the linear mean square estimator (Alencar and da Rocha, 2005).

The modulated signal $s(t)$ is obtained from the following equation:

$$s(t) = A\cos\left(\omega_c t + \theta(t) + \phi\right), \tag{5.66}$$

in which

$$\theta(t) = \Delta_{FM} \cdot \int_{-\infty}^{t} m(t)\mathrm{d}t, \tag{5.67}$$

for FM, and

$$\theta(t) = \Delta_{PM} \cdot m(t)\mathrm{d}t, \tag{5.68}$$

for PM, in which the constant parameters A, ω_c, Δ_{FM} and Δ_{PM} represent, respectively, the carrier amplitude and angular frequency, and the frequency and phase deviation indices. The message signal is $m(t)$, as usual, and is considered to be a zero mean random stationary process. The carrier phase ϕ is also random and is uniformly distributed in the range $[0, 2\pi]$. It is statistically independent of $m(t)$.

The modulating signal, $\theta(t)$, represents the variation in the carrier phase, induced by the message. The FM index, β, and the PM index, α, are defined, respectively, as

$$\beta = \frac{\Delta_{FM}\sigma_M}{\omega_M}, \quad \alpha = \Delta_{PM}\sigma_M, \tag{5.69}$$

in which $\sigma_M = \sqrt{P_M}$, and P_M represents the power of the message signal $m(t)$, and ω_M is the maximum angular frequency of the signal.

The frequency deviation $\sigma_F = \Delta_{FM}\sigma_M$ represents the shift from the original, or spectral, carrier frequency. The modulation index gives an idea of how the modulating signal bandwidth fits into the frequency deviation.

In order to evaluate the spectrum of the angle modulated signal, it is necessary to compute the autocorrelation function of $s(t)$, obtain an estimate of this autocorrelation, for the case of a high modulation index, using the linear mean square estimator (Papoulis, 1981), and finally compute the power spectral density of $s(t)$ as the Fourier transform of the autocorrelation estimate.

Substituting the equation for the modulated signal into (5.66), one obtains

$$R_S(\tau) = \frac{A^2}{2}E[\cos(w_c\tau - \theta(t) + \theta(t + \tau))]. \tag{5.70}$$

For FM, with a low modulation index ($\beta < 0.5$), the autocorrelation function of $s(t)$ can be obtained from (5.70) by expanding the cosine in Taylor series, neglecting the high-order terms, and taking into account that $m(t)$ is a zero mean, stationary process. It follows that (Alencar and da Rocha, 2005)

$$R_S(\tau) = \frac{A^2}{2}\cos(\omega_c\tau)[1 - R_\Theta(0) + R_\Theta(\tau)], \tag{5.71}$$

in which

$$R_\Theta(\tau) = E[\theta(t)\theta(t + \tau)] \tag{5.72}$$

and $R_\Theta(0) = P_\Theta$ is the power of the resultant phase $\theta(t)$.

The power spectral density of $s(t)$ is obtained using the Wiener–Khintchin theorem, through the Fourier transform of (5.71) (Papoulis, 1983):

$$S_S(\omega) = \frac{\pi A^2(1 - P_\Theta)}{2}[\delta(\omega + \omega_0) + \delta(\omega - \omega_0)]$$
$$+ \frac{\Delta_{FM}^2 A^2}{4}\left[\frac{S_M(\omega + \omega_c)}{(\omega + \omega_c)^2} + \frac{S_M(\omega - \omega_c)}{(\omega - \omega_c)^2}\right], \tag{5.73}$$

in which $S_M(\omega)$ represents the power spectral density of the message signal $m(t)$, which has bandwidth ω_M. The modulated signal bandwidth is double the message signal bandwidth $BW = 2\omega_M$.

From (5.73) one can see that the FM spectrum has the shape of the message signal spectrum multiplied by a squared hyperbolic function.

For a high modulation index ($\beta > 5$), the power spectral density of $s(t)$ has the shape of the probability density function of $m(t)$ in (5.67), as the modulation index is increased. This is the result of Woodward's theorem, which will be discussed in the following (Woodward, 1952). The high modulation index also causes a spectral broadening of the modulated carrier.

For most analog television standards, the commercial FM and the audio part of the composite video can be considered as high-modulation-index systems (Lee, 1989). For the high-modulation-index case, one can use Euler's formula and rewrite (5.70) as

$$R_S(\tau) = \frac{A^2}{4} e^{j\omega_c \tau} E[e^{j(-\theta(t)+\theta(t+\tau))}] + \frac{A^2}{4} e^{-j\omega_c \tau} E[e^{j(\theta(t)-\theta(t+\tau))}]. \tag{5.74}$$

The second-order linear mean square estimate of the process $\theta(t+\tau)$ includes its current value $\theta(t)$ and its derivative $\theta'(t)$:

$$\theta(t+\tau) \approx \alpha\,\theta(t) + \beta\,\theta'(t). \tag{5.75}$$

In order to obtain the values of the optimization parameters α and β, the derivative of the error must converge to zero. Thus, the best approximation to the future value of the process, in the mean square sense, is

$$\theta(t+\tau) \approx \frac{R_\Theta(\tau)}{R_\Theta(0)}\theta(t) + \frac{R'_\Theta(\tau)}{R''_\Theta(0)}\theta'(t). \tag{5.76}$$

Considering that the random process is slowly varying, as compared to the spectral frequency of the modulated carrier, leads to the following approximation for the modulated carrier autocorrelation (Alencar, 1989, Papoulis, 1983)

$$R_S(\tau) = \frac{A^2}{4} e^{j\omega_c \tau} E[e^{j\tau\theta'(t)}] + \frac{A^2}{4} e^{-j\omega_c \tau} E[e^{-j\tau\theta'(t)}]. \tag{5.77}$$

But, $\theta'(t) = d\theta(t)/dt = \omega(t)$, in which $\omega(t)$ is the carrier angular frequency deviation at the instantaneous carrier angular frequency, thus

$$R_S(\tau) = \frac{A^2}{4} e^{j\omega_c \tau} E[e^{j\tau\omega(t)}] + \frac{A^2}{4} e^{-j\omega_c \tau} E[e^{-j\tau\omega(t)}]. \tag{5.78}$$

The power spectral density of the modulating signal used in this section is illustrated in Figure 5.19. This signal was originally obtained from a noise generator.

Taking into account that

$$E[e^{j\tau\omega(t)}] = \int_{-\infty}^{\infty} p_\Omega(\omega(t)) e^{j\tau\omega(t)} d\omega(t) \tag{5.79}$$

represents the characteristic function of process $\omega(t) = \theta'(t)$ and that $p_\Omega(\omega(t))$ is its probability density function, which is considered symmetrical, without loss of generality, one can compute the power spectral density for the angle modulated signal.

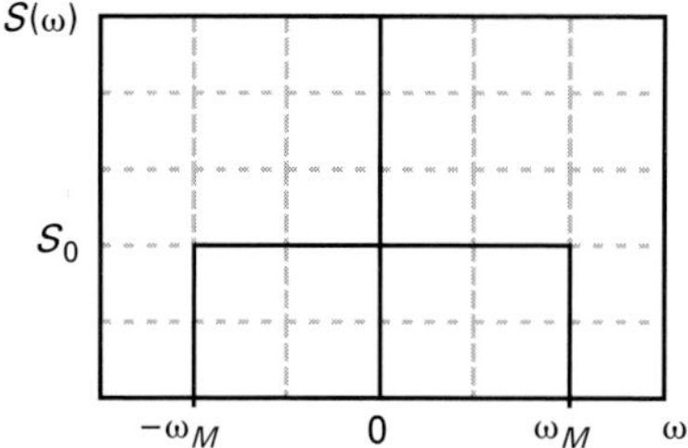

Figure 5.19   Spectrum for the white Gaussian modulating signal

Calculating the Fourier transform of (5.78), it follows that

$$S_S(\omega) = \frac{\pi A^2}{2}[p_\Omega(\omega + \omega_c) + p_\Omega(\omega - \omega_c)]. \tag{5.80}$$

Considering the definition of $\omega(t)$ from (5.67), it is noticed that $\omega(t) = \Delta_{FM} \cdot m(t)$, thus

$$p_\Omega(\omega(t)) = \frac{1}{\Delta_{FM}}p_M\left(\frac{m}{\Delta_{FM}}\right), \tag{5.81}$$

in which $p_M(\cdot)$ is the probability density function of $m(t)$.

Substituting (5.81) into (5.80) gives the formula for the power spectral density for the wideband frequency modulated signal (Alencar and Neto, 1991):

$$S_S(\omega) = \frac{\pi A^2}{2\Delta_{FM}}\left[p_M\left(\frac{w + w_c}{\Delta_{FM}}\right) + p_M\left(\frac{w - w_c}{\Delta_{FM}}\right)\right]. \tag{5.82}$$

Following a similar line of thought one can derive a formula for the power spectral density of the PM signal. The instantaneous angular frequency is given by

$$\omega(t) = \frac{\theta(t)}{dt} = \Delta_{PM}m(t). \tag{5.83}$$

Therefore,

$$S_S(\omega) = \frac{\pi A^2}{2\Delta_{PM}}\left[p_{M'}\left(\frac{w + w_c}{\Delta_{PM}}\right) + p_{M'}\left(\frac{w - w_c}{\Delta_{PM}}\right)\right], \tag{5.84}$$

in which $p_{M'}(\cdot)$ is the probability density function of the derivative of the message signal $m(t)$.

Figures 5.20 and 5.21 illustrate the changes in the modulated carrier spectrum as the modulation index is increased for a Gaussian modulating signal. One can see that for a high modulation index the frequency deviation is given by $\Delta_{FM}\sqrt{P_M}$ and the bandwidth is approximated by BW $= 2\Delta_{FM}\sqrt{P_M}$, in order to include most of the modulated carrier power. As previously derived, the bandwidth for a narrowband FM is BW $= 2\omega_M$. Interpolating between both values gives a formula that covers the whole range of modulation indices β:

$$\text{BW} = 2\omega_M + 2\Delta_{FM}\sqrt{P_M} = 2\left(\frac{\Delta_{FM}\sqrt{P_M}}{\omega_M} + 1\right)\omega_M = 2(\beta + 1)\omega_M. \tag{5.85}$$

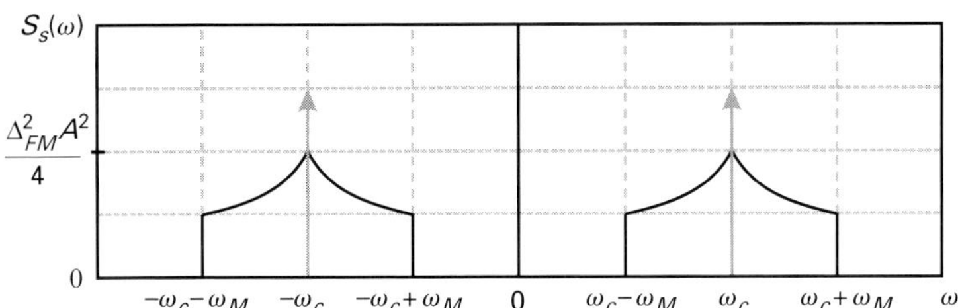

Figure 5.20   Spectrum of an FM signal with a low modulation index for a Gaussian modulating signal

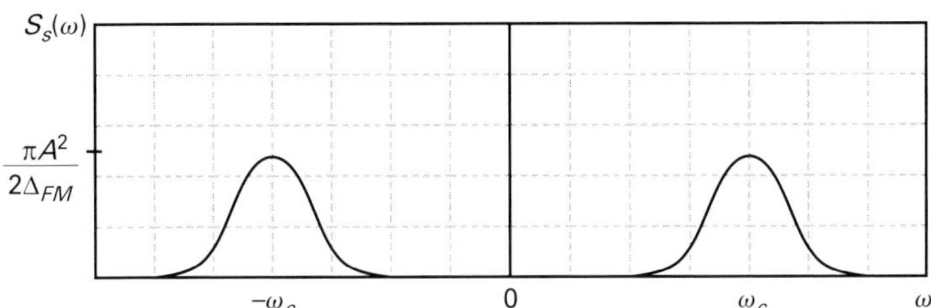

Figure 5.21   Spectrum of an FM signal for a Gaussian modulating signal, with a high modulation index

This is the well-known Carson rule, whose heuristic deduction first appeared in 1922 (Carson, 1922). Using this formula it is possible to compute the bandwidth occupied by the analog television audio signal, for example, based on the maximum frequency of the modulating signal.

5.4.2   Angle modulation with digital signal

Angle modulation with digital signal is described by the following equations, for the case of phase shift keying (PSK), which is used in most digital television systems. The modulated carrier is written as

$$s(t) = A\cos(\omega_c t + \Delta_{PM} m(t) + \phi), \tag{5.86}$$

in which

$$m(t) = \sum_{j=-\infty}^{\infty} m_j p(t - jT_b). \tag{5.87}$$

Figure 5.22 shows the constellation diagram of a PSK modulation scheme, with parameters $\Delta_{PM} = \pi/4$, $\phi = 0$ and symbols $m_j = \{0, 1, 2, 3, 4, 5, 6, 7\}$. This type of modulation is also used for satellite transmission, because its carrier features a constant envelope, with the constellation symbols positioned on a circle. This gives a certain advantage for the scheme, regarding multiplicative noise fading channels or non-uniform amplification. The power of the modulated carrier is $P_S = A^2/2$.

An increase in the transmission rate, by the addition of new symbols, makes the PSK scheme more susceptible to noise, increasing the probability of error, because the

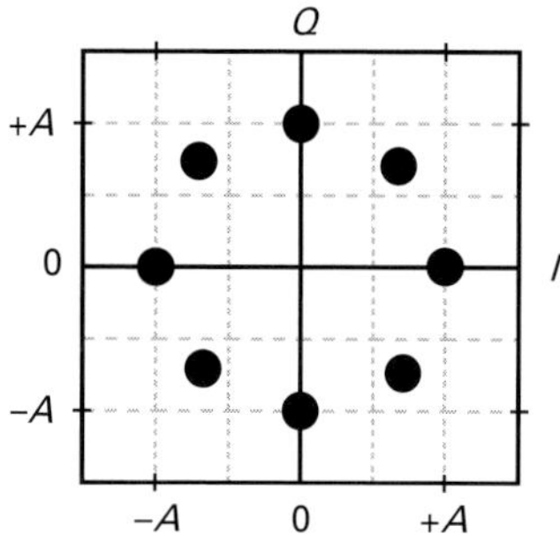

Figure 5.22   Constellation for a PSK signal

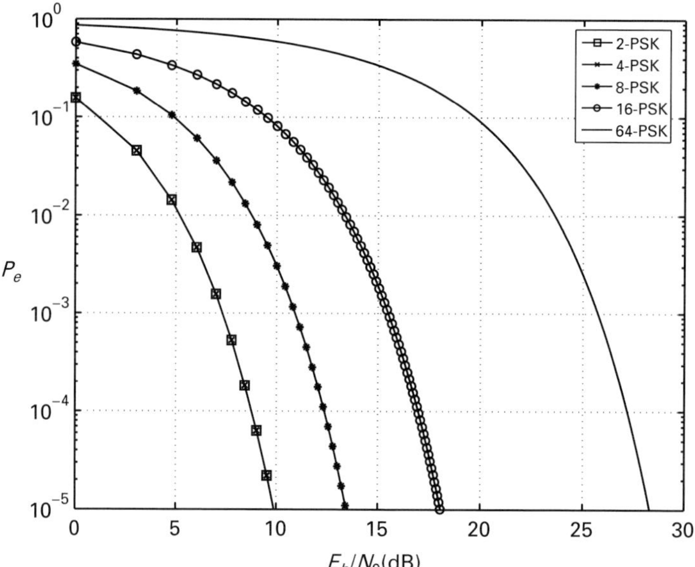

Figure 5.23   BEP for M-PSK

symbols become closer in the signal space. This modulation technique is not as efficient as QUAM for allocating symbols on the constellation diagram.

The probability of error for the coherent binary PSK is (Haykin, 1988)

$$P_e = \frac{1}{2}\mathrm{erfc}\left(\sqrt{\frac{E_b}{N_0}}\right),\tag{5.88}$$

in which, E_b is the pulse energy and N_0 represents the noise power spectral density. Figure 5.23 shows the BEP for the PSK modulation scheme.

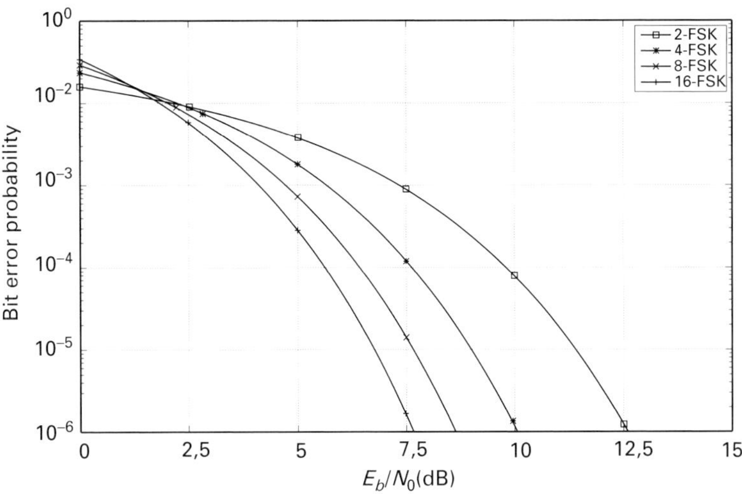

Figure 5.24   BEP for M-FSK

For coherent 2-FSK, the probability of error has a 3 dB penalty, compared to PSK (Haykin, 1988)

$$P_e = \frac{1}{2}\text{erfc}\left(\sqrt{\frac{E_b}{2N_0}}\right).\tag{5.89}$$

Non-coherent 2-FSK has a probability of error given by (Haykin, 1988)

$$P_e = \frac{1}{2}\exp\left(\frac{E_b}{2N_0}\right).\tag{5.90}$$

For the transmission of multiple symbols, the M-ary BEP for the FSK scheme can be upper bounded by (Benedetto and Biglieri, 1999)

$$P_e \leq \frac{M-1}{2}\text{erfc}\left(\sqrt{\frac{E_b}{N_0}\frac{\log_2 M}{2}}\right).\tag{5.91}$$

Figure 5.24 shows the BEP for an M-FSK, for $M = 2, 4, 8, 16$.

6 Advanced Television Systems Committee standard (ATSC)

6.1 Introduction

The Advanced Television Systems Committee, Inc. (ATSC) was formed in 1982 by the member organizations of the Joint Committee on InterSociety Coordination (JCIC): the Electronic Industries Alliance (EIA), the Institute of Electrical and Electronic Engineers (IEEE), the National Association of Broadcasters (NAB), the National Cable Television Association (NCTA), and the Society of Motion Picture and Television Engineers (SMPTE). Currently, there are approximately 140 members representing the broadcast, broadcast equipment, motion picture, consumer electronics, computer, cable, satellite, and semiconductor industries. ATSC digital television standards include digital high-definition television (HDTV), standard-definition television (SDTV), data broadcasting, multichannel surround-sound audio, and satellite direct-to-home broadcasting (ATSC, 2008).

In 1987, the Federal Communications Commission (FCC) formed a group of television industry entrepreneurs, the Advisory Committee on Advanced Television Service, to help it with the development of new technologies for television, as well as with the planning of public policy for organizing the exploration of possible applications of the research.

At first, 23 different systems were proposed to the Advisory Committee. These systems ranged from improved systems, which worked within the parameters of the NTSC system, to enhanced systems, which added additional information to the signal, and to HDTV systems, which presented substantially higher resolution, a wider picture aspect ratio, and improved sound.

The Advanced Television Test Center, funded by the broadcasting and consumer electronics industries, conducted transmission performance testing and subjective tests using expert viewers. Cable Television Laboratories, a research and development consortium of cable television system operators, conducted an extensive series of cable transmission tests as well. The Advanced Television Evaluation Laboratory within the Canadian Communications Research Centre conducted assessment tests using non-expert viewers.

It was only in early 1990s that a company, General Instrument, managed to develop a totally digitalized HDTV system. By the end of 1990, three more systems with this feature had been proposed. Together with two other non-digital systems, all four digital ones were tested by independent laboratories, according to an ATSC document, between July 1991 and October 1992, by means of processes defined by the Advisory Committee (ATSC, 1995b).

Broadcasting and subjective tests were conducted with professional viewers, prepared to engage in critical analysis, and also some laymen. The result was evaluated by a special commission, composed of those who requested the test, which suggested that new experiments be carried out, so that a single system be created, gathering each system's most efficient features and enhancing those that had left something to be desired.

In February 1993, a special panel of the Advisory Committee convened to review the results of the tests, in order to decide on a new transmission standard for terrestrial broadcast television to be recommended by the Advisory Committee to the FCC. The special panel determined that there would be no further consideration of analog technology, and that based upon analysis of the transmission system performance, an all-digital approach was both feasible and desirable.

The special panel recommended that the proponents of the four all-digital systems be authorized to implement certain modifications they had proposed, and that supplementary tests of these improvements be conducted. The Advisory Committee adopted the recommendation of the special panel, but also required a proposal from the remaining proponents for a single system that incorporated the best elements of the four all-digital systems.

In May 1993, the companies that owned the selected systems formed the Grand Alliance, which, managed by the Advisory Committee, produced a prototype to be presented to the FCC as a new standard of television terrestrial broadcasting. The ATSC is an organization composed of representatives of companies in the field of electro-electronics, telecommunications and computers, television stations, movie producers, technical associations and educational institutions, aimed at the development of standards for the improvement of television systems, among which are those related to HDTV. Along with the actions of the Advisory Committee, the ATSC documented the standards specified by the Grand Alliance and gathered several specifications of SDTV which, in addition to HDTV, constituted the digital television system of the ATSC standard.

Then, in November 1995, a specific commission of the FCC for this issue recommended the adoption of the system as standard for terrestrial television broadcasting on North American territory. However, the homologation of the ATSC as a new broadcasting standard only came into effect one year later, leaving it up to the manufacturers and the television stations to choose their HDTV and SDTV specifications, from among the 18 existing at that time.

The North American system has been continually improved, thanks mainly to the contributions of computer companies connected to the ATSC Forum, which has more than 200 members from various countries around the globe led by the USA.

Specific standards for direct-to-home and open satellite broadcasting were created. Nevertheless, the system is still deficient in the reception of signals on mobile receivers. Probably this has happened because it was developed when mobile telephony had not yet reached the importance it did in the 1990s. So, its developers focused their efforts on the creation of standards that provided, above all, high-quality images to the viewers. In these conditions, mobile reception was not considered, hence the modulation standard applied to the system, VSB, is inadequate for this application.

The ATSC signal is more susceptible to changes in radio propagation conditions than DVB-T and ISDB-T, in part because ATSC is not able to dynamically change its error correction modes, code rates, interleaver mode, and randomizer. If it could adapt, the signal could be more robust even if the modulation itself did not change. It also lacks a hierarchical modulation system, which would allow the SDTV signal to be received, in place of HDTV, in areas where signal power density is low. An additional modulation mode, enhanced-VSB (E-VSB), has been introduced to circumvent this problem.

The ATSC was adopted by the USA and Canada, its partner in the North American Free Trade Agreement (NAFTA), and a few other countries, mainly in the Americas. For commercial interests, the South Koreans, traditionally North American allies in Asia and with socioeconomic indicators close to those of developing countries, also implemented the ATSC system. Taiwan at first chose the ATSC, but, at the end of 2000, the government reevaluated this choice due to demands of the radio broadcasting organizations, which argued in favor of digital video broadcasting (DVB), which is more efficient in some respects. Tests with the European standards were carried out, which led to conjectures on the integration of some features of this standard with the ATSC.

The communications authorities in China have come to the conclusion that the best choice would be to adopt the system supported by the broadcasters, although financial resources had already been invested in the implementation of the ATSC standard. A similar situation occurred in Argentina, despite the fact that the amount invested had been smaller, as only two stations in the country had acquired equipment with ATSC technology.

The introduction of digital television has heralded a new era in television broadcasting (Richer *et al.*, 2006) The impact of digital television is more significant than the simple conversion of the analog system into the digital system (Whitaker, 2001). Digital television offers levels of quality and flexibility that are completely inaccessible on analog broadcasting.

Analog television systems are, by nature, limited to a narrow band of performance that offers few choices. The adoption of a digital system generates significant gains in performance, quality, and a wide range of services. Digital television distributes programs using the same 6 MHz band. In terms of performance, the capacity to provide high-definition images with high-quality sound is essential for the future of broadcasting, once consumers have been offered such improvements along with other forms of delivery.

An important element proposed by the ATSC system is the flexibility for the expansion of the functions built on technical bases, such as the digital television ATSC standard (A/53) (ATSC, 1995b) and the standard for the compression of digital audio AC-3 (A/52) (ATSC, 1995a). These ATSC standards, established in 1995, were the first to be launched in the world for digital television and defined the prerequisites for the quality and flexibility that distinguish digital television from all the existing analog television systems.

In the PAL analog system and its competitors NTSC and SECAM, video, audio, and some limited data information (e.g. closed caption) are modulated by an RF carrier in such a way that the receiver has a simple design with current technology. They may be easily demodulated with the reproduction of audio, video, and data components related to the signal. On analog television, the camera, the display, and the broadcasting parameters are strongly connected as part of an end-to-end system, which limits the modification ability of basic features of the system (e.g. image resolution).

The digital television system has made it possible for the capacities mentioned above to be separated. The ATSC standard was the pioneer, proposing an architecture in layers that separates the image format, the encoding for compression, the data transmission, and the broadcasting, as shown in Figure 6.1. This means that additional levels of processing are necessary after an ATSC receiver demodulates the RF signal, before a complete program is put together and presented. This separation is made through the use of information from the system and the services, transmitted as part of the digital signal to generate a collection of program elements (video, audio, and data) that, together, provide the service that the consumer selected.

The audio elements based on the A/52 ATSC standard and the video elements based on the ISO/IEC MPEG-2 standard are distributed in digital form using compression and must be decoded for presentation. The video may be of high definition (HDTV) or of standard definition (SDTV). The audio may be monochannel, stereo, or multichannel. Several forms of data complete the main programs of audio and video (e.g. descriptive texts or commentaries) or may be one or more autonomous services.

The nature of the ATSC standard makes it possible to implement new functions in the system using the same transmission and reception infrastructure. One of the greatest advantages introduced in digital television is, in fact, the integration of processing in the receiving device itself. Historically, in the design of any radio or television broadcasting system, the objective was always the concentration of technical sophistication (when necessary) in the transmission device, the receivers being simple and of low cost. Due to the fact that there are more receivers than transmitters, this approach has certain commercial advantages.

As long as this concept applies, the complexity of the transmitters, the compression of audio, and the video components will demand a significant processing quality in the

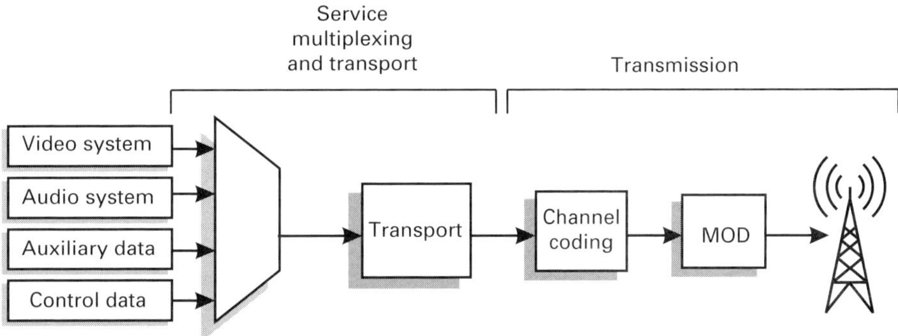

Figure 6.1   Block diagram of the ATSC standard

receiver, which have now become practical owing to the advances in digital television technology.

6.2   Overview of the system

A basic block diagram for the ATSC system is shown in Figure 6.1. This representation is based on the model adopted by the International Telecommunications Union, Radiocommunication Sector (ITU-R). According to the model, the digital television system can be divided into three subsystems:

- source encoding and compression;
- multiplexing of services and transport;
- transmission.

The core standards that describe the ATSC digital television system are the Digital Audio Compression standard (AC-3, E-AC3), and the ATSC Digital Television standard (ATSC, 2008).

Digital Audio Compression (AC-3, E-AC3) document A/52 specifies information on the audio information coding and decoding process. The coded representation specified is suitable for use in digital audio transmission and storage applications, and may convey from one to five full bandwidth audio channels, along with a low-frequency enhancement channel. A wide range of encoded bit rates is supported by this specification. Typical applications of digital audio compression are in satellite or terrestrial audio broadcasting, delivery of audio over metallic or optical cables, and storage of audio on magnetic, optical, semiconductor, or other storage media.

ATSC Digital Television standard document A/53 describes the system characteristics of the advanced television (ATV) system. The document provides detailed specification of the parameters of the system, including the video encoder input scanning formats and the preprocessing and compression parameters of the video encoder, the audio encoder input signal format and the preprocessing and compression parameters of the audio encoder, the service multiplex and transport layer characteristics and normative specifications, and the VSB RF transmission subsystem.

Program and System Information Protocol (PSIP) document A/65 is a collection of tables designed to operate within every transport stream for terrestrial broadcasting of digital television. The purpose of the protocol, described in ATSC document A/65B, is to enable the efficient tuning of programs by specifying the information at the system and event levels for all virtual channels carried in a particular transport stream.

Data Broadcast Standard, document A/90, defines protocols for data transmission within the digital television system. The standard supports data services that are both television program-related and non-program-related. Applications may include enhanced television, webcasting, and streaming video services. Data broadcasting receivers may include PCs, televisions, set-top boxes, and other devices. The standard provides mechanisms for downloading data, delivering datagrams, and streaming data.

Advanced Common Application Platform (ACAP) document A/101 is a suite of documents defining a software layer (middleware) that allows programming content and applications to run on a common receiver. Interactive and enhanced applications need access to common receiver features in a platform-independent manner. This environment provides enhanced and interactive content creators with the specifications necessary to ensure that their applications and data run uniformly on all brands and models of receivers. Manufacturers are thus able to choose hardware platforms and operating systems for receivers, but provide the commonality necessary to support applications made by many content creators.

6.2.1 Source encoding and compression

Source encoding and compression refer to the methods of data compression appropriate to video, audio, and auxiliary digital data. The term auxiliary data encompasses the following, besides referring to independent programs:

- control data;
- data for access control;
- data associated with audio programs and video services.

The function of the encoder is to minimize the number of bits needed to represent the audio and video information.

Figure 6.2 presents an overview of the encoding system. There are two domains within the encoder in which a set of frequencies is listed: the source encoding domain and the channel encoding domain. The source encoding domain, schematically represented by the audio, video, and transport encoders, uses a family of frequencies based on a clock of 27 MHz. A clock is used to generate a sample of 42 bits, which is divided into two elements defined by the specification MPEG-2:

- base clock of 33 bits for program reference;
- extension of the reference clock of 9 bits.

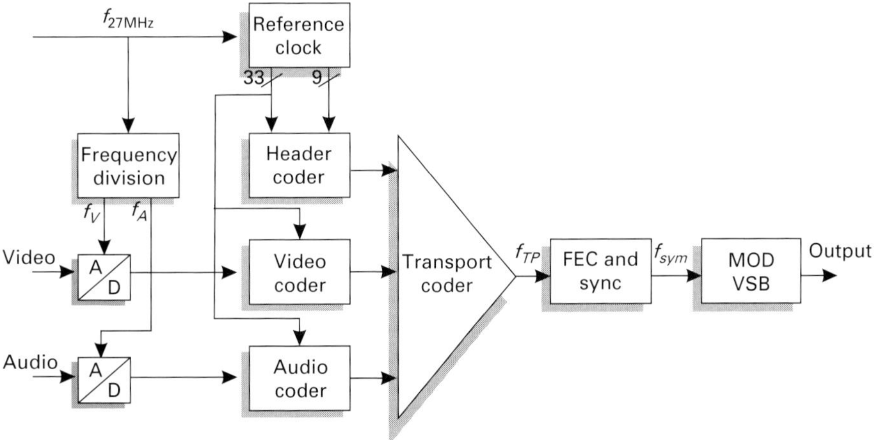

Figure 6.2   Encoding system for digital television

The 33-bit clock is equivalent to a sampling rate of 90 ksamples/s, which is locked on the frequency of 27 MHz and is used by the source encoders for the encoding of audio and video using Presentation Time Stamp (PTS), and their decoding using Decode Time Stamp (DTS). The clocks for the sampling of audio and video, f_A and f_V, respectively, are synchronized at a frequency of 27 MHz. This condition constitutes a prerequisite for the existence of two pairs of integer numbers, (n_A, m_A) and (n_V, m_V), in a way that

$$f_A = \frac{n_A}{m_A} \times 27 \text{ MHz}. \tag{6.1}$$

and

$$f_V = \frac{n_V}{m_V} \times 27 \text{ MHz}. \tag{6.2}$$

6.2.2 Multiplexing of services and transport

Multiplexing of services and transport refers to the division of each bit stream into information packages, to the uniqueness in the identification of each package or type of package, and to the methods appropriate to the interleaving or multiplexing of the streams of video, audio, and data in only one transport mechanism. The structure and the relationships of this one bit stream are transported in the stream of information services and are multiplexed in only one transport mechanism. The digital television system makes use of the MPEG-2 transport syntax for packaging and multiplexing of video, audio, and data signals for the transmission systems.

In the development of the transport mechanism, interoperability between the digital media, such as terrestrial transmission, cable distribution, recorded media, and computer interfaces, is the first consideration.

6.2.3 Transmission system

Transmission refers to the channel encoding and to the modulation. The channel encoder receives the flow of digital bits and adds information that is used by the receiver to reconstruct the signal data received, which, owing to the transmission obstacles, do not precisely represent the transmitted signal. The modulation, on the physical layer, uses the stream of digital information to modulate a carrier for the transmitted signal. The modulation subsystem uses 8-VSB, 16-VSB, 64-QAM, 256-QAM, or QPSK, depending on the type of transmission (broadcasting, cable or satellite).

Figure 6.3 shows the basic block diagram for the transmitter/receiver pair of the ATSC standard. The application encoders/decoders shown in Figure 6.3 refer to the methods of reduction of the bit rate, also known as methods of data compression, appropriate for streams of video, audio, and auxiliary data.

The ATSC receiver must retrieve the bits which represent the video, audio, and other original data of the modulated signal. In particular, the receiver must:

- tune to the selected 6 MHz channel;
- reject adjacent channels and other sources of interference;

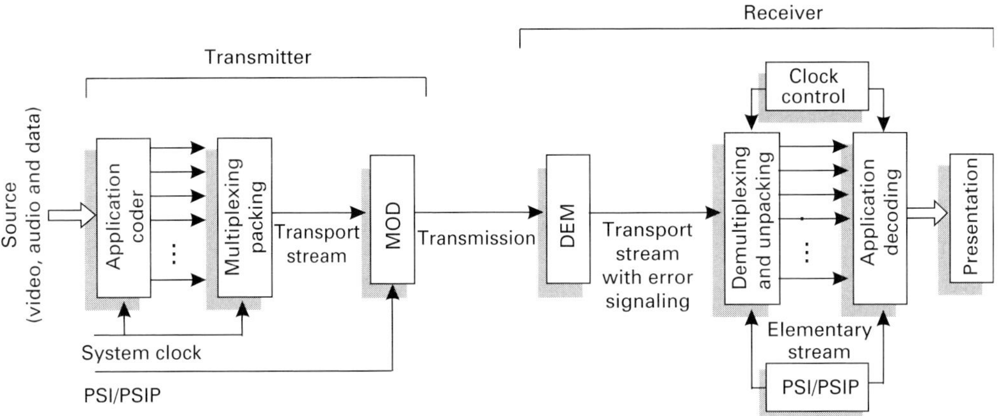

Figure 6.3   Organization of the functionalities in the transmission/receiver pair

- demodulate (and equalize, whenever necessary) the received signal, applying error correction;
- identify the elements of bit flow using the processing on the transport layer;
- select each desired element and send it for adequate processing;
- decode and synchronize each element;
- present the program.

Noise, interference, and multipath are common problems in transmissions, and the receiving circuits are designed to deal with these disturbances. Equalization innovations, automatic control of gain, cancellation of interference, carrier, and retrieving time create products with differentiated performance and improve the signal's coverage.

The decoding of the transport elements that constitute the program is usually considered to be a direct way of implementing the specifications, although there are opportunities for innovation in the efficiency of the circuits or use of power. In particular, innovations in the decoding of video offer opportunities to save memory, for the speed of the circuit, and for complexity. The interface with the user and the new services based on data are other important areas of differentiation of the products.

6.2.4   Basic features of the video system

Table 6.1 shows a list of the main television standards that define the video formats, as well as the compression techniques applicable to the ATSC standard. The image format derives from one or more appropriate video formats.

The video compression algorithms follow the syntax according to recommendations of the main profile of the ISO/IEC 13818-2 (MPEG-2) standard. The acceptable parameters are determined by the upper limits specified in this standard. Table 6.2 shows a list of the compression formats permitted on the ATSC standard (ATSC, 1995b).

Table 6.1. Standardized video formats

Video standard	Active lines	Active samples/line
SMPTE 274M	1 080	1 920
SMPTE 296M	720	1 280
ITU-R BT.601-4	493	720

Table 6.2. Limitations in the compression of formats of the ATSC standard

Vertical dimension	Horizontal dimension	Aspect ratio	Frame rate	Scan sequence
1 080	1 920	16:9, square pixels	1, 2, 4, 5 4, 5	Progressive Interlaced
720	1 280	16:9, square pixels	1, 2, 4, 5, 7, 8	Progressive
480	704	4:3, 16:9	1, 2, 4, 5, 7, 8 4, 5	Progressive Interlaced
480	640	4:3, square pixels	1, 2, 4, 5, 7, 8 4, 5	Progressive Interlaced

Frame rate (the frame rate unit is the frame per second, but it is usual to refer to it in Hz): 1 = 23.976 Hz, 2 = 24 Hz, 4 = 29.97 Hz, 5 = 30 Hz, 7 = 59.94 Hz, 8 = 60 Hz

6.3   Transmission modes

6.3.1   Terrestrial transmission mode (8-VSB)

The terrestrial transmission mode (8-VSB) supports a data transmission rate of approximately 19 Mbit/s on a 6 MHz channel. A block diagram of this mode is shown in Figure 6.4. The input to the transmission subsystem coming from the transport subsystem is a serial data stream made up of MPEG data packages of 188 bytes, representing the 19 Mbit/s.

The input data go through a pseudorandom sequencer and are then processed by a forward error correction (FEC), known as a Reed–Solomon encoder (through which parity bits are added to each package). This process is followed by interleaving one sixth in the data field and trellis coding with a rate of two thirds. The sequencing and the FEC are not applied to the synchronizing bytes of the transport package, which is represented in the transmission by a data segment with a sync signal. Following the sequencing and FEC processes, the data packages are formatted in data frames for transmission, and the segments with synchronizing data added.

Figure 6.5 shows how the data are organized for the transmission. Each data frame includes two data fields, each with 313 data segments. The first segment of each data field is a synchronizing signal, which includes a training sequence used by the equalizer in the receiver. The remaining 312 data segments transport the data equivalent to the transport of 188 bytes, as well as the overload associated with the FEC. These 828 symbols are

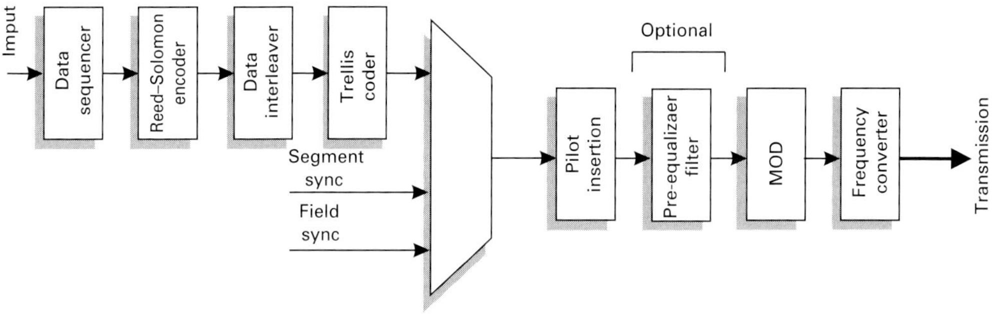

Figure 6.4   Block diagram of the 8-VSB transmitter

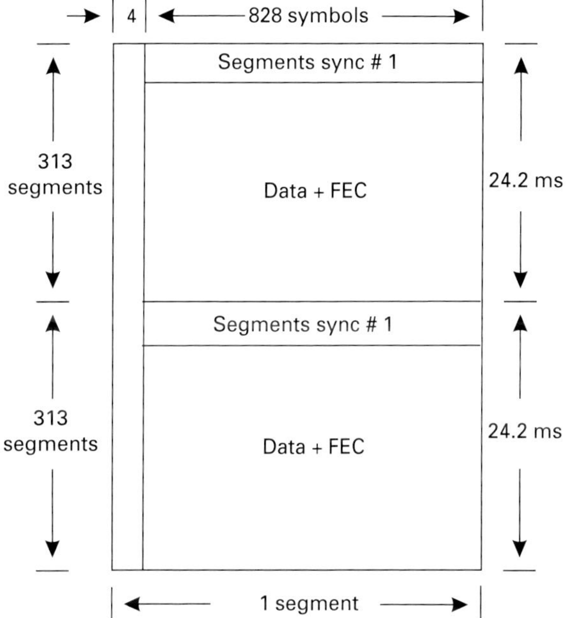

Figure 6.5   The 8-VSB data frame

transmitted as eight-level signals and, therefore, transport 3 bits per symbol. Then, 2484 (828×3) data bits are transported in each data segment, which goes exactly with the prerequisite for sending protected transport packages:

- 187 data bytes $+$ 20 Reed–Solomon parity bytes = 207 bytes;
- 207 bytes $\times$ 8 bits/bytes = 1656 bits;
- the two-thirds rate in the trellis coding requires $3/2 \times 1656$ bits = 2484 bits.

The exact symbol rate is

$$S_R = \frac{4,5}{286} \times 684 = 10.76 \text{ MHz.} \tag{6.3}$$

The frequency of the data segment is

$$f_{seg} = \frac{S_R}{832} = 12.94 \times 10^3 \text{ data segment/s.} \tag{6.4}$$

The frame rate (per second) is represented by

$$f_{quadro} = \frac{f_{seg}}{626} = 20.66 \text{ frame/s.} \tag{6.5}$$

The symbol rate, S_R, and the transport rate, T_R, are fixed in the frequency. The eight-level symbols, combined with the binary data segment, sync, and the sync signals of the data field are used to modulate the carrier. Before the transmission, however, the biggest part of the lower band is removed. The resulting spectrum is, therefore, reduced, except for the sidebands in which a nominal response to a co-sinusoidal square wave results in regions with transitions of 620 kHz. The VSB nominal transmission spectrum is shown in Figure 6.6. It includes a small pilot signal in the suppressed frequency, 310 kHz of the lower side band.

Protection against errors and channel synchronization

The entire useful data load is transported with the same priority. A data pseudorandom sequencer is used in all input data to pseudorandomize the useful load of each package (except for the data field with sync, segments with sync, and Reed–Solomon parity bytes). The data sequencer uses the XOR operation in all bytes with a pseudorandom binary sequence (PRBS) of 16 bits of maximum length. The PRBS is generated in a 16-bit shift-register, which has nine feedback inputs. Eight of the outputs of the shift-register are selected as fixed interleaving bytes. The polynomial generator for the pseudorandom code and its initialization are presented in Figure 6.7.

Reed–Solomon encoding

The Reed–Solomon code used in the 8-VSB transmission subsystem gives a data block size of 187 bytes, with 20 parity bytes added for error correction. A 207-byte Reed–Solomon block is transmitted per data segment. In the creation of bytes from a serial bit stream, the most significant bit (MSB) is the first serial bit. The 20 Reed–Solomon parity bytes are sent at the end of the data segment. The Reed–Solomon encoding/decoding

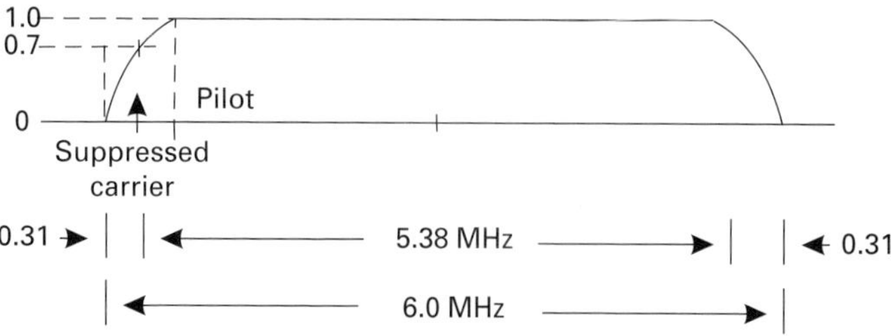

Figure 6.6   Nominal occupation of the 8-VSB channel

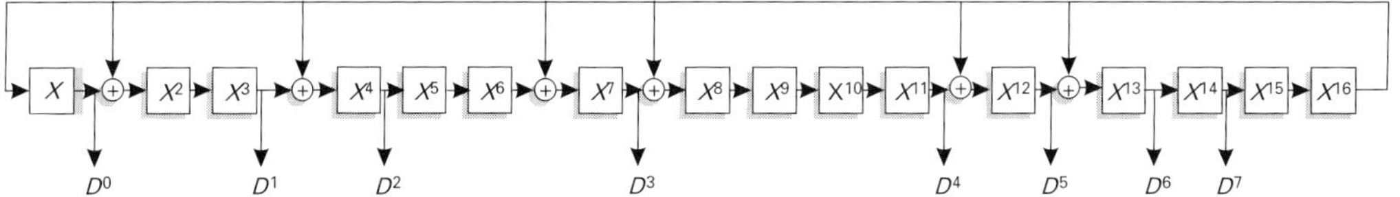

Figure 6.7   Data scrambling polynomial

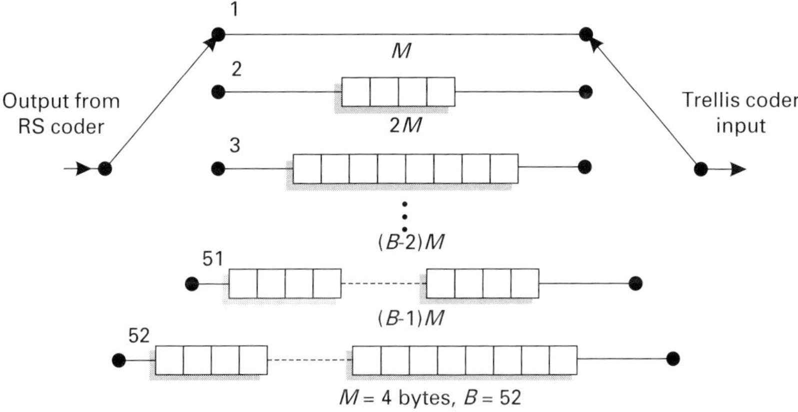

Figure 6.8   Convolutional interleaving scheme

process is expressed as the total number of bytes in a package transmitted to the payload bytes of the current application, in which the Reed–Solomon bytes used are the overload.

Interleaving

The VSB transmission system makes use of a convolutional interleaver of 52 data segments. The interleaving is provided for a depth of about one sixth of a data field (4 ms). Only the data bytes are interleaved. The interleaver for the first data byte of the data field is synchronized. The interleaving between segments is made to improve the trellis coding process. The convolutional interleaver stage is shown in Figure 6.8.

Trellis coding

The 8-VSB transmission subsystem uses a trellis code with a two thirds (2/3) rate. An input bit is encoded in two bits in the output using a half-rate convolutional code, whereas the other bit is pre-encoded. The waveform used with the trellis code is a one-dimensional constellation of eight levels (3 bits). A trellis code is used between interleaved segments.

This requires 12 identical trellis encoders as well as the interleaving operation in the data symbols. The interleaving is accompanied by the encoding of the symbols (0, 12, 24, 36, …) as a group, the symbols (1, 13, 25, 37, …) as a second group, and the symbols (2, 14, 26, 38, …) as a third group, and so forth up to a total of 12 groups.

In the creation of serial bits from parallel bytes, the MSB is sent first: (7, 6, 5, 4, 3, 2, 1, 0). The MSB bit is pre-encoded (7, 5, 3, 1) and the least significant bit (LSB) is fed back to the convolutional encoder (6, 4, 2, 0). The trellis coder uses the four-state feedback encoders, as shown in Figure 6.9. The pre-encoding and the mapping of symbols are also shown in this figure. The trellis code and the pre-encoder between interleaved segments, which feed the mapping shown in Figure 6.9, are illustrated in Figure 6.10. As shown in the figure, data bytes are fed from the byte interleaver to the trellis coder and the pre-encoder. Then, they are processed as full bytes by 12 encoders. Each byte produces four symbols from only one encoder.

The multiplexed output shown in Figure 6.10 advances four symbols in each limit-segment. The estate of the trellis coder does not advance. The data that go out of the

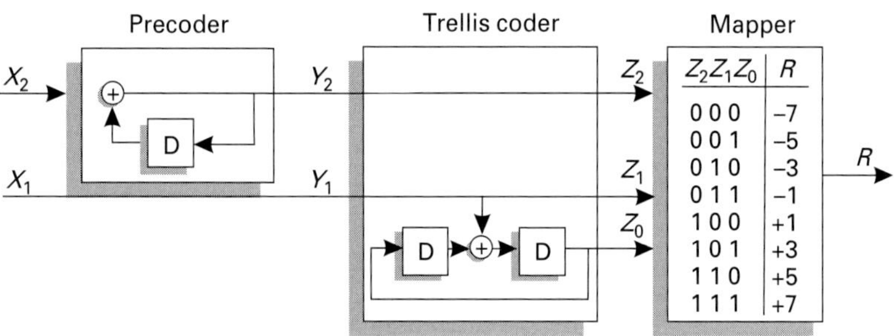

Figure 6.9   Pre-encoder, encoder, and mapping of 8-VSB symbols

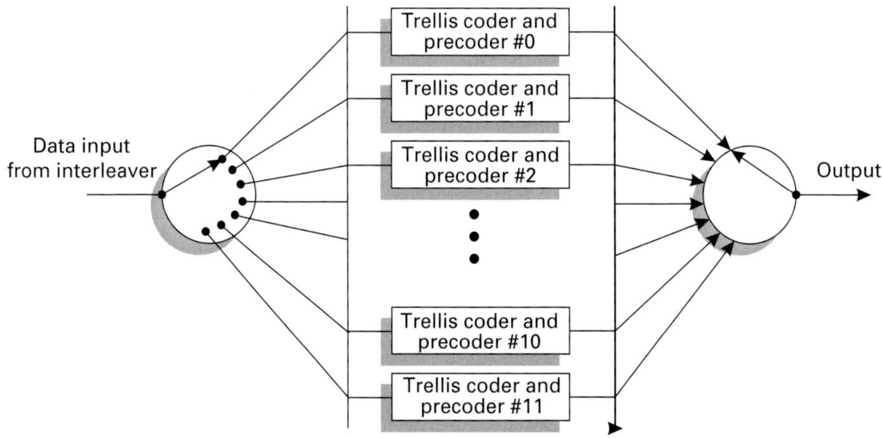

Figure 6.10   Interleaver for the trellis code

Table 6.3. Interleaved partial sequences for trellis encoding

Segment	Block 0	Block 1	...	Block 68
0	D0 D1 D2 ... D11	D0 D1 D2 ... D11	...	D0 D1 D2 ... D11
1	D4 D5 D6 ... D3	D4 D5 D6 ... D3	...	D4 D5 D6 ... D3
2	D8 D9 D10 ... D7	D8 D9 D10 ... D7	...	D8 D9 D10 ... D7

multiplexer follow the regular order of the encoders from 0 until 11 from the first segment of the frame. In the second segment, the order changes and the symbols are read from encoders 4–11, and then from encoders 0–3. The third segment reads from encoders 8–11, and then from encoders 0–7. These three patterns are repeated through all the 312 segments of the frame. Table 6.3 shows the interleaving sequence of the first three data segments of the frame. After the sync data segment is inserted, the order of the data symbols is such that the symbols of each encoder occur at each 12-symbol interval.

A complete conversion of parallel bytes into serial bits needs 828 bytes to produce 6624 bits. The data symbols of two bits sent in the MSB order are created. In this way, a complete conversion operation yields 3312 data symbols, which correspond to the four segments of 828 data symbols. A total of 3312 divided by 12 trellis encoders results in 276 symbols per trellis coder, and 276 symbols per byte resulting in 69 bytes per trellis coder.

The conversion begins with the first segment of a field and involves groups of four segments until the end of the field. A total of 312 segments per field, divided by 4, result in 78 conversion operations per field.

During the sync segment, the input to the four encoders is bypassed, and the encoding cycle is done without any input. The input is stored until the next multiplexing cycle, and then the correct encoder is fed.

Modulation

The mapping of the outputs of the trellis decoder to the levels of nominal signals of the signal $(-7, -5, -3, -1, +1, +3, +5, +7)$ is shown in Figure 6.9. The nominal levels of the sync data segment and the sync data field are -5 and $+5$. The value 1.25 is added to all nominal values after the mapping of bits into symbols, in order to create a small pilot frequency. The frequency of the pilot is the same as that of the suppressed carrier. The power of this pilot is 11.3 dB lower than the average power of the data signal.

The VSB modulator receives 10.7 Msymbols/s, of eight levels, encoded in trellis with a data signal composed of a pilot and added sync frames. The performance of the digital television system depends on the response of a linear phase filter with raised Nyquist cosine in a transmitter connected to the receiver. The response of the system filter is essentially flat throughout the band, except in the transition regions at the end of each band, as shown in Figure 6.11.

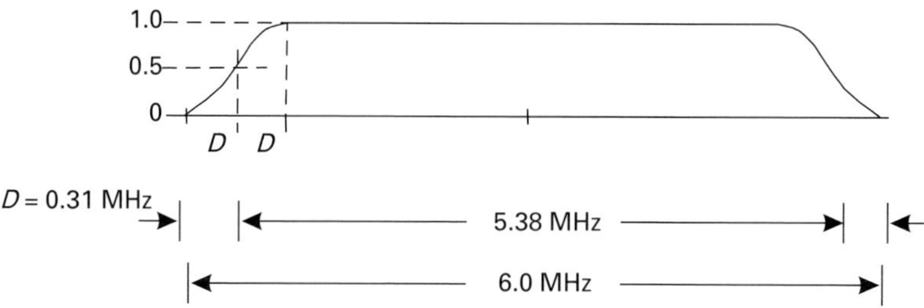

Figure 6.11   Nominal response of the 8-VSB channel (with filter)

6.3.2   Terrestrial transmission mode (16-VSB)

The 16-VSB transmission mode is less robust than the 8-VSB scheme (a threshold SNR of 28.3 dB bit), due to its high-rate of 38 Mbits/s (ATSC, 1995b). Several parts of this mode are similar, to the 8-VSB system. A pilot, a sync data segment, and a sync data field are used to provide a robust operation. The power of the pilot in this mode is 11.3 dB lower than the power of the signal, and the symbols, the segments, the data fields, and all the rates are the same. The data frame definitions are also the same. The preliminary difference is the number of levels transmitted (16 against 8), the use of the trellis encoding, and the filtering used in the NTSC for the rejection of interference in the terrestrial system. The RF spectrum of this high-rate mode using transmitter modems looks identical to the 8-VSB system.

A typical data segment presents 16 levels of data, which result in the duplicated transmission rate. Each portion of 828 data symbols represents 187 data bytes and 20 Reed–Solomon bytes, followed by a second group of 187 data bytes and 20 Reed–Solomon bytes (before the convolutional interleaving).

Figure 6.12 shows a functional block diagram of a transmitter with high rates. It is identical to the 8-VSB system, except for a trellis encoding which is replaced by a mapping that converts the data into multilevel symbols. The interleaver is convolutional with 26 data segments. The interleaving is provided for a depth of about one twelfth (1/12) of a data field (2 ms). Only the data bytes are interleaved. Figure 6.13 shows the mapping of the interleaver outputs to nominal levels $(-15, -13, -11, \ldots, +11, +13, +15)$. The nominal levels of the sync data segment and the sync data field are -9 and $+9$. The value 2.5 is added to all of these nominal levels after the mapping of bits into symbols with the objective of creating a small pilot frequency with the same phase of the suppressed carrier. The modulation method of this mode is identical to the 8-VSB mode, except for the fact that the number of levels transmitted is 16 instead of 8.

6.3.3   Satellite transmission mode

The satellite transmission mode uses of the QPSK modulation technique as a standard, with the choice to use either 8-PSK or 16-QAM. The decision on which of these optional types of transmission to use is up to the manufacturer, as long as the product is compatible with the standard (ATSC, 1999).

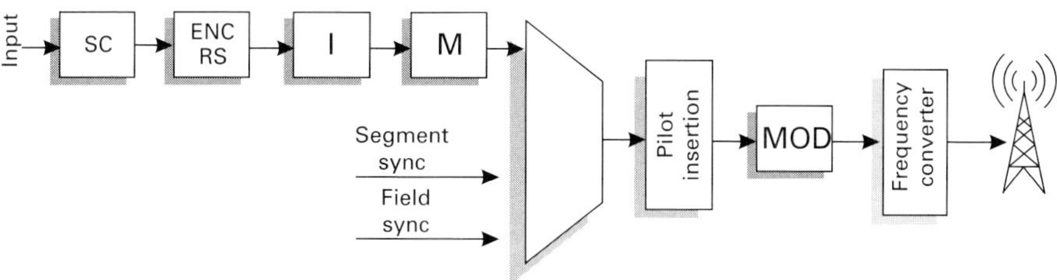

Figure 6.12   16-VSB transmitter functional block

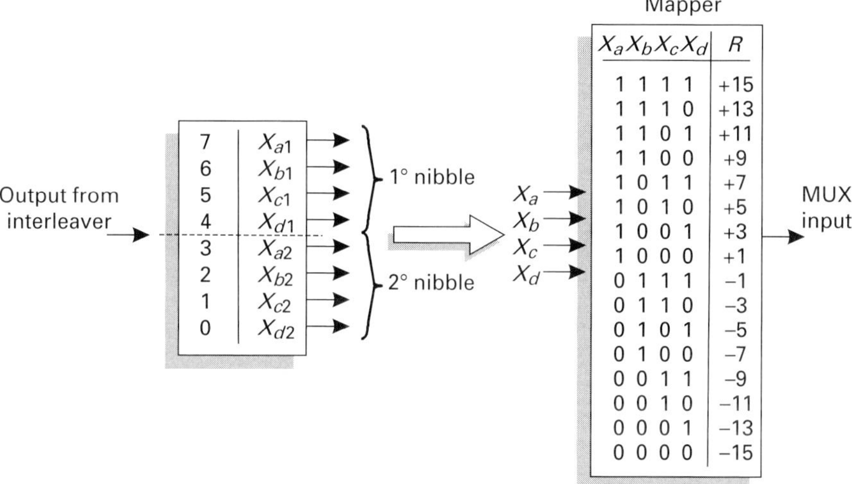

Figure 6.13   16-VSB mapping table

The block diagram of a simple system, as shown in Figure 6.14, describes a source of digital data that can be represented by a video encoder/multiplexer (or decoder/demultiplexer) for ATSC applications, but may also represent a variety of other sources that produce a flow of digital data. The focus of this transmission mode is the segment between the dotted lines and includes the modulator and the demodulator. Only the modulation parameters are specified. The standard does not prevent the combination of equipment outside of the dotted lines with the modulator or the demodulator, but adjusts a logical border between their functions.

The modulator receives a stream of data and does some operations to generate an intermediate frequency (IF) carrier appropriate for satellite transmission. The data go through an FEC. It is then shuffled, mapped into a QPSK constellation, 8-PSK or 16-QAM, and suffers a frequency conversion and other operations to generate the IF carrier (ATSC, 1999).

The selection of the modulation technique and the FEC affects the bandwidth of the signal produced by the modulator.

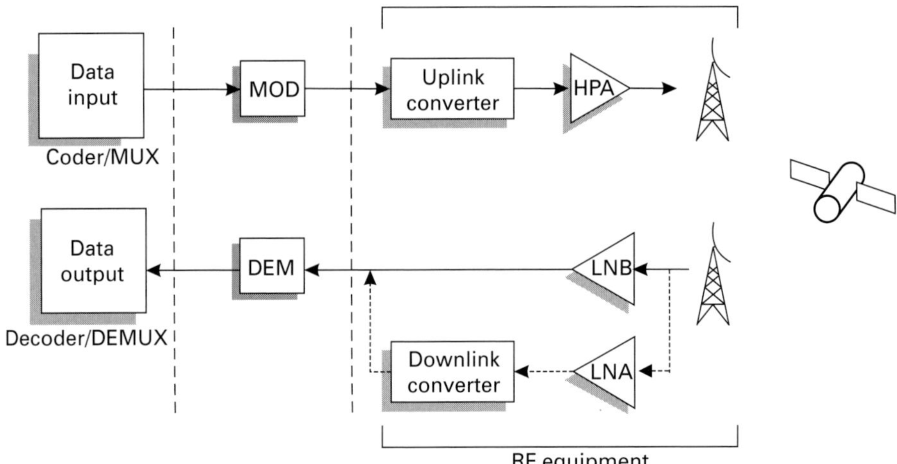

Figure 6.14 Satellite transmission mode block diagram. LNA is low-noise amplifier; LNB is low-noise block converter; HPA is high-power amplifier

There are two types of encoding in the standard. An external Reed–Solomon code is connected with an internal convolutional/trellis code to produce error correction that is more powerful than either of the codes alone. The Reed–Solomon code rate is fixed (188/204), but the convolutional/trellis code rate is selectable, offering the opportunity to change the bandwidth. For example, by choosing a higher internal encoding rate, 7/8 instead of 1/2, the bandwidth used for a given information transmission is reduced.

The main measure in relation to the energy is the SNR (E_b/N_0), and the main performance parameter is the bit error rate (BER). For digital video, a BER of approximately 10^{-10} is necessary to guarantee high quality. Thus the SNR and BER are useful parameters when choosing which modulation scheme and encoding to adopt.

6.4   Standard for interactivity

The DTV Application Software Environment (DASE) standard defines a software layer (middleware) which allows the content of the program and the applications to run on a common receiver. There is a need for the interactive and improved applications to have access to a common receiver, independent of the platform. The standard provides the content developers with the specifications needed to guarantee that their applications and data will run uniformly on all receiver brands and models. The manufactures can choose the hardware platforms and the operational systems for the receivers, but they must provide the common base necessary to accommodate applications developed by multiple content developers.

The DASE standard specifies a standard format of application content and an environment of applications that accommodates hypermedia documents (declarative applications) and/or compiled programs (procedural applications). The DASE declarative applications employ several technologies used on the Internet (web), such as XHTML, Cascading Style Sheets (CSS), ECMAScript, and Document Object Model (DOM), as well as types of single-media content, such as steady images (JPEG, PNG), animated images (MNG), outline and bitmap sources (TrueDoc), and streaming and non-streaming audio and video.

The DASE procedural applications use compiled Java programs with J2ME base, Java Media Framework, and TV Java technologies, as well as (in the user's interface) HAVi Organization, W3C Document Object Model, and Application Programming Interfaces (APIs) specific to the DASE.

The hybrid DASE applications accept a content combination of both declarative and procedural applications, so that it obtains the benefits resulting from this functionality synthesis. In particular, the hybrid DASE applications use the types of content defined by the application via content decoders subject to download.

The DASE application environment provides a standard user's agent (browser) for declarative application content and a Java Virtual Machine and APIs for procedural applications.

The DASE system defines itself independently of the subjacent operational system, and of the data transmission system. For the use of specific data transmission systems, such as the A/90 data diffusion ATSC standard (ATSC, 2000a), DASE defines a binding with the transmission system. Other bindings with the transmission system are contemplated for future standardization.

The DASE standard evolves by means of multiple levels or versions that unfold the previous levels. The first instance of the DASE standard, known as Level 1, provides applications based on the local interactivity. The second instance, Level 2, must accommodate remote interactivity for types of content specified by the DASE. The third instance, Level 3, accommodates interaction by the network and types of general content.

The DASE system was explicitly designed to accept functional interoperability with improved interactive content, specifically Advanced Television Enhancement Forum (ATVET) content, now standardized as SMPTE Declarative Data Essence, Level 1 (DDE-1). In addition, the DASE declarative content format was submitted to the SMPTE for standardization as Declarative Data Essence, Level 2 (DDE-2).

The DASE standard is divided into eight sections:

- Part 1: introduction, architecture, and common facilities;
- Part 2: declarative applications and environment;
- Part 3: procedural applications and environment;
- Part 4: application programming interface;
- Part 5: portable resources;
- Part 6: safety;
- Part 7: application transmission;
- Part 8: compliance.

6.4.1   Java TV

Java TV (Fernandes *et al.*, 2004) is an extension of the Java platform that accepts the production of content for interactive television. Java TV's main aim is to make the development of portable interactive applications possible. These applications are independent of the transmission network technology.

Java TV consists of a Java virtual machine (JVM) and several libraries of reusable and specific codes for interactive digital television. The JVM is hosted and executed in the set-top box (STB) itself. Java was developed in the J2ME environment, which was developed to operate on devices with resource restrictions. In this context, certain characteristics of the Java platform have been excluded, as they are considered unnecessary or inadequate for such environments. Nevertheless, J2ME does not define specific television functionalities. Such functionalities are included in the Java TV.

Java TV allows advanced levels of interactivity, quality graphs, and local processing in the STB itself. These facilities offer an ample spectrum of possibilities to content designers, even in the absence of a return channel. That is, electronic programming guides (EPGs) can offer a general overview of the programs available, which accepts the change, by the user, to the desired channel. By means of synchronization mechanisms, specific applications may be associated with a given television program. In addition to this, isolated applications are executed independently of the television program.

On Java TV, traditional and interactive television programs are featured as a set of individual services. A service is a collection of content for presentation in an STB. For instance, a service may represent a conventional television program with synchronized audio and video, or an interactive television program, which has audio, video, data, and applications associated with it.

The API Java TV provides an abstraction that makes it possible for applications to obtain information about the various services available independently of the hardware and of the adopted protocols. In this way, an application may be reused in a variety of environments.

7 Digital video broadcasting (DVB)

with Jeronimo Silva Rocha, Paulo Ribeiro Lins Júnior and
Mozart Grizi Correia Pontes

7.1 Introduction

The Digital Video Broadcasting (DVB) Project was created in 1993 as a consortium composed of more than 300 members, including equipment manufacturers, network operators, software developers, and regulatory agencies; it started its activities in Europe, but since then has become worldwide (Reimers, 2004). This project developed several technical solutions for coding and transmission of digital television.

For several reasons, but mainly because of market pressures, which demanded satellite, cable, and Satellite Master Antenna Television (SMATV) solutions, the development of a terrestrial transmission system was started in 1994/1995.

The word broadcasting is generally used to describe the transport of media content from a single point of origin to multiple receivers, regardless of the physical layer used for this transport. The specification for transport of Internet Protocol (IP) data in DVB networks, the interactivity solutions and return channels, the software environment, called Multimedia Home Platform (MHP), and the specifications that allow a point–multipoint distribution of different types of data to mobile devices are part of the solution.

DVB is focused on the private customer, who uses the offered services at home, but DVB broadcasting capabilities are also used for professional applications. In addition, DVB makes possible the reception of digital signals by portable and mobile devices.

The DVB standard has been adopted by the European Union and other countries, such as Australia, New Zealand, Malaysia, Hong Kong, Singapore, India, and South Africa. The United Kingdom is the country where the adoption of DVB first consolidated (Reimers, 2006).

The DVB standard allows several different settings for the transmission layer, each of them featuring a different capacity–robustness relation. The main transmission standards which the DVB comprises are: DVB-T (broadcasting terrestrial transmission), DVB-C (cable transmission), DVB-H (transmission to portable devices), DVB-S (satellite transmission), DVB-MC (microwave transmission, operating in frequencies up to 10 GHz) and DVB-MS (microwave transmission, operating in frequencies above 10 GHz).

7.1.1 Application fields of the DVB technologies

A very important aspect of DVB is the technical work commercially led by the requirements of the member organizations. Therefore, the application fields of the DVB technologies have developed significantly. The main goals of the first activities, in 1991 and 1992, can be summarized as follows (Reimers, 2006):

(1) Digital television offering the transmission of HDTV images by terrestrial broadcasting networks.
(2) Transmission of programs with SDTV quality using narrowband channels, which implies a larger number of programs with the existing channel allocation.
(3) Transmission to low-cost receivers (including mobile devices), equipped with internal reception antennas or even short shaft antennas, providing a stable reception for a certain number of television programs.
(4) Television receivers in vehicles (cars, trains or buses) with good transmission quality. In other words, DVB must provide stable reception even for moving vehicles.
(5) Maintaining the digital technology's typical characteristics, such as stability of reception in a defined area and the possibility of a simple distribution through the telecommunications networks as a new type of service.

By the mid-1990s, the transmission of DVB by satellite (DVB-S) had been very successful. The digital television industry then began to extend the standard in order to fit the diverse needs of many different countries. These new requirements can be described by the following:

(1) The ability to transmit multiple television TV programs over a single transmission channel, regardless of whether the transmission is accomplished by satellite, by cable networks or by terrestrial transmissions.
(2) Support of the transmission of radio programs and data transmission for entertainment and business purposes.
(3) The ability for the user to choose the image and audio quality, including the choice of HDTV.
(4) For use in commercial services, specification of a safe encryption method, which ensures that there is no unauthorized access to these services.
(5) Introduction of the DVB standards for the interaction channel between the user and the network operator or contents provider.
(6) An open and interoperable software platform for various types of services, including broadcasting services (with or without interactivity), or even Internet access by the television receiver.
(7) The possibility of installing receivers in all types of environment, from traditional television sets to mobile televisions and devices that receive the television signal.
(8) Concerning the transmission techniques, the incorporation of typical characteristics for use with digital technologies, creating alternatives, for example, the integration with personal computers.

In 2000, the DVB committee revised its goals, and redefined its vision for the following years: "The vision of the DVB is to construct a contents environment that combines the stability and interoperability of the broadcasting world, with the power, innovation and multiplicity of services of the Internet world". Since then, several solutions have been developed to fit the requirements of the member organizations (Reimers, 2006).

7.2   System features

The MPEG-2 standard, which was developed by the Moving Pictures Experts Group (MPEG) was chosen by the DVB Project for audio and video source coding and for the multiplexing of different signals. The reason for this decision was the fact that MPEG-2 promised to become a solution that could be accepted worldwide, which has become true over the years. This makes possible the utilization of embedded systems produced by several companies, especially in the decoders.

7.2.1   Baseband processing – MPEG-2

The international standard ISO 13818 is composed of multiple parts, three of which describe the systems, audio, and video of MPEG-2. It is a set of generic standards (Mendes and Fasolo, 2002, Fernandes *et al.*, 2004), and the use of the existing documentation in order to begin a DVB service becomes rather complicated. Hence, it is necessary to define subsets in the form of implementation guides, so that the specific services can be accomplished. These guides include syntax restrictions and some parameter values described by the MPEG-2 standard, as well as recommendations for preferable values for the use in DVB applications. The complexity of documentation can be shown by the fact that the integrated receiver decoders (IRDs) are classified taking into consideration five factors:

- video frame rate of 25 fps (frames per second) or 30 fps;
- resolution (standard-definition television (SDTV) or high-definition television (HDTV));
- presence of digital interface or base line;
- video encoding MPEG-2 or H.264/AVC;
- several audio formats, selected from a list of options.

The MPEG-2 video compressing method is based on asymmetric algorithms in which the coding costs are much higher than the decoding ones. The algorithms are very flexible, which permits coding the images in different levels of resolution (quality). The part of the MPEG-2 that deals with video is generic, comprising many algorithms and tools. The use of different subsets of MPEG in a disorganized way could make interoperability of systems impossible. For that reason, a hierarchical structure of profiles and levels was created in such a way that the interoperability of systems is guaranteed even if they are working on different levels.

For audio coding, DVB comprises the use of the MPEG-2 bit stream. Since 2004, the use of more efficient source coders for audio and video has become a necessity. Consequently, the DVB group has offered support to H.264/AVC for the video content and optional support to the MPEG-4 high-efficiency AAC audio coder.

MPEG-2 is a solution for traditional broadcasting networks. However, DVB services can be distributed over IP networks. In this case, especially in real-time applications, H.264/AVC is a solution for the video coding and MPEG-4 AAC is used for the audio coding.

For transmission of several signals, a multiplexing and transport subsystem becomes necessary; their function is to an receive elementary sequence of bits, generated by the encoders of the different applications' subsystems (video, audio, data) and, through a multiplexing process, to generate at the output a sequence of packets, whose format is defined by the MPEG-2 systems (Fernandes *et al.*, 2004). The packet coding can be accomplished in two different ways: program flow or transport flow. In the program flow, the packets are of varying size, and are usually large. This kind of flow is used for systems with low error probability, which is not the case for television systems.

In transport flow, the packets have a fixed size of 188 bytes, which is more suitable for error correction, simplifies the implementation of electronic circuits and algorithms and decreases the processing time. The transport packet sequence is multiplexed with other sequences of the same type before being sent to the transmission subsystem. At the receiver, that packet sequence is demultiplexed and the elementary sequence of bits handed to the respective decoder. Using information from the transport packets header, it is possible to accomplish operations such as receiver synchronization, error detection, and signaling.

Before the multiplexing process and formatting into transport packets, the elementary bit sequence can either pass through a process of organization into packetized elementary stream (PES) segments of variable or fixed size. The main goal of the PES segmentation is to provide synchronization of elementary bit sequences of the same program. It is compulsory for the audio and video sequences to pass through this stage. The generation of PES segments can be accomplished directly by the transport and multiplexing subsystem, or even by the encoder of the elementary bit sequence generating application.

Audio and video multiplexing into transport packets is achieved by a simple identifier field in its header. This field is called the packet identifier (PID) and its use allows, for example, the channel capacity to be allocated dynamically for bursts of a given elementary sequence generator subsystem.

DVB services comprise a variety of programs transported through a large number of transmission channels. As the receiver is able to tune such channels and DVB customers can surf through the television guide, efficient navigation schemes need to be provided as a part of the DVB data stream. The service information (SI) is described in the DVB-SI document.

The specification for the TV-Anytime information transport and signaling in the transport streams is the most recent accomplishment of the DVB Project (Reimers, 2006). This document defines how personal digital recorders (PDRs) can be supported using the phase 1 of the TV-Anytime specification in DVB transport streams. The recording process of the TV-Anytime content is *search, select, acquire*. The user selects the correct content and the PDR records it. A number of interrelated technologies are defined that can be used for several different PDR applications.

Transport stream (TS) processing

The baseband signal transmitted for DVB is the MPEG-2 transport stream (TS). The TS is a continuous packet sequence, each packet being 188 bytes long. The first four bytes contain the TS packet header; the following 184 bytes are used for the payload.

The most significant parts of the header are the synchronization byte (sync) and the PID (Ladebusch and Liss, 2006).

The elementary stream identified by a PID can be audio or video, and can carry data such as program guide, subtitles, and IP packets. The transport mechanism is based on the packing of elementary streams or tables mapped onto the TS packets.

In order to make the access to the services easier for users, DVB has defined tables with SI. One of the most important of these is the network information table (NIT), which describes the modulation parameters of the TS that come and go on the network. It is important, for example, for the handover proceedings in a mobile environment. Another important table is the service description table (SDT), which translates program numbers into service names. The events information table (EIT) can offer almost all information found in a printed guide of a given program, the name of event, timetable, duration, and classification are the main features of the EIT.

7.2.2 DVB transmission

In order to be transmitted through the channel, the baseband signal has to undergo a coding and a modulation process. A Forward Error Correction (FEC) technique is essential to allow the receiver to correct the errors that occur as a result of noise and other disturbances on the transmission channel. In addition, a synchronization method has to be provided. The complete block diagram for a DVB-T encoder is shown in Figure 7.1 (Ladebusch and Liss, 2006).

The DVB-C, DVB-S, and DVB-T coding processes are based on the same fundamental concepts (Reimers, 2006a). Thus, the first four blocks in the diagram are common to the three systems. The inner encoder becomes necessary only in satellite and terrestrial transmissions. Note that the DVB-S specification was developed in the middle of 1993. At that time, the chosen encoder was considered to be more advanced, and long discussions about the economic viability of implementing a Viterbi decoder in the user receiver, for example, took place.

There are some differences between the systems, especially concerning the modulation used. For DVB-C, the modulation is QAM (with 16, 32, 64 or 256 points in the

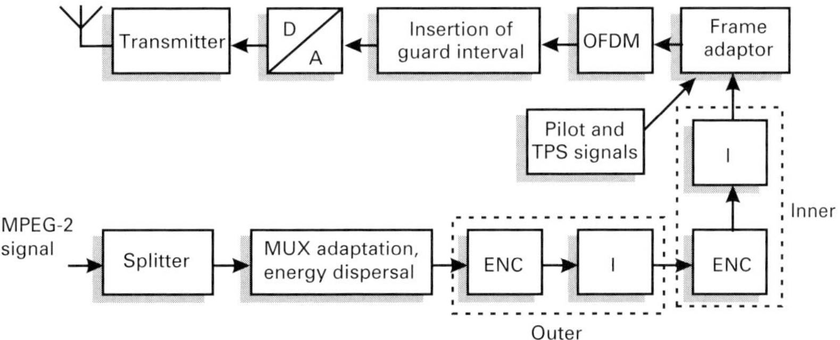

Figure 7.1 Block diagram of the encoder for DVB-T

constellation). DVB-S uses QPSK or PSK and DVB-T orthogonal frequency division multiplexing (OFDM).

To understand the operation of the transmission system, the roles of the main blocks in Figure 7.1 are discussed next (Yamada *et al.*, 2004). Figure 7.1 shows the signal splitter, sync generator and energy dispersion, outer encoder, outer interleaver and the inner interleaver which compose the encoder. An explanation of the subsystems that are common to terrestrial, cable, and satellite transmission is given in the following.

Signal splitter

In DVB-T, the incoming digital signal is split into two streams with different digital contents to form the hierarchical structure. Thus, for example, one stream can contain a standard digital television signal and the other one can be designated for the mobile service (Yamada *et al.*, 2004).

The signal splitter block accomplishes the stream division of the original data. For a single HDTV channel, only one digital stream is used, therefore the digital signal is present in only one of the two signal splitter outputs.

In DVB-T, the bit rate of the incoming digital signal is variable, depending on the system configuration. Thus, subcarrier modulation schemes with fewer symbols and code rates of the inner encoder imply lower bit rates for the input digital signal. As an example, a QPSK modulation (four symbols) leads to a lower input bit rate when compared to a 16-QAM modulation (16 symbols) or 64-QAM (64 symbols). In the same way, a code rate of $\frac{1}{2}$ leads to a transmission data rate lower than that of a $\frac{3}{4}$ rate encoder.

The use of distinct modulation methods and different code rates changes not only the useful bit rate, but also the digital signal immunity to interference. Normally, for the modulated digital signal, the more robust to interference it is, the less the available transmission useful rate is.

Multiplex adaptation and energy dispersion

The data on the baseband interface are combined with the data stream of a pseudorandom noise generator, implemented with a shift register. The aim of this operation is to achieve flat power spectral density. Only the synchronization byte of the TS packets does not take part on this process, in order to keep the reference for the synchronization. After the randomization process, the data packets are organized, as shown in Figure 7.2, and injected into the outer encoder block. The decoders are informed about the start point of the sequence through a reverse synchronization byte (Ladebusch and Liss, 2006).

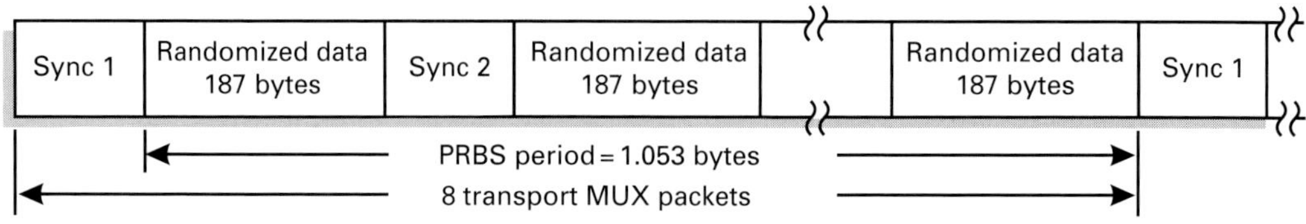

Figure 7.2   Randomized packages

Outer encoder

The main function of this encoder is to allow the receiver to detect and correct errors that occur in the demodulated and regenerated digital signal. It is implemented with a block code oriented to the byte. For each block, i.e. for each TS packet, error correction bits are computed. The result is a correction block of bytes which is added to the TS packet. The block coder uses a Reed–Solomon code (255, 239). That means that 16 correction bytes $(255 - 239 = 16)$ are added to the 239 information bytes. As long as the TS packet is 188 bytes long, the first 51 bytes are set to zero and are not transmitted (they are used for the parity check). Hence, a (204,188) Reed–Solomon code is created.

Figure 7.3 illustrates the data packets at the encoder output, already protected by the 16 parity bytes.

Outer interleaver

The outer interleaver rearranges bytes in an order that makes the burst error correction easier. This rearrangement allows the receiver to uniformly spread the errors caused by burst interferences (impulsive noise, for example) which occurs in a transmission. After the byte interleaving of depth equal to 12 $(I = 12)$, the packets are structured as shown in Figure 7.4.

The interleaver is a byte-wise convolutional type. The convolutional interleaving process is based on the Forney approach, which is compatible with the type III Ramsey approach for $I = 12$. This process does not introduce alterations on the input signal bit rate (Yamada *et al.*, 2004).

Inner encoder

For DVB-S and DVB-T, an inner encoder is needed because of the channel impairments. The role of this encoder is complementary to the Reed–Solomon coding, i.e. it allows the receiver to detect and correct errors introduced by interference present on the transmission channel. A convolutional encoder with basic code rate $\frac{1}{2}$ is used to accomplish this task. For each individual bit in the data input, two bits are computed at the output. The data stream is fed one bit at a time by a shift register and two bit streams are obtained as a result of the combination of several shift-register derivations.

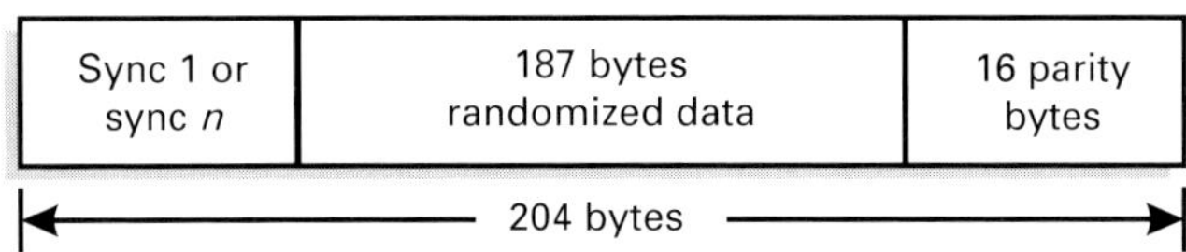

Figure 7.3   Reed–Solomon packages

Figure 7.4   Data structures after interleaving

The high amount of redundancy that results from the $\frac{1}{2}$ code rate can be reduced with the help of a technique that simply avoids transmitting some output bits through the channel. The DVB specifies the following code rates: $\frac{1}{2}$, $\frac{2}{3}$, $\frac{3}{4}$, $\frac{5}{6}$, and $\frac{7}{8}$.

7.3   Transmission standards

7.3.1   DVB-T

From all the DVB solutions for content transmission, DVB-T has been given the most attention (ETSI ETS 300 744, 1997). Its excellent performance has led to its adoption around the world. Since 2005, DVB-T has been used in Australia, France, Germany, Italy, the Netherlands, Singapore, Spain, Sweden, Taiwan, the United Kingdom, and several other countries. It seems to be safe to say that DVB-T is currently in operation in the whole African continent, in Australia, in all Europe and in most parts of Asia. The following subsystems present specific features for terrestrial transmission:

Inner interleaver

The role of the inner interleaver is to deal with the frequency selectivity effect in a channel, caused by the multipath effect. It is designed in such a way that provides the best performance for a specific memory size and complexity and consists of a combination of one bit and one symbol interleaved. The bit interleaver combines 126 successive bits into an interlaced block. Subsequently, the symbol interleaver of the pseudorandom sequence changes the symbol sequence.

This interleaver is composed of three distinct functional blocks, which are: the demultiplexer, the bit interleaver, and the symbol interleaver. A brief summary of each block is given as follows:

- Demultiplexer: this splits the input stream into two output streams for the QPSK modulation, or four output streams for the 16-QAM, or even six streams for the 64-QAM modulation.
- Bit interleaver: the interleavers, of which there are two (for QPSK), four (for 16-QAM) or six (for 64-QAM), mix up the bits contained in blocks of 126 bits each. Each OFDM symbol has 1512 useful subcarriers in the 2K mode and 6048 useful carriers in the 8K mode. Therefore, 12 sets of blocks are required to transmit a symbol in the 2K mode, and 48 of such subsets in the 8K mode.
- Symbol interleaver: the signals assembled in "v" bits (2, 4, or 6 bits) are allocated in sequence to form an OFDM symbol, in which each of the 1512 subcarriers in the 2K mode, or 6048 for the 8K mode, assumes one of the "2v" possible states. The interleaving of the "v" bits is accomplished inside the symbols in which they are inserted.

Generally, the interleaving provided by the inner interleaver block has a similar function to that performed by the outer interleaver, i.e., to prevent the burst errors introduced in the transmission decreasing the quality of signal reception.

Symbol mapping

The symbol mapping block translates the input signal into a frame structure and inserts into this structure the signals responsible for the transmission of: (a) the receiver configuration (or transmission parameter signals, TPS) and (b) the information transported by the pilots to the frame synchronization, frequency synchronization, channel estimation, and identification of the transmission mode.

The symbol mapping consists of the following functional blocks:

Mapper The mapper establishes a relation between the "v" bits (2, 4, or 6 bits) coming from the symbol interleaver block and the phase/amplitude of the OFDM subcarriers. Thus, the sequences in Figure 7.5 use a Gray code to map the "v" bits into the respective states of the modulations QPSK, 16-QAM and 64-QAM.

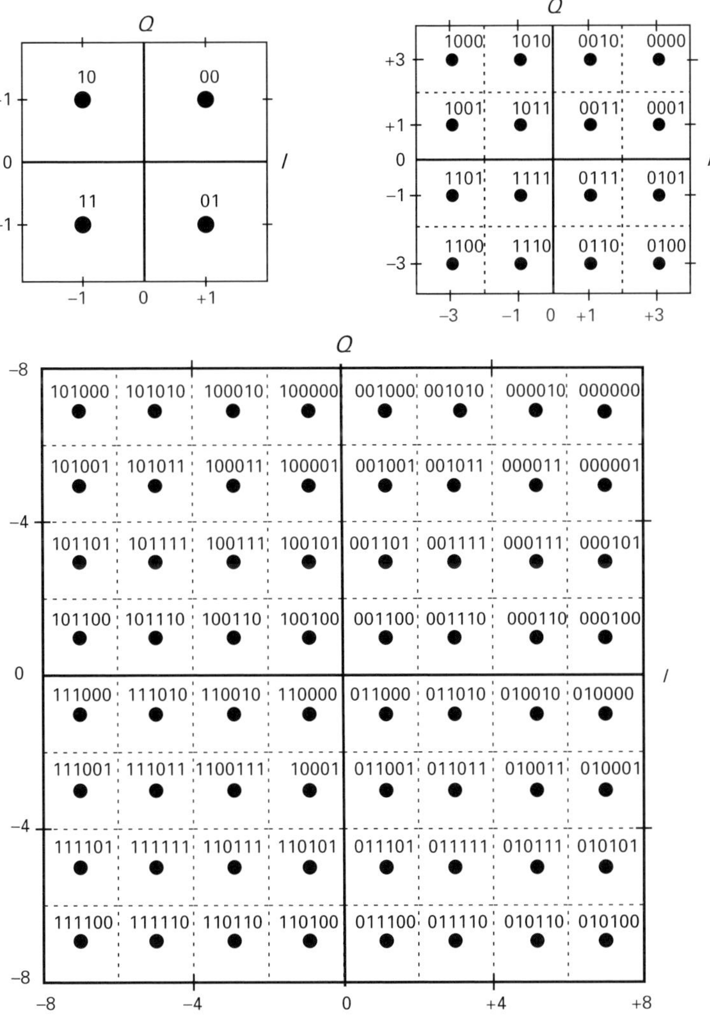

Figure 7.5　　QAM mapping

Table 7.1. Frame adaptation parameters

Parameter	8K mode	2K mode
Carrier number K	6 817	1 705
Number of lower carrier	0	0
Number of upper carrier K_{MAX}	6 816	1 704
T_U duration	1194.667μs	298.6667μs
Separation between carriers $1/T_U$	0.837 054 kHz	3 348 214 kHz
Separation between the upper and lower carriers	5.71 MHz	5.71 MHz

In the hierarchical modulation, two digital streams with different modulations are mapped into a single data stream.

Frame Adapter The pilot signals and the TPS are added to the mapped signals to compose the architecture of the orthogonal frequency division multiplexing (OFDM) frame. Table 7.1 shows the values of the main parameters of the coded OFDM (COFDM) symbol. The values in the table correspond to the 2K and 8K modes. They do not depend on the chosen modulation method, or on the convolutional FEC code rate value, or on the chosen safety band value. They are numbers adopted by the group that defined the DVB-T standard.

COFDM transmission

The terrestrial channel is different from the satellite and cable channels, and is exposed to multipath effects, as a result of the surface topography and the presence of buildings. During DVB-T development, the mobile reception was not an exigency, but the system can handle the signal variation due to the movement of transmitters/receivers.

The system that best fits the needs for such channel conditions is COFDM. COFDM manages the multipath problem very well, and offers a high level of frequency economy, allowing better utilization of the available bandwidth (Ladebusch and Liss, 2006).

The OFDM part describes how the data stream is allocated to carriers with individual frequencies, and the initial C indicates the use of an individual coding for each carrier. The COFDM technique is based on the utilization of several carriers which transport the signal in frequency division multiplexing (FDM) sub-channels with 6, 7, or 8 MHz. In this way, each carrier transports a limited bit rate. The interference between the carriers is avoided by making them mutually orthogonal. This occurs when the separation between the carriers is the inverse of the period over which the receiver will accomplish the signal demodulation (Reimers, 1997). In the DVB standard, 1705 (in the 2K mode) or 6817 (in the 8K mode) carriers are used. To increase the immunity to external interference, several coding techniques are employed, including a pseudorandom exchange of the useful load among the carriers.

For a given configuration of the above-mentioned parameters, the bits are assembled to form a word. Each word modulates a carrier during the time of one symbol (T_U). The set of words of all the carriers in a given time interval T_U is called a COFDM symbol. Thus, the COFDM symbol defined for DVB-T consists of any of the 6817 carriers of the

8K mode or the 1705 carriers of the 2K mode. In the DVB-T system, the OFDM symbols are combined to form a transmission frame, which consists of 68 consecutive symbols. Four consecutive transmission frames form a superframe (Ladebusch and Liss, 2006).

The beginning of each COFDM symbol is preceded by an interval called the guard band. The aim of this interval is to guarantee immunity from echoes and reflections, so that if, for example, an echo is in the guard band, it does not affect the decoding of useful data transmitted on the symbols. The guard band is a cyclic continuation of the symbol and its length in relation to the duration of a symbol can assume four different values: $\frac{1}{4}$, $\frac{1}{8}$, $\frac{1}{16}$, or $\frac{1}{32}$ of symbol time.

To achieve synchronism and phase control, pilot signals are inserted. These pilots are transmitted with an amplitude $1/0.75$ larger than the amplitude of the other carriers, with the aim of making them particularly robust to the transmission effects. The two end carriers (0 and 1704 for the 2K mode, and 0 and 6816 for the 8K mode) play this role. Another 43 carriers in the 2K mode and 175 for the 8K mode are used as a continuous pilot. For the remaining carriers, some words are employed in predefined sequences to act as dispersed pilot signals, which are used to estimate the transmission carrier (and neighboring carriers) characteristics (Yamada *et al.*, 2004). Some carriers are used to send a TPS control signal which identifies the transmission channel parameters, such as modulation type and number of carriers.

Digital to analog converter

All the operations to obtain the OFDM signal in the preceding blocks take place in the digital signal domain. There is, therefore, at the output of the guard band insertion block, a digital signal that represents the modulated signal in OFDM, already inserted digitally in the guard band. It is necessary to convert this signal to the analog domain, to confine it in the 6 MHz band. This conversion is accomplished by a digital to analog converter. After this conversion, the signal is ready to occupy the range between 41 MHz and 47 MHz of the spectrum.

Transmitter

The transmitter circuit is composed of two blocks: the up-converter and the power amplifier. The first is responsible for transferring the signal to the desired television channel frequency. The second amplifies the signal to be transmitted by antenna.

7.3.2   DVB-H

In 2000, the EU-sponsored Motivate (Mobile Television and Innovative Receivers) Project came to the conclusion that mobile reception with DVB-T was possible, but this would require dedicated broadcasting networks, with different architectures and more robustness as compared to the traditional broadcasting networks for fixed receivers.

In 2002, research on the use of spacial diversity began to achieve digital television transmissions for automobiles. In the same year, the DVB community started to take into consideration the possibility of creating technical specifications for digital television transmissions to mobile devices.

In the same year, the DVB Project started to define a set of commercial requirements for a system that comprised support for mobile devices. This technical specification has been given the name of Digital Video Broadcasting – Handheld (DVB-H) and it was finally published in Europe as European Telecommunications Standards Institute (ETSI) Standard EN 302 304, in November 2004 (Faria *et al.*, 2006).

DVB-H System

Compared with DVB-T, DVB-H has two additional features: a time slicing block and an FEC subsystem. The time slicing significantly reduces the average power at the receiver interface – a 90–95% reduction. The use of these two features is compulsory in DVB-H.

The FEC for encapsulated multiprotocol data (MPE-FEC) provides an improvement of the signal-to-noise ratio in mobile channels, and enhances robustness in relation to burst interference. The use of MPE-FEC is optional on DVB-H.

It is important to point out that neither the time slicing nor the MPE-FEC elements, implemented on the link layer, have contact with the DVB-T physical layer. That means that existing DVB-T receivers are not disturbed by DVB-H signals. The DVB-H payload includes IP datagrams or some other network datagram encapsulated in MPE sections. In view of the data rates suggested for individual services and the small displays of mobile terminals, the most popular audio and video coding schemes are not suitable for DVB-H. One solution is to switch MPEG-2 video to H.264/AVC or another standard of high-efficiency video encoding.

The DVB-H physical layer has four extensions compared with the DVB-T physical layer. First, the receiver configuration (TPS) includes two additional bits to denote the presence of DVB-H services and the eventual use of MPE-FEC. Furthermore, a new model for OFDM, the 4K (only the 2K and 8K models had been defind was adopted with the aim of negotiating the mobility and the size of the cell in a single frequency network (SFN), allowing reception with only one antenna even at high speeds. This gives additional flexibility to the network project. The 4K mode is optional on DVB-H and operates with the 2K and 8K modes. All the modulation schemes used on DVB-T, i.e. QPSK, 16-QAM and 64-QAM, with hierarchical and non-hierarchical modes, are also available on DVB-H. In addition, a new way of using the DVB-T symbol interleaver was defined. For the 2K and 4K modes, one can choose to interleave the bits in four or two OFDM symbols, respectively (instead of the interleaver, which interleaves one OFDM symbol). This approach allows these modes to be robust against the impulsive noise equivalent to the 8K mode, and also increases the robustness in relation to the mobile environment. Finally, the fourth addition to the DVB-T physical layer is the 5 MHz channel bandwidth, which is for use in bands not designated for broadcasting.

The DVB-H receiver structure is illustrated in Figure 7.6. It comprises a DVB-T demodulator, a time slicing module, an optional MPE-FEC module, and a DVB-H terminal. The DVB-T demodulator retrieves the MPEG-2 transport stream of the received signal. It offers three modes of transmission: 8K, 4K, and 2K with the corresponding signaling. The time slicing module controls the receiver in order to decode only the stream of bits of the required service, switching to idle otherwise. This implies a reduction in receiver power consumption. The MPE-FEC module provides, along with the traditional

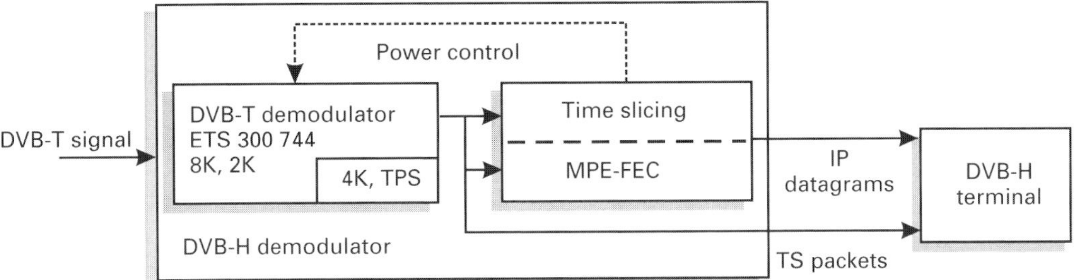

Figure 7.6   Structure of the DVB-H receiver

error correction on transmitted bits, a complementary feature, which changes both the encapsulating process (MPE) and the coding (FEC), allowing the receiver to handle particularly difficult reception conditions.

7.3.3   DVB-S2

At first, in 1993, the main goal of DVB-S was direct-to-home (DTH) television service transmission around the world. Since then, DVB-S has also been used in point-to-point television emission and services transmission, taking the audio and video material originated from studios and/or captured in outdoor transmission directly to television stations, without the need of a fixed local network. In 1998, DVB produced a second standard for satellite applications, extending the DVB-S features to include higher-order modulations, such as 8-PSK and 16-QAM (Morello and Mignone, 2006), besides the QPSK modulation used until then.

In the last decade, television stations, as well as customers, have demanded a larger transmission capacity and innovative services by satellite, forcing the DVB to define, in 2003, the second generation of high-bandwidth services by satellite, named DVB-S2. DVB-S2 is structured as a toolkit that allows the implementation of several services by satellite, such as television and radio transmission, Internet access, and interactivity.

This standard is defined according to three main concepts: better transmission performance, full flexibility, and a reasonable complexity at the receiver. In order to have the best relation between complexity and performance, which is 30% better than that of DVB-S (Morello and Mignone, 2006), DVB-S2 uses more sophisticated modulation and coding schemes. Low parity density codes are adopted together with QPSK, 8-PSK, 16-APSK, 32-APSK modulations, so that the system works satisfactorily on the non-linear satellite channel. For point-to-point interactive applications, the adoption of adaptive modulation and coding allows the optimization of the transmission parameters for each user individually, taking into account individual reception conditions. The frame structure allows much flexibility in a versatile system as well as synchronization in the worst cases of configuration (low signal-to-noise ratio).

The replacement of DVB-S by DVB-S2 is under way. A large number of DVB-S decoders are currently in use, and contribute to the success of the transmission by satellite around the world. New applications use more and more the satellite transmission, like HDTV and Internet services, and for these new applications, DVB-S2 stands out as the best choice.

7.3.4   DVB and Internet

In 2004, the DVB Project approved the first edition of the *DVB IP Handbook*, to give DVB some features to support the first commercial distributions of television services through the Internet.

The aim of the *DVB IP Handbook* is to specify technologies at the interface between a network based on the Internet Protocol (IP) and a digital television receiver, allowing the distribution of television services on IP networks (IPTV) and the mass production of IPTV receivers by the industry. The specified technologies, implemented in a DVB IP receiver, allow the end user to buy one of these receivers, connect it to a large-bandwidth network and receive DVB services over IP-based networks (Stienstra, 2006).

Normally, as mentioned, DVB develops technical specifications based on commercial needs. However, in the beginning, when TV transmission over the Internet was virtually non-existent, the detailed needs were not known. Hence, the *DVB IP Handbook* was created to take into account the support of future services that would eventually become available, based on customer demand.

The first edition of the *DVB IP Handbook* deals with three basic types of services: live transmission (Live Media Broadcast, LMB), transmission with tools (that allows the user, for example, to pause or forward and audio/video stream) and video on demand.

Architecture

The general architecture of the broadcast system of DVB services over IP-based networks is illustrated in Figure 7.7.

In this architecture, four different hosts can be recognized: the content provider, the service provider, the network host, and the local host. Communication between these hosts is accomplished in five layers: the application layer, the session layer, the transport and network layers, and the physical layer. Although on the IPTV service most of the transport data run from the content provider host to the local host, it is important to point out that the communication is bidirectional, as an IP network requires.

On the application layer, there is end-to-end communication between the content provider and the local host. That communication is mainly responsible for the transmission of the digital rights management (DRM).

The service provider shown in Figure 7.7 is the organizer entity and the television service provider to the final user at the local host. This activity can include the function of Internet service provider, which in general also allows user connectivity.

The content provider is a set of entities providing contents for digital TV services, such as movie studios, broadcasting stations, news services providers, and game developers.

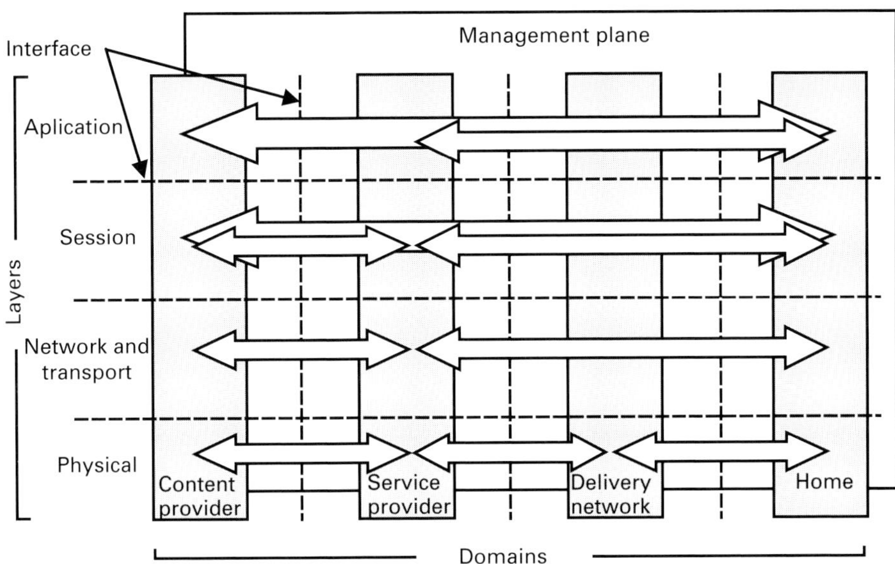

Figure 7.7 Architecture of DVB services over IP networks

The delivery of data from the network host to the local host uses an IP network, through a bidirectional cable, xDSL access network, satellite or wireless systems.

The *DVB IP Handbook* standardization focuses on the local host, mainly because of its DVB IP interface nature. The logic of the DVB standardization is to specify only the minimum needed to guarantee interoperability. In this way, the receiver implementation can vary from manufacturer to manufacturer. There is also a certain level of independence in the implementation of the networking systems and the service providers by communications companies.

7.4   Interactivity

E-commerce, interactive programs, Internet access, are some of the new services that can be made available with the deployment of digital television. The return or interactive channel allows the receivers reverse communication. New services, such as home shopping, e-mail, and remote teaching are made possible by interactivity.

Adding interactivity to the DVB infrastructure requires the system extension to provide the means of communication between the end user and the interactive-service provider. The interactive service provider can be the broadcast service provider or not, as illustrated in Figure 7.8. In this case it uses DVB channels of large bandwidth deliver information from the interactive service user with typical rates of 20 Mbit/s per channel in terrestrial transmission networks, and up to 38 Mbit/s per channel in cable or satellite broadcast networks. Table 7.2 lists the DVB specifications of the return channels in addition to the protocol-independent network specification that has to be used by each of them.

Table 7.2. Specifications of the DVB interactive channel

Interactive channel	Acronym	Standard
PSTN (fixed telephony)	DVB-RPC	ETS 300 801
DECT	DVB-RPD	EN 301 193
GSM	DVB-RPG	EN 301 195
GPRS	DVB-RPGPRS	EN 202 218
CATV	DVB-RCC	ETS 300 800
Satellite	DVB-RCS	EN 301 790
Terrestrial	DVB-RCT	EN 301 958

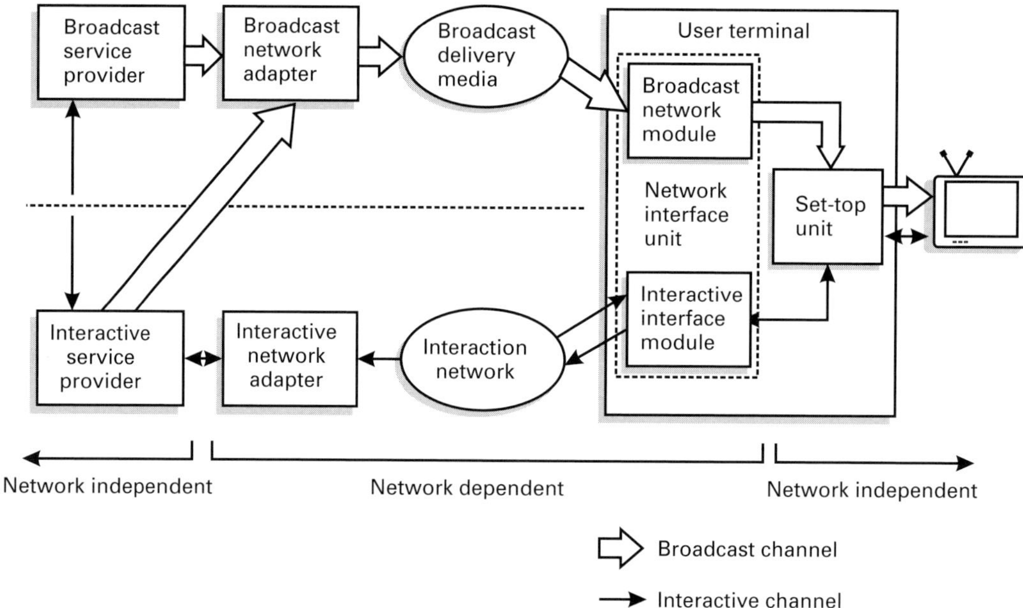

Figure 7.8   General model of an interactive DVB system

7.4.1   DVB-RCT

For DVB-T, the DVB-RCT interactive channel (Santos *et al.*, 2005) is a solution that offers wireless support to real-time applications of interactive services, for the following reasons:

- It has spectral efficiency, low cost and provides multiple access through an OFDM system.
- It can cover large areas, up to a 65 km radius, typically providing several kilobits per second of capacity for each television receiver.
- It can support high traffic peaks, as it was specifically designed to process up to 20 000 interactions per second in each cell's sector.

- It can be implemented with small cells to supply networks of up to a 3.5 km radius, providing the user with a capacity of several megabits per second.
- It is designed for use in any space or spectrum used in bands III, IV, and V, without interference from analog and digital broadcasting services.
- It is a standard able to deal with mobile devices, taking interactivity anywhere the DVB-T signal can be received.
- It can be used by other systems, apart from DVB-T, including 6, 7, and 8 MHz channels.
- It requires a maximum transmission power of 1 W (30 dBm) from the user terminal or set-top box to the base station.

7.5 Middleware

The DVB consortium's specification for the user terminal required its ability to receive and present applications to be independent of both the actual equipment used and the service provider (Herrero *et al.*, 2003). The middleware layer is described in the following.

7.5.1 Multimedia Home Platform (MHP)

For the middleware layer, the DVB Project adopts the MHP platform, whose specification is titled Digital Video Broadcasting – Multimedia Home Platform (DVB-MHP). The MHP platform began to be specified by the DVB project in 1997. However, the first version (MHP 1.0) was officially released in June 2000. One year later a new specification (MHP 1.1) was developed. In June 2003, MHP version 1.1.1 was released.

MHP defines a generic interface between the applications and the set-top box (hardware and operational system), in which the applications are run. Furthermore, MHP defines a model and a life cycle for the applications, as well as the protocols and data distribution mechanisms for pseudointeractive and interactive television environments.

In its 1.1 and 1.1.1 versions, MHP provides additional features, including, for example, the possibility of loading interactive programs by the return channel and the optional support of developed applications.

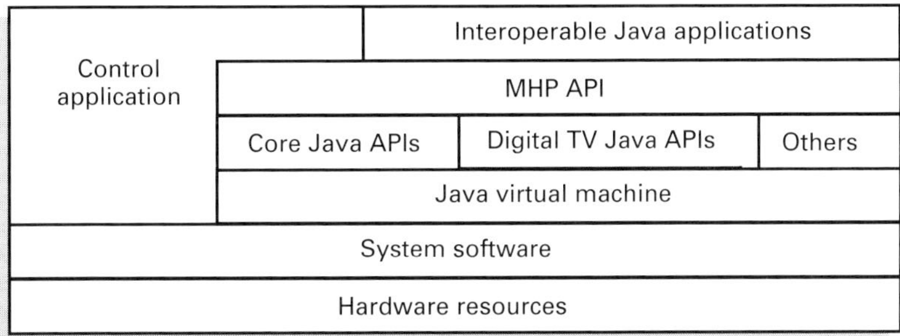

Figure 7.9   Architecture of the MHP platform

In the 1.1 and later versions, the MHP adopts applications models based on procedural and declarative languages. In the procedural model, the MHP supports the execution of Java TV applications, called DVB-J. In the declarative model, the MHP supports the execution of applications developed with technologies related to HTML, named DVB-HTML.

Figure 7.9 illustrates the MHP architecture, which presents hardware and software system resources chosen by the manufacturer according to its product specifications. Regarding the software, MHP works on a virtual Java machine (JVM), which offers several interfaces for application development (API), some of them based on personal Java elements. The total set of APIs generates the API MHP with which the interoperable application, transmitted to the MHP terminal, can be executed.

The MHP is a solution that focuses on the horizontal market. This means that a customer can buy a receiver compatible with the MHP to run all the MHP applications currently on air, regardless of where these applications are originated.

8 Integrated Services Digital Broadcasting for Terrestrial Television Broadcasting (ISDB)

8.1 Introduction

The Integrated Services Digital Broadcasting for Terrestrial Television Broadcasting (ISDB-T) was conceived in Japan to broadcast digital television signals. It allows high-definition television (HDTV) to be accessible for users of the mobile and wireless receivers, with high or low image definition.

The ISDB-T standard can be analyzed as the transformation of a binary signal, which is the digital signal generated by the television operators formatted by MPEG-2, into electromagnetic waves. These waves travel in the air until they reach the ISDB-T receivers, which have the function of executing the reverse process i.e. turning electromagnetic waves into digital signals that the television decoders will understand: the video and audio signals.

The transformation process is complex and attempts to guarantee a perfect signal at the receiver, which implies high quality of sound and image.

8.2 ISDB transmission

The ISDB transmission can be divided into two main phases, called channel encoding and modulation, as shown in Figure 8.1.

The first block of the system is called the channel encoder and has the function of encoding the bits to minimize the destructive effects of the communication channel. The input of this block is fed by an MPEG-2 stream. The first alteration suffered by the signal is structural, i.e. the MPEG-2 stream is rearranged in groups of predefined sizes that remain unchanged until they reach the transmission subsystem.

The method used to arrange the data accepts the implementation of special services, such as hierarchic transmission. The ISDB-T uses a multicarrier modulation method for data transmission, which has advantages over the single-carrier mode. For this reason, at the beginning of the encoding process, the data are arranged in several streams of data, which are introduced in the system's carriers.

Three different configurations related to the number of carriers are permitted in ISDB-T. These different configurations are called transmission modes and determine the number of carriers used in the transmission process, and, consequently, the size of the fast Fourier transform (FFT). Table 8.1 shows the three possible configurations.

Table 8.1. Number of system carriers

Mode	Number of carriers
1	2048
2	4096
3	8192

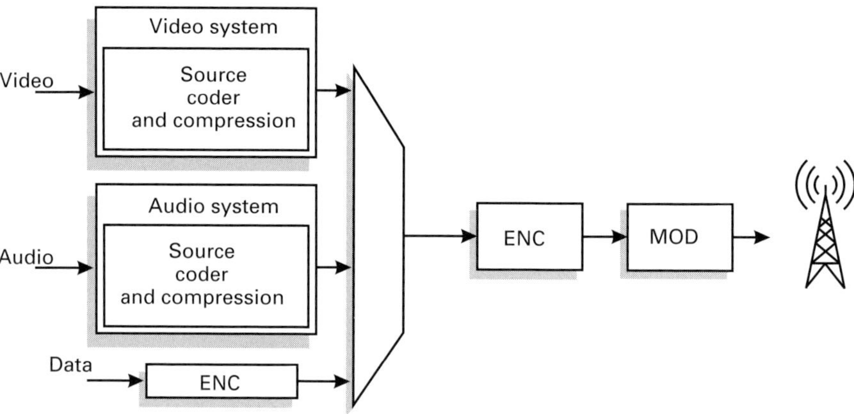

Figure 8.1   Block diagram of the ISDB-T system

As stated before, ISDB-T sees that services, such as the hierarchic transmission, are implemented. Hierarchic transmission enables different receivers to receive and process the signal, besides generating images and sounds according to their characteristics. Two types of receivers receive the signal: wide-band and narrowband receivers.

The hierarchic transmission allows the data to be encoded and modulated in a different way. This feature enables narrow-band receivers to receive only part of the transmitted signal, and, in general, this portion of the signal is better protected than the remaining data. In order to implement these features, a method of gathering carriers is used. Each group of carriers, called a segment, is made up of a fixed number of control data carriers, according to the transmission model chosen.

A detailed description of the segment composition for the three transmission modes is presented in Table 8.2. In this table the channels spread pilot (SP) and continuous pilot (CP) are used by the receiver for synchronization and demodulation, the transmission and multiplexing configuration control (TMCC) provides control information, and the auxiliary channel (AC) is used for transmission of additional information. AC1 is available in equal numbers for both segments, and AC2 is available only for segments that use additional modulation.

The ISDB-T is made up of 13 segments, which are arranged in up to three different groups. These groups are named layers and are configured at the beginning of each

Table 8.2. Parameters of the OFDM segment

	Mode					
	Mode 1		Mode 2		Mode 3	
Bandwidth	$3\,000/7 = 428.57\,\text{kHz}$					
Space between carriers	$250/63 = 3\,968\,\text{kHz}$		$125/63 = 1\,984\,\text{kHz}$		$125/126 = 0.992\,\text{kHz}$	
No. of carriers						
Total	108	108	216	216	432	432
Data	96	96	192	192	432	432
SP	9	0	18	0	36	0
CP	0	1	0	1	0	1
TMCC	1	5	2	10	4	20
AC1	2	2	4	4	8	8
AC2	0	4	0	9	0	19
Modulation scheme	QPSK 16-QAM 64-QAM	DQPSK	QPSK 16-QAM 64-QAM	DQPSK	QPSK 16-QAM 64-QAM	DQPSK
Symbols per frame	204					
Symbol duration	$252\,\mu\text{s}$		$504\,\mu\text{s}$		$1\,008\,\mu\text{s}$	
Guard interval	$63\,\mu\text{s}\,(1/4)$ $31.5\,\mu\text{s}\,(1/8)$ $15.75\,\mu\text{s}\,(1/16)$ $7.875\,\mu\text{s}\,(1/32)$		$126\,\mu\text{s}\,(1/4)$ $63\,\mu\text{s}\,(1/8)$ $31.5\,\mu\text{s}\,(1/16)$ $15.75\,\mu\text{s}\,(1/32)$		$252\,\mu\text{s}\,(1/4)$ $126\,\mu\text{s}\,(1/8)$ $63\,\mu\text{s}\,(1/16)$ $31.5\,\mu\text{s}\,(1/32)$	
Frame duration	$64.26\,\text{ms}\,(1/4)$ $57.83\,\text{ms}\,(1/8)$ $54.62\,\text{ms}\,(1/16)$ $53.01\,\text{ms}\,(1/32)$		$128.52\,\text{ms}\,(1/4)$ $115.67\,\text{ms}\,(1/8)$ $109.24\,\text{ms}\,(1/16)$ $106.03\,\text{ms}\,(1/32)$		$257.04\,\text{ms}\,(1/4)$ $231.37\,\text{ms}\,(1/8)$ $218.48\,\text{ms}\,(1/16)$ $212.06\,\text{ms}(1/32)$	
IFFT sampling frequency	$512/63 = 8.126\,98\,\text{MHz}$					
Inner code	Convolutional encoder (1/2, 2/3, 3/4, 5/6, 7/8)					
Outer code	RS (204,188)					

transmission. It is possible to make a transmission with one, two, or three layers, depending on its requisites.

The segments of each layer have the same configuration, and it is this feature that makes the hierarchic transmission possible. Figure 8.2 shows the allocation of a passband for the partial or complete transmission of the signal. Figure 8.3 shows the layout of the carriers and segments in the frequency band. An example of layer distribution in the frequency band is shown in Figure 8.4.

The developers of this standard were careful to give it a high degree of flexibility. For this, they provided a series of parameters that are configured according to the conditions of the channel. The signal parameters are presented in Table 8.3. Table 8.4 presents the data rate for one segment. The data rate for the system is shown in Table 8.5.

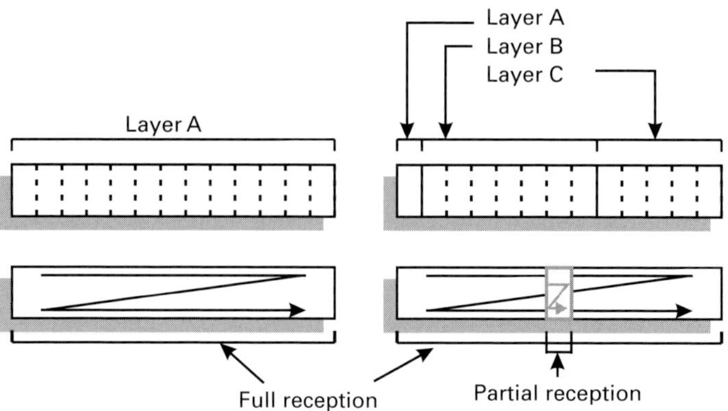

Figure 8.2   Band allocation for complete or partial transmission

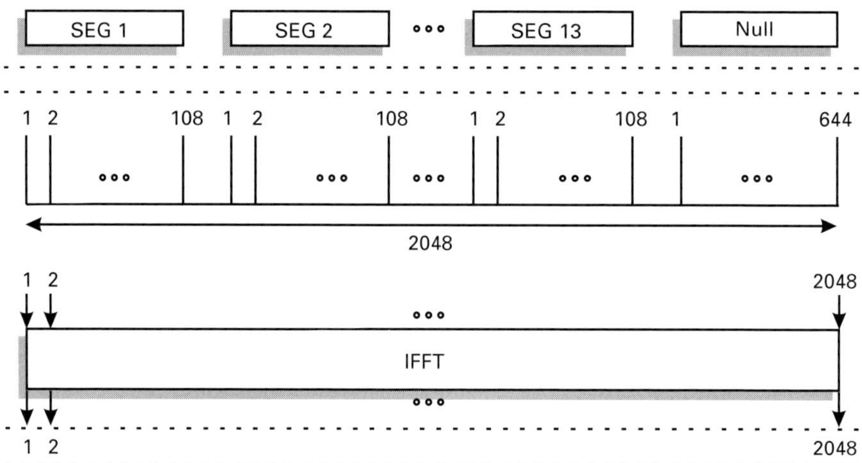

Figure 8.3   Layout of the carriers and segments in the frequency band

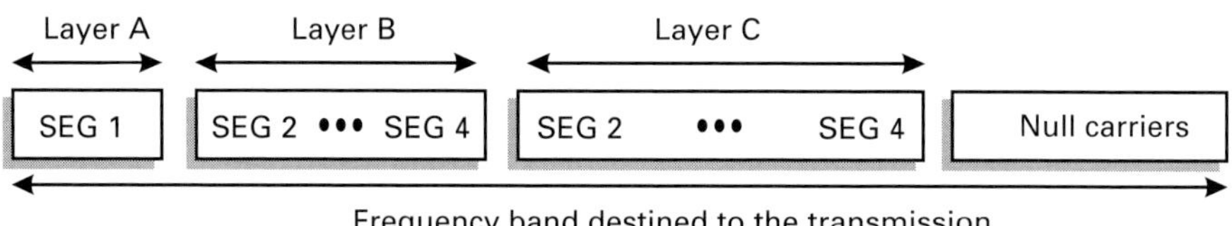

Figure 8.4   Example of layer distribution in the frequency band

8.3   The transmission process

The signal that is inserted in the first encoding block, the (transport stream (TS))
remultiplexer, comes from an outside MPEG-2 multiplexer. The function of the

Table 8.3. Parameters of the transmitted signal

	Mode		
	Mode 1	Mode 2	Mode 3
Number of segments N_s		13	
Bandwidth	$3\,000/7\,(\text{kHz}) \times N_s$ $+250/63\,(\text{kHz})$ $= 5\,575\,\text{MHz}$	$3\,000/7\,(\text{kHz}) \times N_s$ $+ 125/63\,(\text{kHz})$ $= 5\,573\,\text{MHz}$	$3\,000/7\,(\text{kHz}) \times N_s$ $+125/126\,(\text{kHz})$ $= 5\,572\,\text{MHz}$
Number of segments with differential modulation		n_d	
Number of segments with synchronous modulation		$n_s\,(N_s = n_s + n_d)$	
Space between carriers	$250/63 = 3\,968\,\text{kHz}$	$125/63 = 1.984\,\text{kHz}$	$125/126 = 0.992\,\text{kHz}$
No. of carriers			
Total	$108 \times N_s + 1 = 1\,405$	$216 \times N_s + 1 = 2\,809$	$432 \times N_s + 1 = 5\,617$
Data	$96 \times N_s = 1\,248$	$192 \times N_s = 2\,496$	$384 \times N_s + 1 = 4\,992$
SP	$9 \times n_s$	$18 \times n_s$	$36 \times n_s$
CP	$n_d + 1$	$n_d + 1$	$n_d + 1$
TMCC	$n_s + 5 \times n_d$	$2 \times n_s + 10 \times n_d$	$8 \times n_s + 20 \times n_d$
AC1	$2 \times N_s = 26$	$4 \times N_s = 52$	$8 \times N_s = 104$
AC2	$4 \times n_d$	$9 \times n_d$	$19 \times n_d$
Modulation schemes		QPSK, 16-QAM, 64-QAM , DQPSK	
Symbols per frame		204	
Symbol duration	$252\,\mu s$	$504\,\mu s$	$1\,008\ \mu s$
Guard interval	$63\,\mu s\,(1/4)$ $31.5\,\mu s\,(1/8)$ $15.75\,\mu s\,(1/16)$ $7.875\,\mu s\,(1/32)$	$126\,\mu s\,(1/4)$ $63\,\mu s\,(1/8)$ $31.5\,\mu s\,(1/16)$ $15.75\,\mu s\,(1/32)$	$252\,\mu s\,(1/4)$ $126\,\mu s\,(1/8)$ $63\,\mu s\,(1/16)$ $31.5\,\mu s\,(1/32)$
Frame duration	$64.26\,\text{ms}\,(1/4)$ $57.83\,\text{ms}\,(1/8)$ $54.62\,\text{ms}\,(1/16)$ $53.01\,\text{ms}\,(1/32)$	$128.52\,\text{ms}\,(1/4)$ $115.67\,\text{ms}\,(1/8)$ $109.24\,\text{ms}\,(1/16)$ $106.03\,\text{ms}\,(1/32)$	$257.04\,\text{ms}\,(1/4)$ $231.37\,\text{ms}\,(1/8)$ $218.48\,\text{ms}\,(1/16)$ $212.06\,\text{ms}\,(1/32)$
Inner code		Convolutional encoder (1/2, 2/3, 3/4, 5/6, 7/8)	
Outer code		RS (204,188)	

Table 8.4. Data rate for one segment

Modulation	Convolutional encoder rate	Number of transmitted TSPs Mode 1 / 2 / 3	Data rate (kbit/s)			
			IG 1/4	IG 1/8	IG 1/16	IG 1/32
DQPSK	1/2	12 / 24 / 48	280.85	312.06	330.42	340.43
	2/3	16 / 32 / 64	374.47	416.08	440.56	453.91
	3/4	18 / 36 / 72	421.28	468.09	495.63	510.65
QPSK	5/6	20 / 40 / 80	468.09	520.10	550.70	567.39
	7/8	21 / 42 / 84	491.50	546.11	578.23	595.76
16-QAM	1/2	24 / 48 / 96	561.71	624.13	660.84	680.87
	2/3	32 / 64 / 128	748.95	832.17	881.12	907.82
	3/4	36 / 72 / 144	842.57	936.19	991.26	1 021.30
	5/6	40 / 80 / 160	936.19	1 040.21	1 101.40	1 134.78
	7/8	42 / 84 / 168	983.00	1 092.22	1 156.47	1 191.52
64-QAM	1/2	36 / 72 / 144	842.57	936.19	991.26	1 021.30
	2/3	48 / 96 / 192	1 123.43	1 248.26	1 321.68	1 361.74
	3/4	54 / 108 / 216	1 263.86	1 404.29	1 486.90	1 531.95
	5/6	60 / 120 / 240	1 404.29	1 560.32	1 652.11	1 702.17
	7/8	63 / 126 / 252	1 474.50	1 632.34	1 734.71	1 787.28

Table 8.5. Data rate for the system

Modulation	Convolutional encoder rate	Number of transmitted TSPs Mode 1 / 2 / 3	Data rate (Mbit/s)			
			IG 1/4	IG 1/8	IG 1/16	IG 1/32
DQPSK	1/2	156 / 312 / 624	3 651	4 056	4 259	4 425
	2/3	208 / 416 / 832	4 868	5 409	5 727	5 900
	3/4	234 / 468 / 936	5 476	6 085	6 443	6 638
QPSK	5/6	260 / 520 / 1 040	6 085	6 761	7 159	7 376
	7/8	273 / 546 / 1 092	6 389	7 099	7 517	7 744
16-QAM	1/2	312 / 624 / 1 248	7 302	8 113	8 590	8 851
	2/3	416 / 832 / 1 664	9 736	10 818	11 454	11 801
	3/4	468 / 936 / 1 872	10 953	12 170	12 886	13 276
	5/6	520 / 1 040 / 2 080	12 170	13 522	14 318	14 752
	7/8	546 / 1 092 / 2 184	12 779	14 198	15 034	15 498
64-QAM	1/2	468 / 936 / 1 872	10 953	12 170	12 886	13 276
	2/3	624 / 1 248 / 2 496	14 604	16 227	17 181	17 702
	3/4	702 / 1 404 / 2 808	16 430	18 255	19 329	19 915
	5/6	780 / 1 560 / 3 120	18 255	20 284	21 477	22 128
	7/8	819 / 1 638 / 3 276	19 168	21 298	22 551	23 234

TS remultiplexer is to arrange the MPEG-2 data stream according to the ARIB STD-B32 standard (Video Coding, Audio Coding and Multiplexing Specifications for Digital Broadcasting). This arrangement consists of the division of the stream in blocks of 188 bytes, which are called MPEG-2 transport stream packets (MPEG-2 TSPs).

The TSPs are sent to be segmented and rearranged into one, two, or three encoding blocks. These blocks have independent coding and ensure different levels of protection for the information that travels through the channel. The blocks, which are layers of the system, are made up of a series of phases and perform functions such as interleaving, encoding, and modulation. There is still a delay adjuster, which is responsible for adjusting the delay between the layers, even if these have distinct configurations and different delays.

The layers data also go through modulating blocks, which map the data into constellations of digital modulation. After the mapping, the data are regrouped into one single stream and sent to the frequency interleaver. Its function is to shuffle the carriers to prevent the burst fading from jeopardizing the information that travels in the channel. The time interleaver is the next stage, followed by the frame composer and by the inverse fast Fourier transform (IFFT).

At this stage, the signal is almost ready to be broadcast, except for the insertion of a guard interval, the transformation of the samples into a continuous signal, and the increase of the frequency of this continuous signal to the broadcasting band defined in the standard, according to the current rules and regulations.

8.4 The channel encoding

The channel encoding process was presented, in a general way, in Chapter 4, and is approached in detail in this section. The block diagram of this process is shown in Figure 8.5.

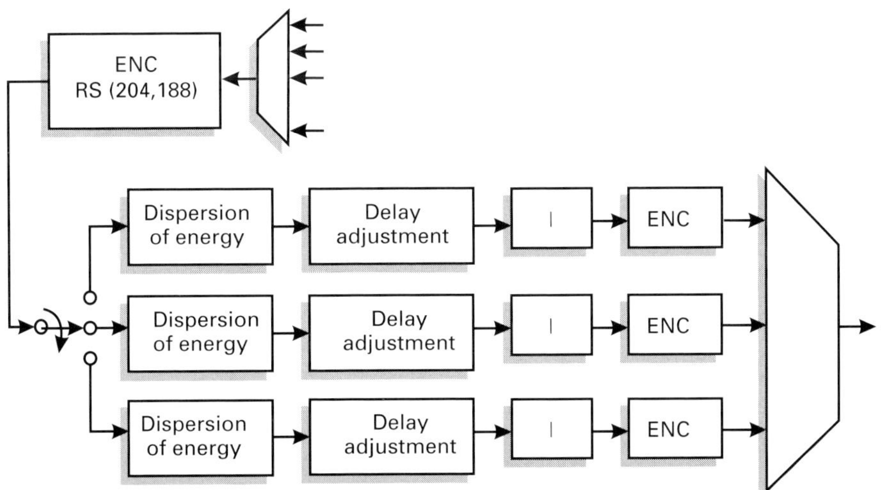

Figure 8.5   Block diagram of the channel encoding

Table 8.6. Dimensions of the multiplex frame

Mode	$\delta = 1/4$	$\delta = 1/8$	$\delta = 1/16$	$\delta = 1/32$
1	1 280	1 152	1 088	1 056
2	2 560	2 304	2 176	2 112
3	5 120	4 608	4 352	4 224

8.4.1 Remultiplexing of the transport stream

The TS remultiplexer block, as the name implies, has the function of formatting the MPEG-2 data stream that comes from the outer multiplexer and passes it on to the outer encoder. The process begins with the breaking down of the input stream into elementary units of 188 bytes named MPEG-2 TSPs. Then they receive another 16 null bytes, and the total block of 204 bytes is called a transmission TSP. The generation rate of MPEG-2 packages is four times higher than the IFFT frequency so that synchronism is maintained.

The output of the TS remultiplexer is formed through a stream of multiplex frames, which are made up of transmission TSPs. The size of the multiplex frame is not fixed and is defined by the configuration of the transmission mode and by the rate of the guard interval (δ). Table 8.6 lists the parameters and numbers of TSPs of the frames that make up the multiplex stream. This table was generated from

$$T = 2^{N-1}(1 + \delta), \tag{8.1}$$

in which N is the dimension of the FFT and δ is the dimension of the guard interval.

The multiplex frame also includes TSPNULL. These TSPs do not contain information, their sole function being that of grafting the multiplex frame. This procedure is carried out when the number of TSPs required by a given layer is higher than the number of TSPs made available by the TS remultiplexer. The TSPNULLs are appropriately identified and excluded in the reception process. Figure 8.6 shows a multiplex frame for the configuration of the transmission mode 1 and guard interval $\delta = 1/8$.

8.4.2 Outer coding

The external encoding scheme employed by ISDB-T uses Reed–Solomon (RS) code and has the function of protecting the data at the beginning of the process of channel encoding. RS(204,188) is used in each transmission TSP that reaches the encoder, independently of the layer to which it belongs. This type of coding is known as a shortened scheme, and is named as such, as 225 bytes are generated from the 188 made available at the beginning of the encoder, which later results in 204 being processed and 51 being eliminated.

This coding scheme uses the Galois field (2^8), with the polynomial

$$p(x) = x^8 + x^4 + x^2 + 1, \tag{8.2}$$

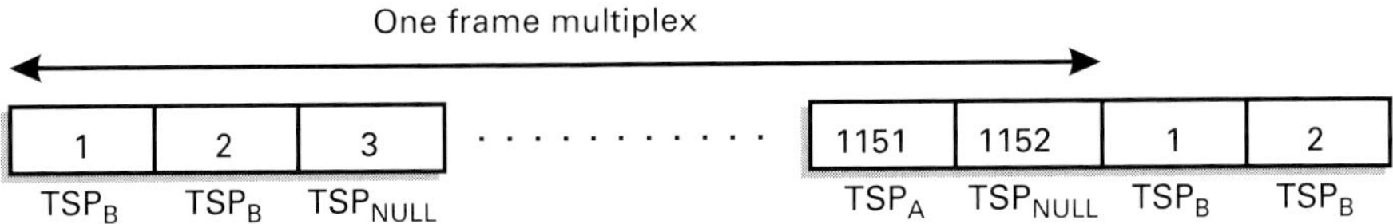

Figure 8.6   Example of the remultiplexed transportation stream

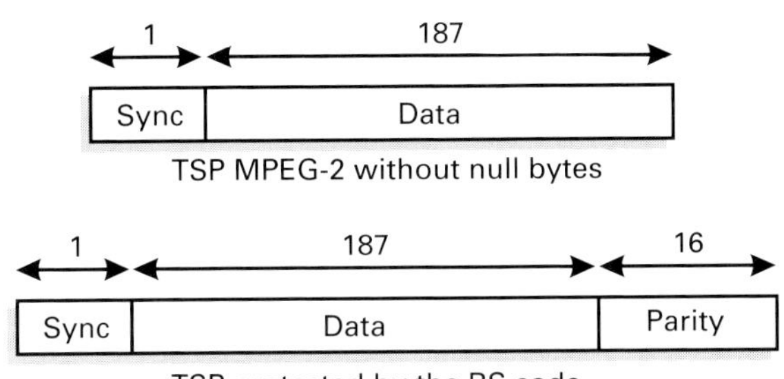

Figure 8.7   MPEG-2 TSP and TSP protected by the RS code

and generates the codes according to the polynomial

$$g(x) = (x - 1)(x - \alpha)(x - \alpha^2) \cdots (x - \alpha^{15}). \tag{8.3}$$

It is important to mention that $\alpha = 02_{HEX}$.

The transmission TSP that reaches the RS encoder, as described previously, contains 16 null bytes. The remainder is the MPEG-2 TSP itself, which contains 188 bytes, being one byte for synchronism and 187 for data. In the encoding process, a new TSP is generated from the transmission TSP, which excludes the 16 null bytes and aggregates 16 parity bytes following the data field, preserving the size of 204 bytes. Figure 8.7 shows this structure.

8.4.3   Division of the transport stream in hierarchic layers

The dividing block has the function of separating the different streams according to the layers to which they belong. In this way, depending on the initial configuration, the transmission TSPs are led to one, two, or three layers. A similar and inverted process is executed in the receiver. This process gathers the data that come from the different layers and passes them on to the RS(204,188) external decoder.

8.4.4   Power dispersion

The process of power dispersion happens independently in each of the layers and has the function of preventing a sequence of successive zeros or ones being transmitted, in order to guarantee an adequate binary transition. This process is executed with an XOR

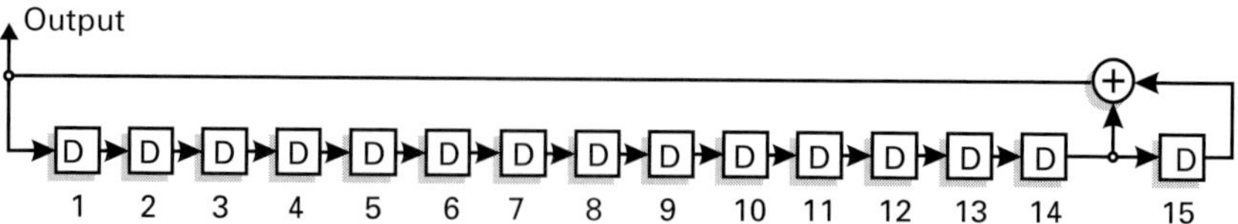

Figure 8.8   Flowchart of the shift-register

binary operation, between the data of the referred layer and a pseudorandom sequence generated by the shift-register shown in Figure 8.8.

The generator polynomial is

$$g(x) = X^{15} + X^{14} + 1, \tag{8.4}$$

and its initial state is "1 0 0 1 0 1 0 1 0 0 0 0 0 0 0." This initial state is composed from left to right in ascending order, and has been chosen to produce an optimal sequence. For a sequence to be considered optimal, it must not repeat a state in its maximal length.

8.4.5   Delay correction

A characteristic of hierarchic transmission is that it is responsible for the configuration of different parameters (convolutional encoder rate, modulation scheme, among others) in the layers in an independent manner. These different configurations imply data with distinct sizes and cause different delays in the receiver.

To prevent different delays between the layers, a scheme of delay correction in the data transmission process is executed. The adjustment is made by inserting different delays in the layers according to the scheme that will be used, as well as by the rate of the code that the internal encoder will use. Table 8.7 shows the delays in the TSPs for several configuration possibilities.

Another characteristic that is intrinsic to hierarchic transmission is the different bit transmission rates achieved. Once the code rates used in the convolutional encoder are different, so will be the number of bits added for protection.

8.4.6   The byte interleaver

The byte interleaver is responsible for mixing the bytes that form the transmission TSP. Each transmission TSP consists of 204 mixed bytes as a way to prevent possible problems caused by the transmission channel.

This is the first interleaver that acts upon the data before it is transmitted, and, just like the power disperser, works independently by layers. It is made up of twelve arms, each of which inserts a different delay, as they have different size buffers. The exception is the first arm that does not insert any delay, as can be observed in Figure 8.9, which shows the byte interleaver scheme.

Table 8.7. Delays in TSPs

Modulation	Code rate	Mode 1	Mode 2	Mode 3
QPSK	1/2	$12N-11$	$24N-11$	$48N-11$
	2/3	$16N-11$	$32N-11$	$64N-11$
	3/4	$18N-11$	$36N-11$	$72N-11$
	5/6	$20N-11$	$40N-11$	$80N-11$
	7/8	$21N-11$	$42N-11$	$84N-11$
	1/2	$24N-11$	$48N-11$	$96N-11$
16-QAM	2/3	$32N-11$	$64N-11$	$128N-11$
	3/4	$36N-11$	$72N-11$	$144N-11$
	5/6	$40N-11$	$80N-11$	$160N-11$
	7/8	$42N-11$	$84N-11$	$168N-11$
	1/2	$36N-11$	$72N-11$	$144N-11$
64-QAM	2/3	$48N-11$	$96N-11$	$192N-11$
	3/4	$54N-11$	$108N-11$	$216N-11$
	5/6	$60N-11$	$120N-11$	$240N-11$
	7/8	$63N-11$	$126N-11$	$252N-11$

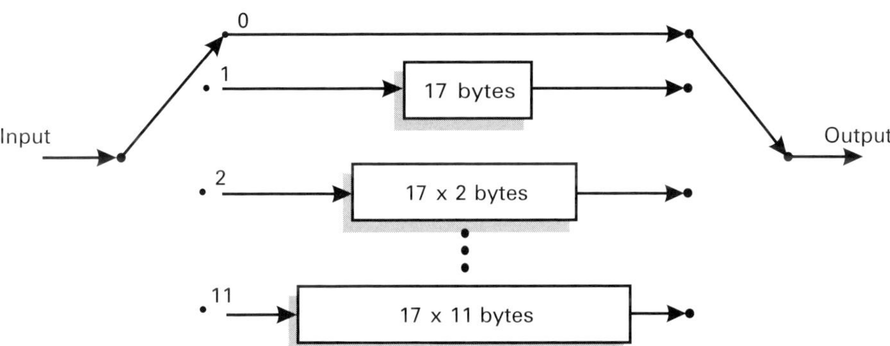

Figure 8.9   Flowchart of the byte interleaver

The sizes of the arm buffers are multiples of 17 bytes in the following order: 17×0, $17 \times 1, 17 \times 2, \ldots, 17 \times 11$. The data are arranged in increasing order in the interleaver and feed the arms also in increasing order. Once the last arm has been reached, the cycle is complete, and the bytes are inserted into arm zero again.

8.4.7 Inner coding

The internal encoding is executed by a convolutional encoder, which is also responsible for protecting the data, inserting redundant bits according to the flowchart shown in Figure 8.10. This encoding scheme provides the generation of codes with different rates. In this way, the rates $\frac{2}{3}$, $\frac{3}{4}$, $\frac{5}{6}$, and $\frac{7}{8}$ are originated from the output of the convolutional encoder, which generates a $\frac{1}{2}$ rate code.

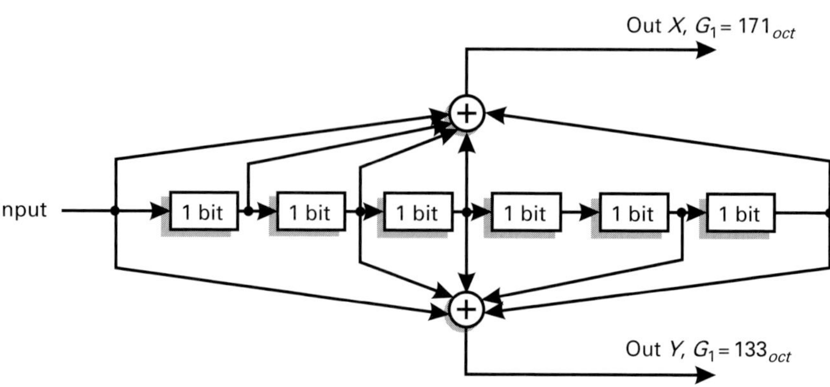

Figure 8.10   Flowchart of the convolutional encoder of rate = 1/2 and $k = 7$

The encoder diagram used has a constraint length $k = 7$, and generates a code with rate $\frac{1}{2}$. This block contains two outputs, each formed by the junction of different stages of the shift register, which are defined by $G_1 = 171_{oct}$ and $G_2 = 133_{oct}$.

This coding scheme permits the production of codes with different rates. Therefore, the rates $\frac{2}{3}$, $\frac{3}{4}$, $\frac{5}{6}$, and $\frac{7}{8}$ are generated from the output of the rate $\frac{1}{2}$ convolutional encoder. For such an event to take place, a bit suppression scheme that is called puncturing is used, as shown in Table 8.8.

8.5   Modulation

The modulation subsystem is responsible for transforming the coded information into electromagnetic waves to be transmitted by the antenna. The ISDB-T modulator consists of several blocks, which perform different operations. The sequence of these blocks is shown in Figure 8.11.

8.5.1   OFDM

ISDB-T uses the technique of orthogonal frequency division multiplexing (OFDM) as a transmission scheme. The subcarriers form a set of functions that are orthogonal to each other, i.e., the integral of the product between any two of these functions within the interval of a symbol is null.

This orthogonality ensures that the intersymbol interference in the frequencies of the subcarriers is null. Figure 8.12 illustrates the effect of the orthogonality. The orthogonality allows the OFDM signal to occupy the narrowest possible band, which, in turn, makes the signal fit into a 6 MHz passband channel.

In Figure 8.12, the frequency of the first subcarrier (f_0) equals the inverse of the duration of the symbol ($1/T_U$). In the 8K mode, f_0 is 837 054 Hz and, in the 2K mode, it is 3 348.214 Hz. The values calculated for f_0 result from the need to maintain the orthogonality between the subcarriers.

Table 8.8. Convolutional encoder, rate 1/2

Code rate	Suppression pattern	Transmitted sequence
1/2	X:1, Y:1	X_1, Y_1
2/3	X:10, Y:11	X_1, Y_1, Y_2
3/4	X:101, Y:110	X_1, Y_1, Y_2, X_3
5/6	X:10101, Y:11010	$X_1, Y_1, Y_2, X_3, Y_4, X_5$
7/8	X:10000101, Y:1111010	$X_1, Y_1, Y_2, X_3, Y_4, X_5, Y_6, Y_7$

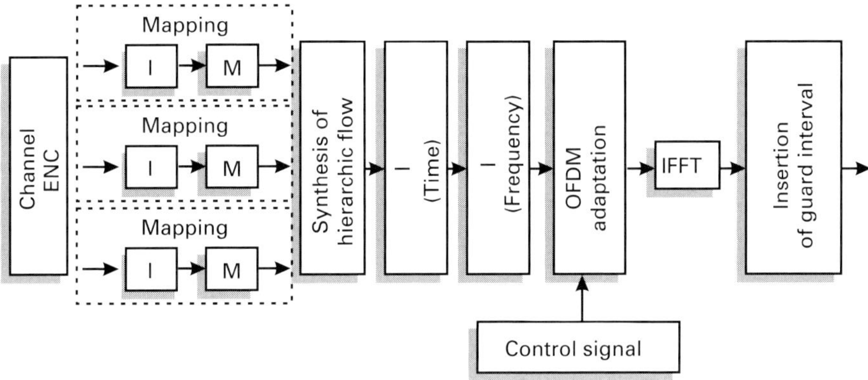

Figure 8.11   Block diagram of the modulation process

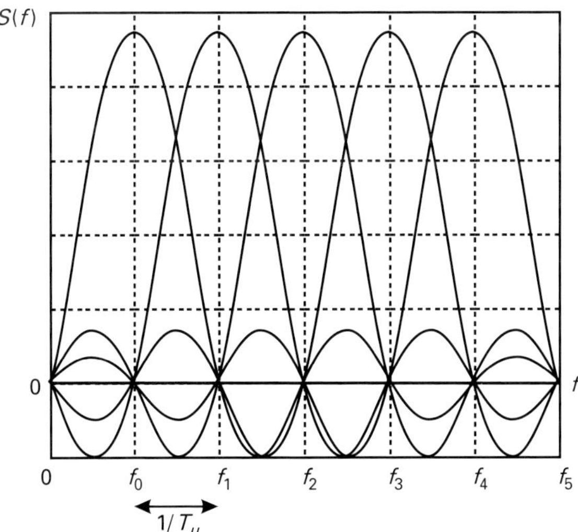

Figure 8.12   Orthogonality between the carriers in the OFDM system

Each subcarrier is modulated in QPSK, 16-QAM, or 64-QAM by one of the "v" bit sets mapped by the following block mapper. For every set of "v" bits (2, 4, or 6 bits) there is a given state of phase/amplitude of the subcarrier. In a symbol (T_U), the states of the subcarriers remain unchanged. In the next symbol, they acquire new states due to the new sets of "v" bits that are found at the input of the modulators of each subcarrier.

It is worth mentioning that the state of the subcarriers, within the transmission of a symbol, possesses information on the frequency spectrum isolated from the OFDM signal. To convert this information to the time domain, the IFFT is used. All the operations are digitally performed by means of digital processors. The OFDM signal obtained is in digital format and ready to be injected in the next block, in which the guard interval is inserted.

Guard interval

The guard interval (Δ) is a time interval added before the transmission of each symbol. During this interval, no information is transmitted. Its function is to eliminate or decrease the interference between the symbols when the interfering signal is an echo of the main signal, but the value of the delay experienced by this echo is smaller than the value of the guard interval. These echoes are produced by reflections of the main signal in obstacles existing in the space between the transmitter and the receiver.

Figure 8.13 illustrates the transmission of two symbols headed by their respective guard intervals. The relationship between the guard interval and the useful time (T_U) is called the guard ratio (k):

$$k = \frac{\Delta}{T_U}.$$ (8.5)

8.5.2   Mapping

The first block of the modulation process is the mapping. Its first function is to execute an interleaving of bits, and the second is to map them into a constellation of digital modulation.

The interleaving process must shuffle the bits according to the rules of each map used. Consequently, a bit delay is introduced in each segment, which varies from mapping to mapping. This difference is fixed by the insertion of a correcting delay, according to Table 8.9, in a way that the total delay is uniform and equal to two OFDM symbols. Notice that N represents the number of used segments in each layer.

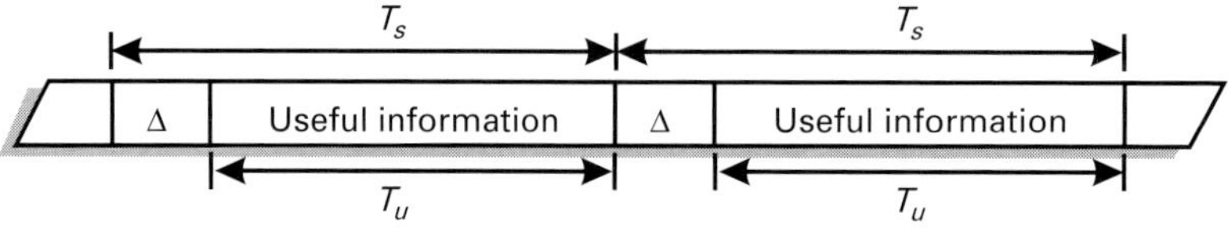

Figure 8.13   OFDM guard interval

Table 8.9. Bit delay values

Modulation scheme	Delay adjust in bits		
	Mode 1	Mode 2	Mode 3
QPSK / DQPSK	$384 \times N - 240$	$768 \times N - 240$	$1\,536 \times N - 240$
16-QAM	$768 \times N - 480$	$1\,536 \times N - 480$	$3\,072 \times N - 480$
64-QAM	$1\,536 \times N - 720$	$3\,072 \times N - 720$	$4\,608 \times N - 720$

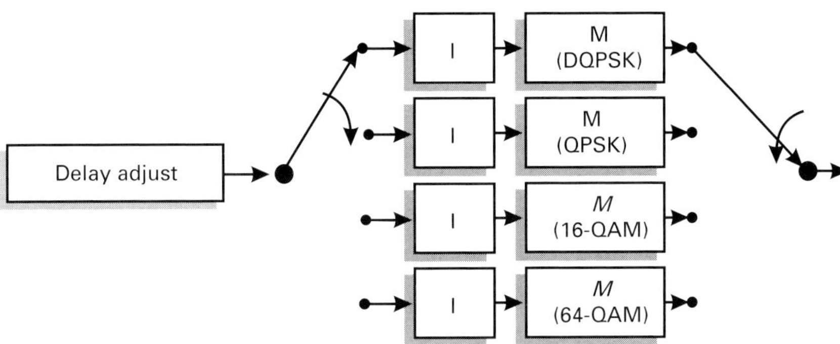

Figure 8.14   Mapping scheme

The constellations used in the ISDB-T and their respective interleaving blocks are presented in the next subsections, while a schematic mapping flowchart is shown in Figure 8.14.

Differential quadrature phase shift keying (DQPSK)

This is the only incoherent modulation scheme used in the data. The $\pi/4$ DQPSK has the capacity to transmit its information in the phase difference between the current and the previous symbol. This mapping scheme uses eight possible pseudosymbols to carry the information of two bits and, at every transmission, only four of them are used, and in the following transmission, only the other four symbols are used.

The possible phase transitions between symbols are $\pm\pi/4$ and $\pm3\pi/4$, as can be observed in the constellation shown in Figure 8.15. Thus, there are two distinct constellations in the same map, each with four symbols, used at different times and lapsed by $\pm\pi/4$. This feature makes it possible to state that the DQPSK mapping consists of eight pseudosymbols.

The complete scheme proposed for the DQPSK mapping is presented in Figure 8.16. Its function is to turn a serial sequence of bits into two distinct streams, identified as I and Q, which are the phase and quadrature channels. After the separation, the interleaving takes place, which is the introduction of a delay of 120 bits in the Q channel in relation to the I channel, both being transmitted to the next block.

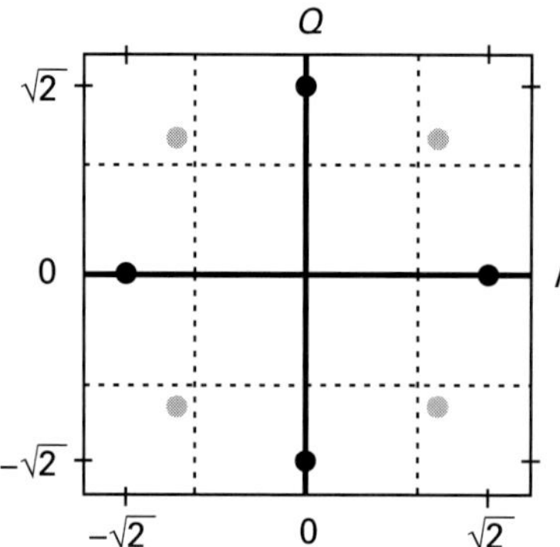

Figure 8.15   DQPSK constellation

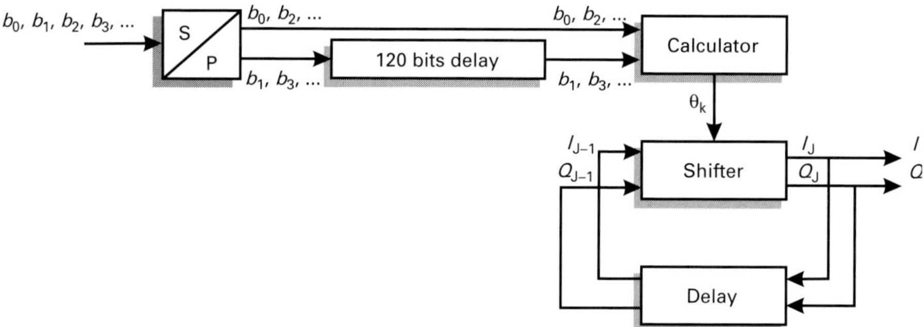

Figure 8.16   Bit interleaving and DQPSK modulator

The mapping process begins with the input of bits in the phase calculator. It is responsible for converting two bits, one from the I channel and the other from the Q channel, into one phase, which is inserted into a phase shifter. The function of this stage is to add the current phase to the previous one, obtaining, as a result, the different phases in time, each of which is related to the input bits and is ready to be forwarded to the next block.

Quadrature phase shift keying (QPSK)

The QPSK mapping scheme is similar to the DQPSK, from which it is derived. The serial sequence of bits coming from the internal encoder is separated into two parallel channels, the I channel and the Q channel, as in the DQPSK. Also identical is the interleaving process, which executes a delay of 120 bits in the Q channel in relation to the I channel. The difference between the two systems is that QPSK encodes the bits in phases with reference to the zero phase.

QPSK is a coherent modulation scheme, as the information transmitted is contained within the phase of the symbol that refers to the initial point, i.e. zero, and no longer in the previous phase. Its name suggests the number of symbols available in its mapping, which are four; they can be seen in Figure 8.17. The process of channel division and interleaving is shown in Figure 8.18.

16-quadrature amplitude modulation (16-QAM)

The 16-QAM mapping scheme has a bit sequence separation process and an interleaving process slightly different from the schemes previously described. Whilst in DQPSK and QPSK the sequence is separated into two distinct streams, 16-QAM needs four distinct streams, as each symbol of this constellation is formed by four bits. Consequently, the interleaving scheme for 16-QAM mapping is adapted to insert a different delay in each of the information streams, as can be seen in Figure 8.19.

16-QAM is also considered a coherent modulation type, made up of 16 symbols. Unlike QPSK, which only saves information on the phase of the symbol, 16-QAM stores the information about phase and amplitude simultaneously. The map of this constellation is presented in Figure 8.20, and it is possible to confirm that there are symbols with distinct amplitudes and equal phases, as well as distinct phases and equal amplitudes.

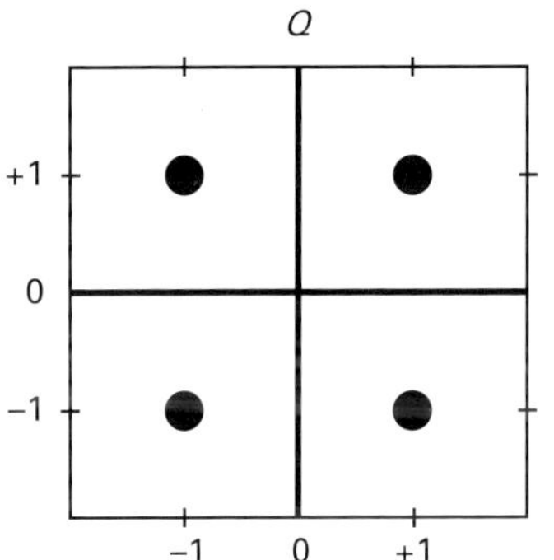

Figure 8.17   QPSK constellation

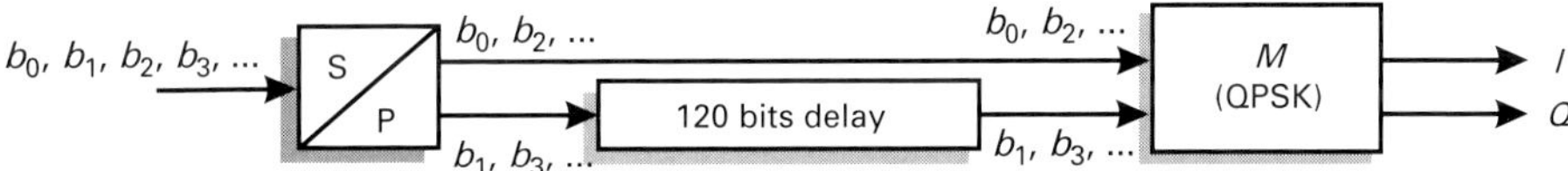

Figure 8.18   Bit interleaving and QPSK modulator

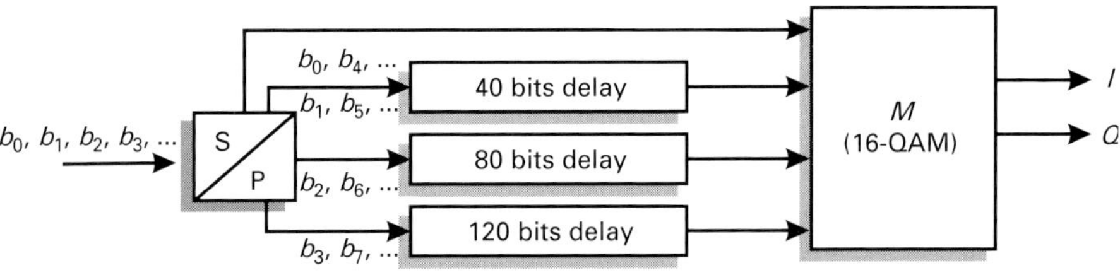

Figure 8.19   Bit interleaving and 16-QAM modulator

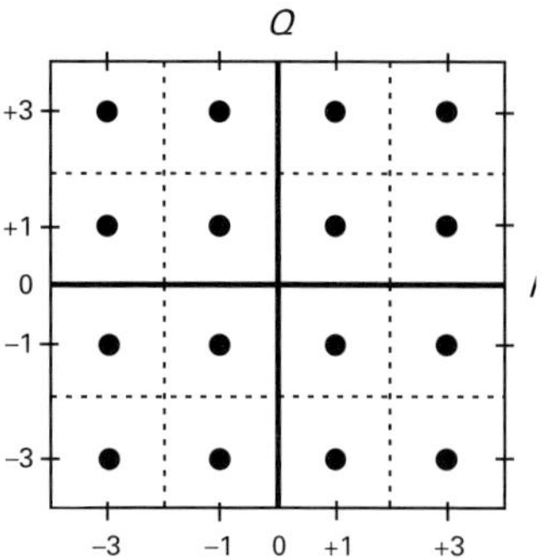

Figure 8.20   16-QAM constellation

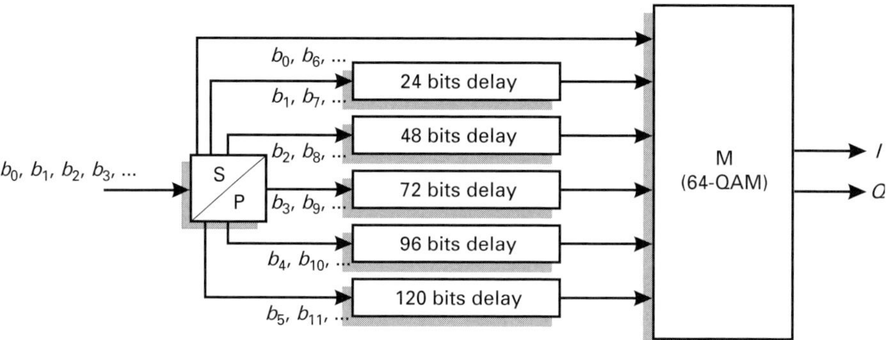

Figure 8.21   Bit interleaving and 64-QAM modulator

64-quadrature amplitude modulation (64-QAM)

The 64-QAM scheme is similar to 16-QAM one. The difference between them is only the number of bits contained in each symbol and, consequently, the number of symbols of the constellation. The separation process of the serial stream of bits and the mapping process also adapt to support this change.

In a constellation of 64 symbols, each symbol needs 6 bits to represent it, and, consequently, the serial sequence is divided into six independent streams. In the interleaving, the delays are inserted proportionately in the streams, as shown in Figure 8.21. The mapping process of 64-QAM also encodes its information in the phase and in the amplitude of its symbols. Figure 8.22 shows the 64-QAM constellation.

Normalization of the modulation level

When working with modulations of different levels, one must be careful to normalize the maximum symbol energy from different constellations. In this way, the normalization factor is multiplied by the symbol of each constellation, which is in the $Z = I + jQ$ format. These factors are shown in Table 8.10.

Table 8.10. Normalization factor

Modulation scheme	Normalization factor
$\pi/4$ DQPSK	$Z/\sqrt{2}$
QPSK	$Z/\sqrt{2}$
16-QAM	$Z/\sqrt{10}$
64-QAM	$Z/\sqrt{42}$

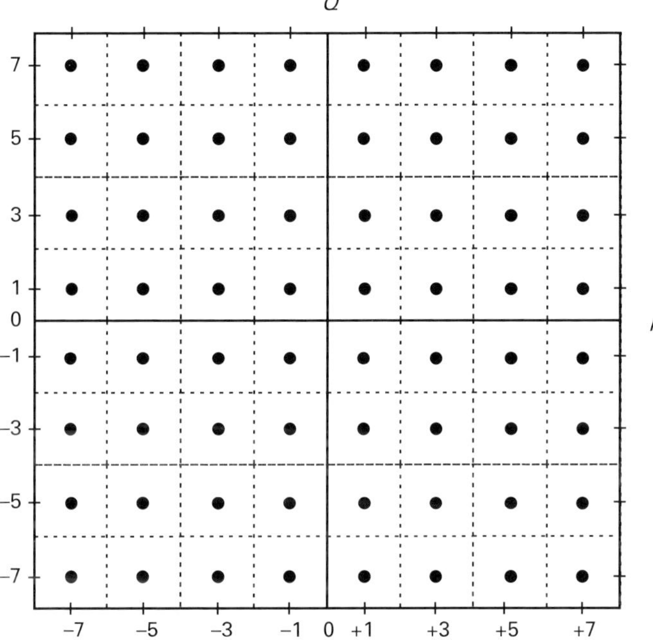

Figure 8.22   64-QAM constellation

8.5.3   Time interleaving

The time interleaver inserts delays between the segment carriers in such a way that more than one sequence of data is not transmitted at that same time, even if they are in different carriers. This action is taken in order to prevent burst fading from corrupting the signal. Burst fading acts upon a sequence of data, thereby it has a seriously damaging effect.

One corrupted symbol is easily recovered by the internal and external codes, but a lost sequence of symbols implies the loss of a piece of information, for the codes will not be able to recover it. The interleaver spreads the data in the time domain when it inserts different delays, and it is in this way that the protection against burst fading occurs, as the symbols will no longer be corrupted in sequence.

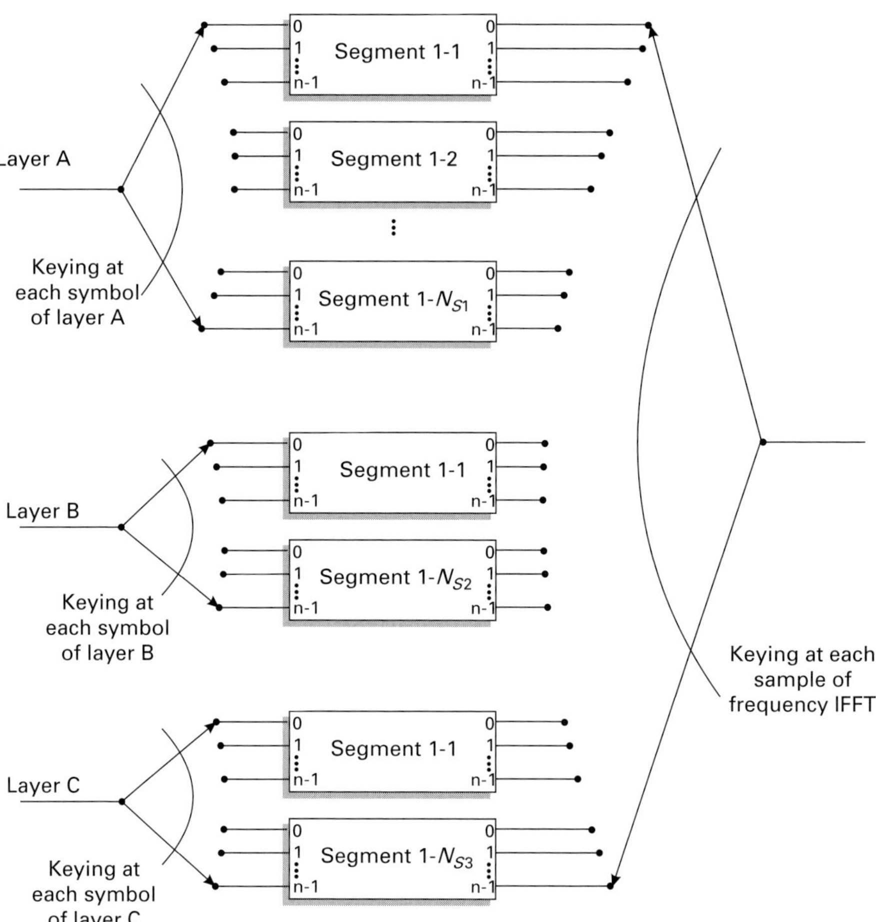

Figure 8.23 Time interleaving

The delays are inserted in the carriers by the time interleaver sequentially in each segment. The disposition of the segments, as well as the order for the insertion of the delays, is shown in Figure 8.23. It is important to remember that, at every new segment, the sequence is restarted.

The dimension of the delay inserted is defined at the beginning of the transmission by means of the variable I:

$$m_i = [(i \cdot 5) \mod 96] \cdot I, \tag{8.6}$$

which may assume preestablished values, as can be seen in Table 8.11. The decision is taken according to the transmission mode adopted and the degree of protection that must be chosen to fight the effects of the burst fading. Consequently the bigger the value of I, the better protected is the transmitted signal.

Table 8.11. Delay adjustment made by the interleaver

	Mode 1			Mode 2			Mode 3	
I	Number of adjust symbols	Number of delayed symbols	I	Number of adjust symbols	Number of delayed symbols	I	Number of adjust symbols	Number of delayed symbols
0	0	0	0	0	0	0	0	0
4	28	2	2	14	1	1	109	1
8	56	4	4	28	2	2	14	1
16	112	8	8	56	4	4	28	2

8.5.4   Frequency interleaving

The frequency interleaver is in charge of executing some shuffling in the carriers of the same segment, in order to give a random aspect to the frequencies. This is another protection for the signal, intended to reduce the destructive effects of the noise with memory (noise that attacks several symbols in series) in the signal transmitted. This block is composed of sub-blocks, that execute the functions of data interleaving, shuffling of frequencies, and rotation of carriers within the same segment. The flowchart of the frequency interleaver is shown in Figure 8.24.

There are certain peculiarities to the frequency interleaver that must be considered. The first one is that no data interleaving is executed in the layer used for partial reception. Another peculiarity is that the shuffling in the carriers must be done in an independent manner when one of the layers uses a coherent modulating scheme and the other uses an incoherent one. This is due to the fact that the format of the OFDM transmission frame is different for these types of modulation. The shuffling of carriers is used in the segments independently of the transmission mode. Actually, each transmission mode specifies a shuffling table for the carriers of a segment.

8.5.5   Frame composer

The frame composer is responsible for ordering the symbols into OFDM frames, which will later be transmitted. There are two basic frame structures selected from the type of modulation used in the layer. These two structures also receive the continuous pilot (CP) signal, the transmission multiplexing configuration control (TMCC) channel, and the auxiliary AC1 and/or AC2 channels.

Configuration for differential modulation

The transmission frame for differential modulation and transmission mode 1 is presented in Figure 8.25. In ISDB-T there is only one type of differential modulation for the data, which is the DQPSK. The symbols delivered by the interleavers are arranged in the carriers and represented by $S_{i,j}$, in which i represents the number of the carrier in question, and j the transmission position within each carrier. Note that the frame has 204 symbols

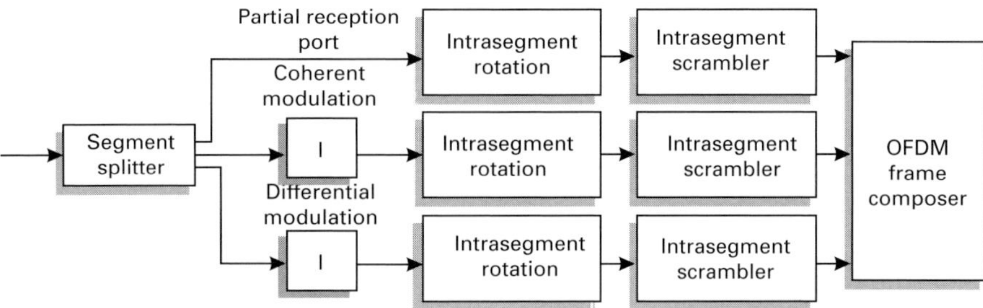

Figure 8.24   Frequency interleaving

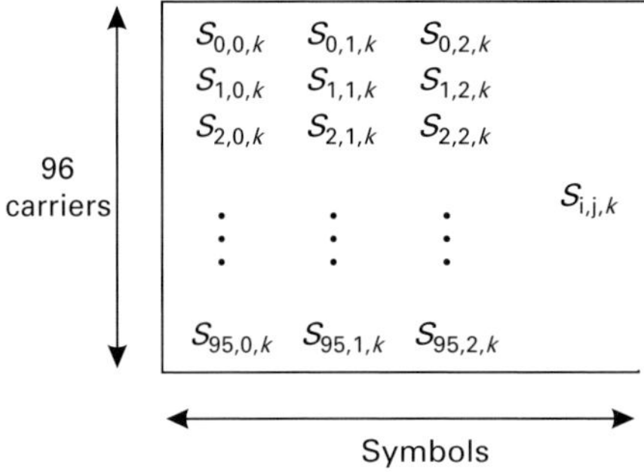

Figure 8.25   Configuration of a segment for differential modulation in mode 1

per carrier, and, for transmission mode 1, the total number of carriers is equal to 108, comprising 96 data carriers and 12 control carriers. In transmission mode 2 there are 216 carriers, and in mode 3 there are 432.

The control signals are inserted in the frame in a particular manner between the segments. The TMCC and AC channels are inserted in distinct carriers to lessen the effects caused by the multipath.

The CP signal is transmitted in the carrier furthest to the left of the spectrum, i.e. the lowest-frequency carrier. Its function is to provide the receiver with a reference for the execution of differential demodulation. The scattered pilot (SC) is always introduced in the 0 carrier in each segment and is responsible for transmitting a sequence for a reference of synchronism.

Configuration for synchronous modulation

The synchronous modulation schemes used for data transmission in the ISDB-T standard are: QPSK, 16-QAM, and 64-QAM. Unlike the frame for differential modulation, the SP signals are inserted in the synchronous frame at every 12 carriers, and at every four symbols within the same carrier. This arrangement ensures the synchronism of information in the receiver. The TMCC and AC1 channels are also inserted in distinct carriers in the segments. The AC2 channel is available only in the frame that uses differential modulation.

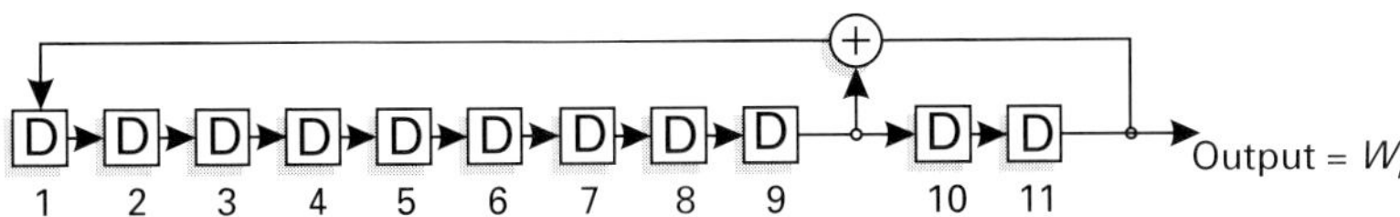

Figure 8.26   Generator of the pseudo-random sequence.

The SP pilot-signal

The SP is one of the ISDB-T standard pilot signals, and is formed by a pseudorandom sequence. This sequence is generated by a shift register, as shown in Figure 8.26. In spite of the SP signals of all segments being generated by the same register, they are different. This happens because each segment has a different initial state value, which provides distinct sequences.

The SP signal has the function of ensuring synchronism for the segments that use synchronous modulation. The modulation scheme used in these channels is BPSK.

The CP pilot-signal

Like the SP signal, the CP pilot-signal is BPSK modulated. Its function is to serve as a reference for the segments that use differential modulation. This signal is constant in time and defined according to the value of W_i.

The TMCC signal

The TMCC signal is a pilot-signal modulated by the DBPSK scheme, and uses the reference signal W_i. This signal is responsible for carrying information on the configuration of each segment in such a way that the receiver knows how to demodulate and decode it.

9 International System for Digital Television (ISDTV)

with Mylène Christine Queiroz de Farias and Marcelo Menezes de Carvalho

9.1 Introduction

Television is one of the most popular communication media in Brazil, where over 94.6% of households have at least one television set. This corresponds to more than 80 million television sets. Besides providing free access to entertainment and culture, broadcast television acts as an unifying factor of fundamental importance to a country of the size of Brazil.

After 50 years of existence, Brazilian television is undergoing significant changes with the arrival of the digital television (DTV) technology. DTV not only represents progress in terms of communications technology, but also a new concept in terms of accessing media information. Within the same bandwidth currently used by analog television channels, DTV is able to transmit different types of programs that may include high-definition television (HDTV) programs, several standard-definition television (SDTV) programs, several audio programs or even several multimedia (lower-definition) programs. One of the major novelties of DTV is the advent of interactivity between users and broadcasters/content providers. With DTV, users are able to participate in polls, play games, surf the Web and send and receive emails. In summary, besides providing a better quality of image and sound, and a large variety of services, the technology of DTV transmission allows better utilization of the spectrum, which translates into more channels in the same geographic area.

Effective action to implement DTV in Brazil started in the late 1990s with the work of the Brazilian Commission of Communications, a body of the National Telecommunications Agency (Anatel). From November 1998 to May 2000, extensive field and laboratory tests were carried out with the three DTV standards available at that time (Alencar, 2005b):

(1) the Advanced Television System Committee (ATSC) standard developed by the USA (ATSC, 1995c, 2000b, Richer *et al.*, 2006);
(2) the Digital Video Broadcasting – Terrestrial (DVB-T) standard developed by the European Union (Reimers, 2006, ETSI, 1997a, 2001, EISI, 1997b);
(3) the Integrated Services Digital Broadcasting – Terrestrial (ISDB-T) standard developed by Japan (Asami and Sasaki, 2006, Takada and Saito, 2006).

As well as comparing the performance of the available standards, these tests provided insights on the technologies appropriate for the unique environmental, economic, and social conditions of Brazil.

In November 2003, the project for the development of the Brazilian Digital Television System (SBTVD), which came later to be known as International System for Digital Television (ISDTV), was officially launched. The main objective of the ISDTV project was to define the reference model for the Brazilian DTV standard, which includes not only the technology itself, but also the ways of exploiting the rights to transmit, and the transition model from analog to digital. The ISDTV, also called ISDB-Tb, project aims to provide the population with access to DTV technology, promoting social inclusion and a democratic access to information (Farias *et al.*, 2008).

A total of 105 institutions were involved in the ISDTV project, including industry, universities, research centers, and broadcasting companies. In February 2006, the report containing the recommendations for the ISDTV standard was released (CPqD, 2006).

In June 2006, the Brazilian president, Mr. Luiz Inácio Lula da Silva, signed the decree that officially defined the transition period from analog to digital television. According to the decree, ISDTV permits digital transmission of standard- and high-definition video, simultaneous transmission for fixed, mobile and portable devices, and interactivity. The innovations of ISDTV are:

- the use of the more advanced H.264 standard for digital video coding, as opposed to the MPEG-2 standard used in the ATSC, DBV-T, and ISDB-T standards;
- Ginga middleware, which has been specifically designed for ISDTV;
- the use of WiMAX technology as the communications platform for the interactivity channel.

The decree established that ISDTV must use the same technology as that used by the Japanese standard ISDB-T for coding and modulating digital television signals, which means that the signals are transmitted with the band segmented transmission (BST) technique and orthogonal frequency division multiplexing (OFDM). The Japanese model was chosen because it was considered to be the most advanced of the three available standards (ISDB, DVB, and ATSC), supporting the transmission of high and standard definition to fixed and mobile receivers. In the field and laboratory tests performed in Brazil, the modulation and coding scheme of ISDB-T had the best performance (Bedicks *et al.*, 2006).

The transmission of ISDTV signals started in the city of São Paulo on December 2, 2007. A photograph of the inauguration ceremony in São Paulo is displayed in Figure 9.1. In this picture, President Luiz Inácio Lula da Silva is addressing the audience.

ISDTV maintains the same characteristics as analog television: (1) it is broadcast in the VHF/UHF band, free of charge and (2) each broadcast company is assigned a 6 MHz channel. ISDTV roll-out plans mandate that all state capitals must be covered by the end of 2009, with the rest of the Brazilian cities following before the end of 2013. During this period, analog and digital signals are to be transmitted simultaneously. Initially, users do not need to change their television sets, as an adapter (set-top box) can be used to convert the digital signals transmitted so they can be displayed on an analog television.

In this chapter, a description of the ISDTV is given, considering the reference model for DTV systems defined by the International Telecommunications Union (ITU) (ITU-T,

Figure 9.1   ISDTV inauguration ceremony. (Source: Ricardo Stuckert/Agência Brasil)

1996). For each subsystem of the model, a description of the corresponding technology adopted for ISDTV is given.

9.2   Reference model for a terrestrial DTV system

A terrestrial DTV system is defined as a platform capable of transmitting and receiving audio, video, and data using a broadcasting channel in VHF and UHF band frequencies. The ITU has defined a reference model for terrestrial DTV systems that is common to all standards currently available (ITU-T, 1996). The reference model gives a general view of the main structures that compose a DTV system, describing how it works and how these structures interact with each other. According to this model, a terrestrial DTV system can be divided into the following subsystems:

- source coding: this is composed of audio, video, and data coding at the transmission side, and audio, video, and data decoding at the receiver side;
- transport layer: this is composed of the multiplexing stage at the transmission side and the demultiplexing stage at the receiver side;
- transmission and reception: this comprises channel coding and modulation at the transmission side and channel demodulation and decoding at the receiver side;
- middleware: this is the software layer, which allows applications to be run at the receiver side;
- interactivity channel: this consists of the downstream and return channels, including the underlying communications technology and infrastructure.

ISDTV is fully compliant with the reference model defined by the ITU (ITU-T, 1996). A block diagram of the ISDTV standard is presented in Figure 9.2. Table 9.1 shows a summary of the technical specifications for the stages of ISDTV. The following sections describe each of the subsystems of ISDTV and, in particular, the technologies chosen for the stages of each subsystem.

Table 9.1. ISDTV technical overview

Stage	ISDTV choice
Video coding	H.264
Audio coding	MPEG-2 AAC
Middleware and data coding	Ginga (NCL + Java)
Transport layer	MPEG-2 Systems
Modulation	BST-OFDM

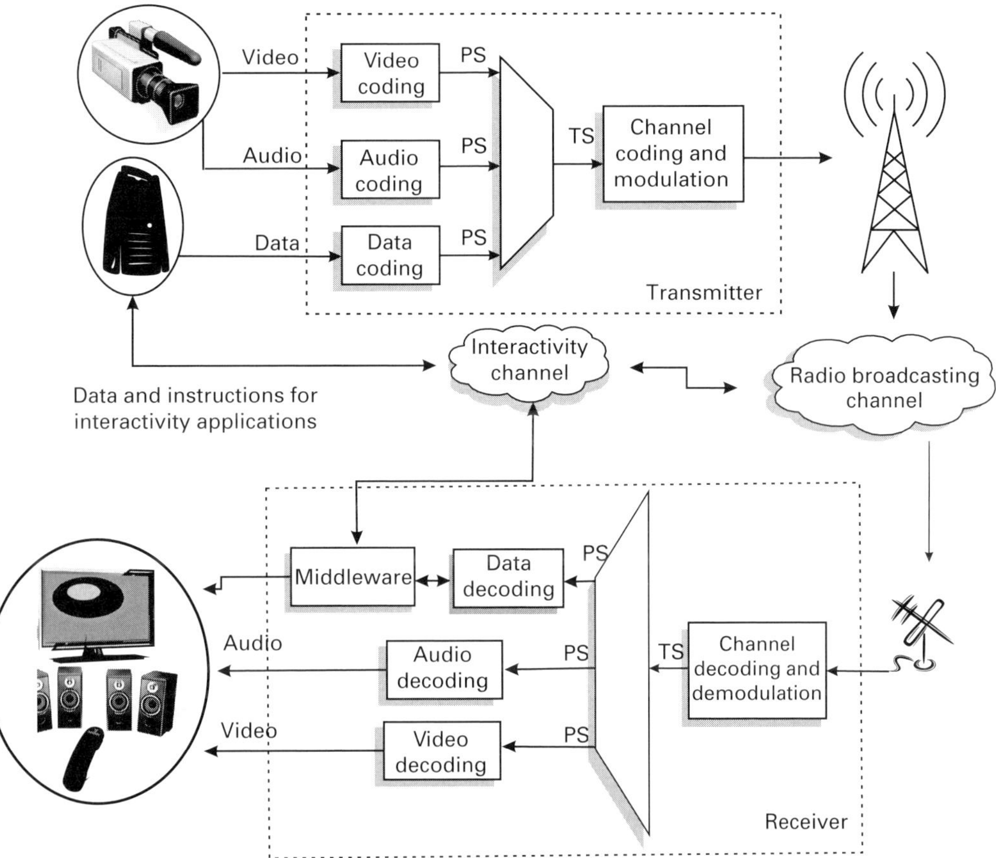

Figure 9.2 Block diagram of the terrestrial DTV reference model

9.3 Source coding

The source coding stage is responsible for the compression of the audio and video signals with the goal of reducing the transmission bit rate. Source coding exploits the inherent redundancy in the source signals to reduce the number of data (compression). This section presents three stages that are part of the source coding subsystem: audio coding, video coding, and data coding.

9.3.1   Video coding

The bit rate for an uncompressed digital video signal can reach up to 270 Mbits/s for standard definition (SD) and 1.5 Gbit/s for high definition (HD). Since the bit rate available for terrestrial digital video broadcasting is around 19 Mbits/s (for a 6 MHz channel), the video coding stage is of great importance. It has the goal of compressing the video signal, reducing the transmission bit rate. Compression is achieved by removing the redundant information (spatial, temporal, and statistical) from the video.

The video coding stage is carried out by a video coder (on the transmission side) and by a video decoder (on the reception side). The video coder receives the uncompressed digital video signal as input and performs its compression. At its output, it generates a coded video stream that is forwarded to the multiplexer in the transport layer. The video decoder receives the coded video stream from the demultiplexer and performs its decoding (decompression), obtaining the reconstructed video signal as output.

Currently, there are several standards for video compression. The most popular ones were produced by the Motion Picture Experts Group (MPEG) and the Video Coding Experts Group (VCEG). MPEG is a working group of the International Organization for Standardization (ISO) and of the International Electrotechnical Commission (IEC), formally known as ISO/IEC – JTC1/SC29/WG11. Among the standards MPEG has developed are MPEG-1, MPEG-2, and MPEG-4. MPEG-2 is a very popular standard used not only for broadcasting, but also in DVDs (Haskell *et al.*, 1997, ITU-T, 1998). The main advantage of MPEG-2 is the low cost of the receivers due to its popularity and the large scale of production. MPEG-2 is also undoubtedly a very mature technology.

VCEG is a working group of the Telecommunication Standardization Sector of the International Telecommunications Union (ITU-T). Among the standards developed by VCEG are H.261 and H.263. A collaboration between MPEG and VCEG resulted in the development of the H.264, also known as MPEG-4 Part 10 or Advanced Video Coding (AVC) (Richardson, 2003a, ITU, 2003). H.264 represents a major advance in the technology of video compression, giving a considerable bit rate reduction when compared to previous standards (Lambert *et al.*, 2006). For the same quality level, H.264 provides a bit rate of about half that provided by MPEG-2.

ISDTV has adopted H.264 as its video compression standard. H.264 is used to code both SD and HD video, as well as reduced-resolution videos targeted at mobile or portable receivers. The adoption of H.264 is a key innovation in relation to all other DTV standards.

H.264 video compression standard

H.264 is the most recent digital video compression standard (ITU-T, 2003), developed by VCEG and MPEG. Compared to MPEG-2, H.264 provides the same perceptual quality at about one-third to one-half of the MPEG-2 bit rate. It is a flexible standard that can be used in a large variety of applications (from low-bit-rate mobile video to HDTV) in different types of transmission media (terrestrial, satellite, DSL, etc.). Like MPEG-2, H.264 is a hybrid video coding scheme, i.e. it is a block-based algorithm that uses interframe prediction with motion compensation. The performance of H.264 is a consequence of the introduction of new coding techniques that do a better job of reducing the redundancies of the video. As expected, the improvements come at the cost of a higher computational complexity.

The block diagrams of the encoder and decoder of the H.264 standard are displayed in Figures 9.3 and 9.4, respectively. In the encoder, the input frame video is divided into macroblocks. Each macroblock corresponds to one block of 8×8 or 4×4 picture elements (pixels). These frame macroblocks can be encoded in *intraframe* or *interframe* mode. In interframe mode, a macroblock is predicted using motion compensation (temporal prediction). The prediction error, which is the difference between the original and the predicted macroblock, is transformed (discrete cosine transform, DCT), quantized, and entropy coded. In order to reconstruct the same image on the decoder side, the quantized

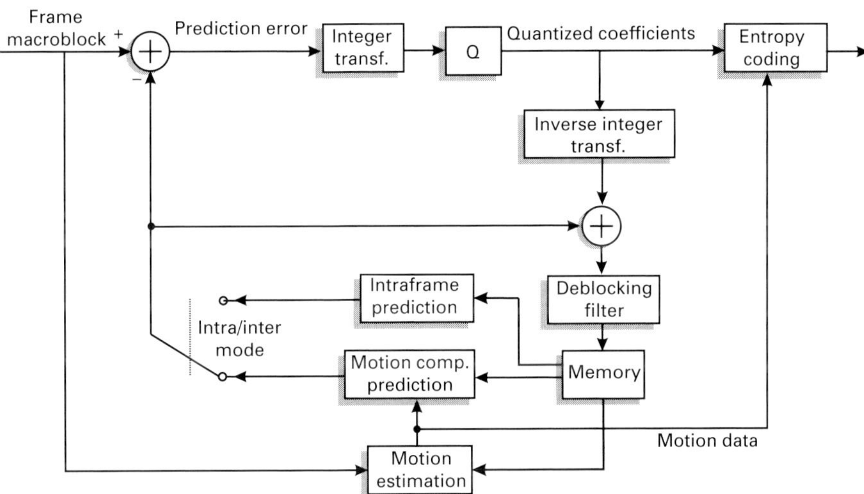

Figure 9.3   Block diagram of the H.264 encoder

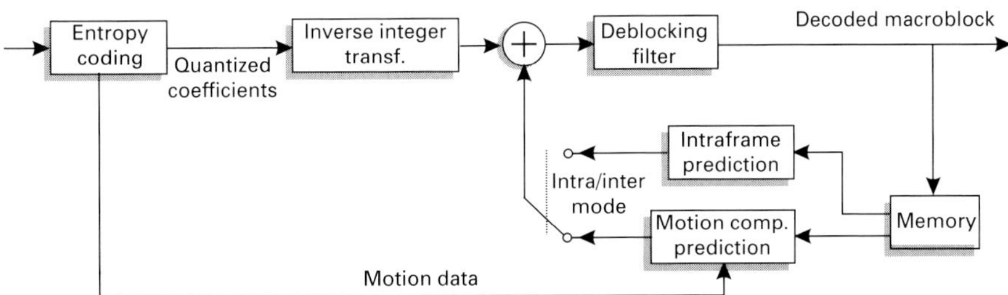

Figure 9.4   Block diagram of the H.264 decoder

coefficients are inverse transformed and added to the prediction signal. The result is the reconstructed macroblock on the decoder side.

In previous algorithms, such as MPEG-2, the frame macroblocks in intraframe mode are coded independently without any reference to previous or future frame macroblocks. At the encoder side, the prediction is set to zero and the original macroblock is transformed, quantized, and entropy coded. One of the innovations introduced by H.264 is the utilization of prediction for the intraframe mode. H.264 improves the performance of the compression algorithm by exploiting the spatial redundancy of frames for which temporal prediction cannot be used (e.g. the first frame of a video) or is inefficient (e.g. at scene changes). The prediction used by H.264 for the intraframe mode is a spatial prediction, which is calculated using the weighted average of the pixels to the left of and above the current macroblock.

Besides the introduction of intraframe prediction, H.264 also brings the following improvements to the prediction algorithm used in the interframe mode:

- use of different macroblock sizes (16×16, 8×8, or 4×4);
- reference to multiple frames;
- type B frames can be used as reference;
- weighted prediction;
- subpixel precision.

MPEG-2 uses a DCT as its transform. In the case of H.264, an integer transform is used instead. The integer transform is designed in such a way that its calculation requires only additions and shifts, which reduces the complexity of the algorithm considerably. The size of the integer transform is 4×4, or 2×2 in some special cases. The smaller block size enhances the performance of the algorithm, giving a better precision.

H.264 has two alternative methods for entropy coding. The first one is a low-complexity technique based on the use of context-adaptive switched sets of variable length codes (CAVLC). The second is the context-based adaptive binary arithmetic coding (CABAC), a more computationally demanding algorithm. Both methods represent improvements in terms of coding efficiency compared to the techniques used in MPEG-2, which employs a fixed variable length code (VLC).

Another innovation of H.264 is the use of a filter to reduce the blocking effect, which improves the perceptual quality of the decoded video. The anti-blocking filter is applied only to the borders of the blocks, except in the borders of the frame. This filter is highly adaptive and several parameters, thresholds, and the local characteristics of the picture control the strength of the filtering process. The dimensions of the block used by the transform and the quantizer determine the parameters of the filter.

9.3.2 Audio coding

According to ITU-T, the use of six streams of audio information is a desirable feature for all new DTV systems (ITU-T, 1996). For such systems, the six streams are divided in the following way: three frontal streams (L – left, C – center, and R – right), two back streams (Ls – left surround, Rs – right surround), and a low-frequency effects (LFE)

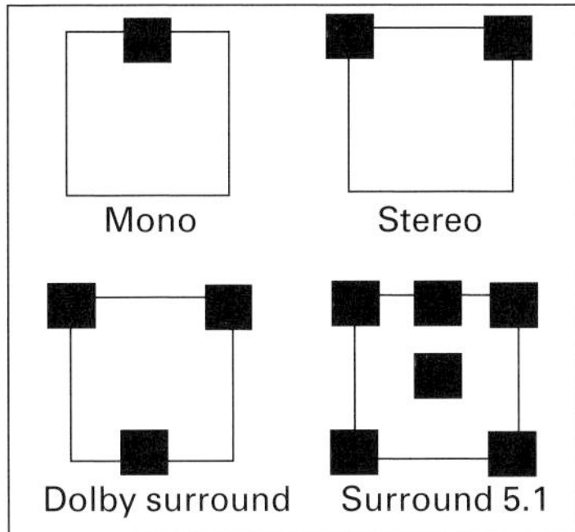

Figure 9.5   Spatial arrangement for mono, stereo, Dolby surround, and surround 5.1 audio

stream, which occupies only a fraction of the bit rate of the other streams. This system is known as multichannel 3/2+1 or 5.1. The different types of spatial arrangements for mono, stereo, and surround multichannel audio are presented in Figure 9.5 (ITU-T, 1996).

Audio signals transmitted without any type of compression require bit rates of 1.41 Mbits/s for stereo sound and 4.23 Mbits/s for multichannel systems 5.1 (for a sampling frequency of 44.1 KHz and a resolution of 16 bits). Although these bit rates are low when compared to the bit rates required for video, a significant reduction can be obtained using audio coding techniques. Compression is achieved by removing redundancy in audio signals. Advanced audio coding techniques include psycho-acoustic models to eliminate perceptually irrelevant data: this is known as perceptual coding. Also common are frequency domain techniques (subband and transform coding), dynamic bit allocation, and window switching.

ISDTV is expected to transmit in stereo and 5.1 multichannel, simultaneously. When necessary, receivers should convert from multichannel to stereo using a down-mixing technique, as described in ISO/IEC 14496-3 (Table 4.70) (ISO/IEC, 2005). Both signals are coded using the MPEG-2 Advanced Audio Coding (AAC) standard. MPEG-2 AAC incorporates developments in the area of audio coding. It is able to deliver CD-quality sound with bit rates around 96 kbit/s and it allows up to 48 audio streams and up to 15 distinct audio programs. In the following, there is a brief description of the MPEG-2 AAC system.

MPEG-2 AAC system

The block diagram of a generic audio coding scheme is displayed in Figure 9.6. It consists of the following stages:

(1) Time/frequency analysis: This stage has the goal of splitting the source spectrum into frequency bands to generate nearly uncorrelated spectral components. There are basically three ways of implementing the analysis: transform coding (TC), subband coding (SBC), and a hybrid of the two. Most of the techniques use an analysis filter bank to decompose the signal into frequency bands.

(2) Perceptual model: Using either the time-domain input signal and/or the output of the analysis stage, the perceptual model calculates the masking threshold values

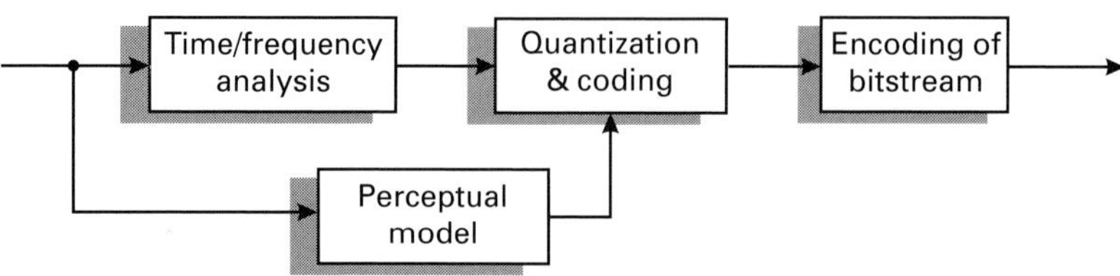

Figure 9.6   Block diagram of a generic audio coding system

(or allowed noise) for each frequency component. In some cases, the frequency components are roughly equivalent to the critical bands of human hearing.

(3) Quantization and coding: At this stage, the frequency components obtained from the time/frequency analysis stage are individually quantized and coded. If the quantization noise can be kept below the masking threshold for each component, the compression result should be perceptually indistinguishable from the original sound. The coding is performed using an entropy code, such as the Huffman coding.

(4) Encoding of bitstream: A frame packetizer is used to assemble the audio bitstream.

The original MPEG-2 audio standard is known as MPEG-2 bidirectional compatible (MPEG-2 BC) (ISO/IEC, 1994). MPEG-2 BC consists of two extensions of the previous MPEG-1 audio coding standard (ISO/IEC, 1992), which can only code stereo audio signals. MPEG-2 BC codes multichannel audio signals with backward and forward compatibility with MPEG-1. It also adds sampling frequencies of 16 k sample/s, 22.05 k sample/s, and 24 k sample/s, improving the coding efficiency for very low bit rates. Neither extension introduces new coding algorithms.

MPEG-2 AAC is a second-generation audio coding algorithm for coding stereo and multichannel audio signals. It is specified as both Part 7 of the MPEG-2 standard (ISO/IEC, 1997) and as Part 3 of the MPEG-4 standard (ISO/IEC, 2005). As such, it can be referred to as MPEG-2 Part 7 and MPEG-4 Part 3, depending on its implementation. MPEG-2 AAC has the same structure as the generic audio coding displayed in Figure 9.6. The block diagram of an MPEG-2 AAC encoder is presented in Figure 9.7. The new features of MPEG-2 AAC include high-resolution filter banks, new prediction techniques, and noiseless coding.

MPEG-2 AAC uses a standard switched modified discrete cosine transform (MDCT) filter bank, with an impulse response of 5.3 ms and a sampling frequency of 48 k sample/s. The maximum number of frequency lines is 1024 and the perceptual model is taken directly from MPEG-1. The time dependence of the quantization noise is controlled by the temporal noise shaping stage, which is a tool designed to control the location (in time) of the quantization noise. On the other hand, the adaptive predictor enhances the compressibility of stationary signals. It bases its prediction on the quantized spectrum of the previous block. Finally, an interactive method is employed for quantization and

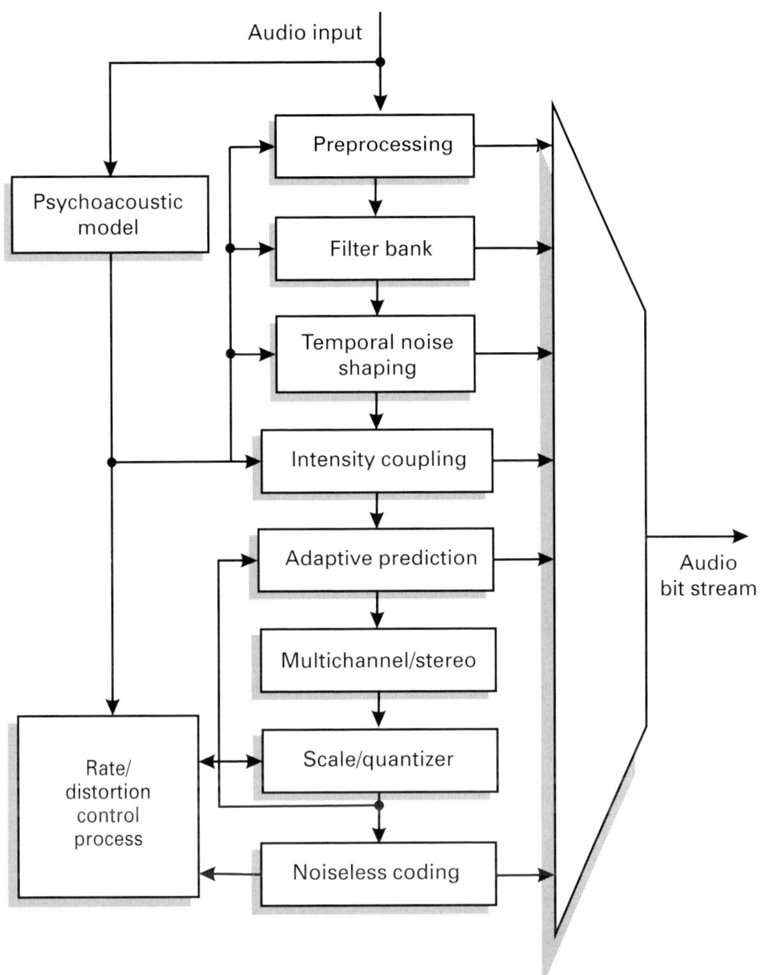

Figure 9.7 Block diagram of the MPEG-2 AAC encoder

noiseless coding to keep the quantization noise in all critical bands below the global masking threshold.

AAC encoding methods can be classified into the following profiles (MPEG-2) or object types (MPEG-4):

(1) Low complexity AAC (LC-AAC): this is the most widely supported profile. It does not use time-domain prediction, which saves in complexity;

(2) Main AAC: this is similar to AAC-LC profile, with the addition of backward prediction;

(3) Sampling-rate-scalable AAC (SRC-AAC): in this a hybrid filter bank is used instead of the switched MDCT. It offers the lowest complexity;

(4) High-efficiency AAC (HE-AAC): this is an extension of the AAC-LC, optimized for low-bit-rate applications.

These different profiles are not necessarily compatible with each other and they vary in complexity and coding efficiency. In some applications, the benefits of the more complex profiles are not worth the effort required to encode/decode. As a result, AAC-LC is currently the profile that is used by most encoders and decoders. However, it has become more popular with its addition to the Nero AAC encoder. The ISDTV audio coding standard allows for the following profiles of AAC:

- LC-AAC, level 2 (LC-AAC@L2);
- LC-AAC, level 4 (LC-AAC@L4);
- HE-AAC, level 2 (HE-AAC v1@L2) for stereo;
- HE-AAC, level 4 (HE-AAC v1@L4) for multichannel.

9.3.3   Data coding

The main goal of the data coding stage is to fragment and organize the data so that it can be multiplexed into the transport stream. A block diagram showing the information flow between the audio coding, transport layer, and interactivity channel is displayed in Figure 9.8.

The type of data coding used by the DTV system depends on the type of application available at the transmission side. The data are transmitted in a datagram format, which is a logical structure that contains all the information about the data (size and content), their destination, and their function. At the output of the transport layer, the data are transmitted in fixed-size packets. The data encoder, shown in Figure 9.8, provides two possibilities for

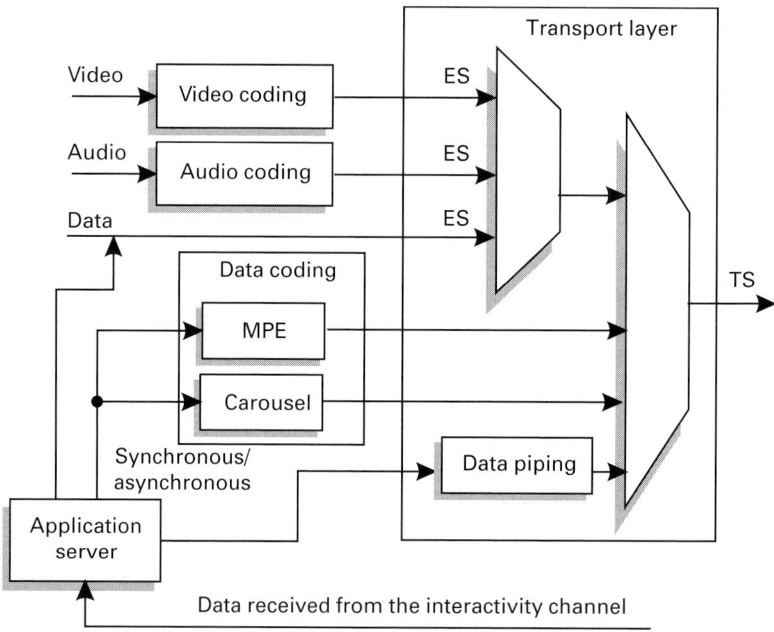

Figure 9.8   Flow diagram of the data coding, transport layer, and interactivity channel

transport mechanisms: the carousel and the multiprotocol encapsulation (MPE). These transport mechanisms are responsible for the fragmentation and reassembly processes. For a higher degree of reliability, they use error correcting codes.

The data carousel is a transport mechanism that allows the periodic transmission of data to a decoder in a broadcasting system. The content of a data carousel is repeated cyclicly, therefore, the decoder at the receiver must wait for the correct time instant when the requested data are being transmitted. Although the content of a carousel can be modified in response to a user request, it is more common that this content be repeated periodically, independently of how many users are connected or what type of data is needed at that moment. In this case, if a user needs a specific set of data, the user must wait until this set is available in the carousel. The data repetition rate in the carousel is chosen to maximize the bandwidth usage.

There are two types of carousel:

- Data carousel, which contains only data that are not specified or identifiable. It is up to the receiver to identify the received data. This type of carousel can be used, for example, for automatic updates of the set-top box software.
- Object carousel, which allows different types of data (images, text files, or application executable files) to be identified from a directory list that contains all the objects in the carousel. This type of carousel can be used for applications that require the identification of the data to be accessed. Examples of such applications include shopping services and the downloading of games and applications.

MPE is a transport mechanism that allows data from any communication protocol to be transmitted in a private table section, using a mapping to a specific structure. This structure is composed of a binary sequence with information corresponding to its own structure (header), as well as available space for the data transport. This type of mechanism is commonly used in computer network traffic in which the TCP/IP protocol carries the information about the addresses (IP) of the origin and the destination, besides the medium access control (MAC) target address. The MPE supports any type of protocol and can be used for unicast, multicast, or broadcast applications.

The way the functions executed by the data encoder are implemented depends on the specifications of the transport layer and the middleware. Therefore, the data encoder does not have its own specifications. It merely conforms to the specifications of these two subsystems.

9.4 Transport layer

Program is a term used to describe a single broadcast service or channel. A program is composed of one or more elementary streams (ES), which are single, digitally coded video or audio. A traditional analog television broadcast comprises three elementary streams: one carrying coded video, one carrying coded stereo audio, and one carrying teletext, data. A DTV broadcast, on the other hand, includes extra elementary streams: video elementary streams carrying alternative resolutions of the original video, audio or teletext, elementary streams carrying alternative choices of language for the audio and captions related to the video, and the necessary data for the synchronized display of this information in the receiver.

The transport layer of a DTV system provides a set of resources that can be used to extend the functionalities of this platform besides broadcasting of audio and video. On the transmission side, the transport layer combines (multiplexes) the several programs in a unique digital signal stream in order to prepare them for transmission. On the receiver side, it decomposes (demultiplexes) the digital stream into several elementary streams that are forwarded to the decoders. The block diagram of the transport layer is shown in Figure 9.9.

In the multiplexing stage of the transport layer there are two levels of multiplexing. In the first level, the elementary streams of audio, video, and data associated with specific programs are packetized and multiplexed. The data associated with programs include subtitles, additional information or any type of data that might be sent together with the TV signal.

The second level of multiplexing involves adding data that are not related to the television content (datacasting). This includes information that is available to all users, like, for example, electronic program guides, traffic conditions, weather forecasts, horoscopes, and a summary of the daily news. At this level, any information related to the structure of the multiplexed signal, of the network, and the broadcasting system, as well as eventual updates for the receiver software, is added. This type of information, which represents the static characteristics of conventional television content, is transmitted within a structure called a section that can be multiplexed and sent in a lower frequency than the frequency used for regular television content.

Another stage of the transport layer is the *control unity and QoS* (quality of service). This stage is coupled to the multiplexer from which it receives information about the data traffic. Its main function is to optimize the bandwidth usage, avoiding idle time of

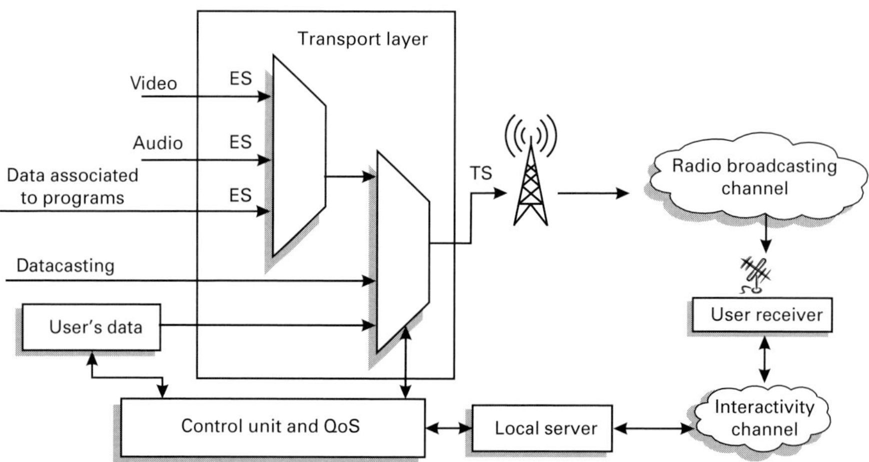

Figure 9.9   Block diagram of the transport layer

the broadcasting channel. The goal is to balance the traffic of the television programs and data (datacasting).

Besides multiplexing and demultiplexing data, the transport layer has the following functions:

- it supports transmission of coded bit streams in environments prone to errors;
- it supports multiplexing of multiple programs into one transport stream, which requires a sophisticated synchronization and the capacity of describing and identifying a network formed by several multiplexed streams, each containing different programs;
- it supports conditional access;
- it buffers administration to guarantee that input data are not lost;
- it supports any information that is not considered conventional television content, including a private data system that characterizes the customization of a commercial standard;
- it supports add-drop capacity, i.e. the capacity to substitute part of the packets corresponding to the original content by other content. This is particularly important for local broadcast transmissions, in which there is a need to insert local programs in the content program.

ISDTV uses MPEG-2 Systems as the standard for the transport layer functions, which is the same standard used by ATSC, DVB, and ISDB. One of the features of MPEG-2 Systems is its great capacity to absorb and adapt to local requirements of a given DTV system. Since it is a widely used system, its adoption provides compatibility between ISDTV and the three most well-known DTV systems. Next, a brief description of the MPEG-2 Systems is provided.

9.4.1 MPEG-2 Systems Layer

The ITU-T H.222 Recommendation (ITU-T, 2000b), also known as MPEG-2 Systems Layer, provides a set of tools that can be employed in the conception of the transport layer for a DTV system. This set of tools consists of general functionalities, some or all of which can be chosen. The restrictions to the semantics can be specified to reflect the local requirements and necessities, extending the original functionalities described in the ITU-T H.222 Recommendation.

The working principles of the multiplexing operations in MPEG-2 Systems are illustrated in Figure 9.10. The first stage of the block diagram is the packetizer, which converts each elementary stream into a packetized elementary stream (PES). A PES packet is formed by a header and a payload. The payload consists of a series of data bytes taken sequentially from the elementary stream. PES packets have variable lengths, with a maximum size of 64 kbytes. After the packetizer, the next step consists of combining (multiplexing) the video PES, audio PES, and data PES into a single digital signal stream. MPEG-2 Systems specification defines two alternative multiplexes: the program stream (PS) and the transport stream (TS).

The PS is based on MPEG-1 multiplex and can only accommodate a single program. It is intended for storage and retrieval of program material from digital storage media in

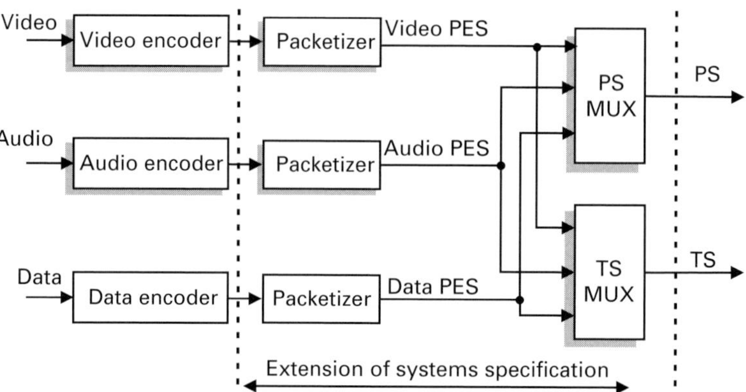

Figure 9.10   Block diagram of MPEG-2 Systems Layer

error-free environments. It is rather susceptible to errors for two reasons: (1) The program stream is composed of a succession of relatively long packets of variable length, so a loss or corruption of a single packet may represent the loss or corruption of an entire video frame. (2) Since the packets are of variable size, the decoder cannot predict where a packet begins or end and, therefore, where a new packet will start. The information about the length of the packet is located inside its header. As a result, if the header is corrupted the decoder will lose synchronism with the stream, resulting in the loss of at least one packet.

The TS is intended for multiprogram applications, such as broadcasting. A single TS can accommodate several independent programs. A TS is composed of a series of packets of 188 bytes, which are called "transport packets". The use of short, fixed-length packets reduces the susceptibility to errors. Error protection can easily be added using error protection codes, such as Reed–Solomon. As a consequence, TSs are used for transmissions over communications channels. On the other hand, the TS is a more sophisticated multiplex than the PS and, consequently, it is more difficult to create and demultiplex.

If a PS is being generated, a PS multiplex (PS MUX) is used. In this case, PES packets are organized in packs, which are made of a pack header, an optional system header and several PES packets. As previously pointed out, there is no constraint on the length of a pack, except that a pack header must occur every 0.7 seconds within the program stream. The pack-header contains important timing information. The system header contains a summary of the characteristics of the program stream, such as its maximum data rate, the number of contributing video and audio elementary streams, and further timing information.

A TS multiplex (TS MUX) is used to generate a transport stream. The length of a TS packet is always 188 bytes. It is composed of a 4-byte header, an adaptation field or a payload or both. PES packets from the various elementary streams are divided among the payload parts of several transport packets. Since it is unlikely that a PES packet fills the payload of an integer number of transport packets, the excess space is deliberately wasted by including an adaptation field of the appropriate length. A careful choice of the packet size can avoid waste. The use of long PES packet lengths, for example, ensures that a greater proportion of the transport packets are completely filled. There are no constraints on the order in which transport packets appear within the multiplex, except that the chronological order of packets belonging to the same elementary streams must be preserved. The resulting transport packets are output sequentially to form an MPEG-2 TS.

9.5 Transmission and reception

The main objectives of the transmission and reception subsystem are:

- to maximize the total bit rate of transmission, with the goal of transmitting the maximum amount of information through the broadcasting channel (limited by the total bandwidth allocated to the television channel), which also means maximizing the usage of the bandwidth (spectrum);
- to maximize the robustness of the TS, which is subject to degradation introduced by the broadcasting channel, as, for example, attenuation, obstructions, multipath, noise, and interferences;
- to minimize the power required for transmission;
- to keep the frequency spectrum of the transmitted signal contained in the appropriate bandwidth.

A simplified block diagram of the transmission and reception subsystems is presented in Figure 9.11. The transmission subsystem is divided in the following stages:

- Channel encoder: this adds redundant information to the TS with the goal of making it more robust to errors in transmission. This redundant information is used in the receiver to correct possible errors introduced by the transmission channel.

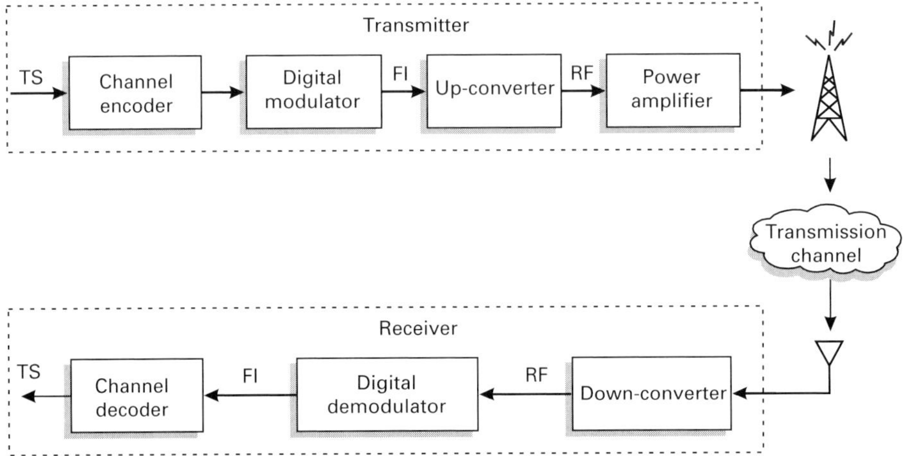

Figure 9.11 Simplified block diagram of the transmission and reception subsystem

- Digital modulator: this processes the coded signal so that it is possible to broadcast it in air with limited bandwidth and as robust to error channels as possible.
- Up-converter: this converts the modulated signal from an intermediate frequency (IF) to the required radiofrequency (RF), in the UHF or VHF band.
- Power amplifier: this amplifies the power signal from the RF level to the power required to cover the designated area. The power amplifier must be as linear as possible to avoid distortion in the amplified signal and to restrict the signal bandwidth to the bandwidth of the channel, minimizing interference in adjacent channels.

The reception subsystem is divided in the following stages:

- Down-converter: this receives the RF signal captured by the antenna and converts it from the RF (in the UHF or VHF band) to an IF. This stage also includes filters and amplifiers to eliminate interference from other channels and to adjust the signal to the level required by the digital demodulator.
- Digital demodulator: this is responsible for the recovery of the TS from the IF signal.
- Channel decoder and estimator: this is responsible for removing the redundant information and correcting errors (possibly introduced by the channel) and compensating for the distortion present in the signal.

ISDTV adopted the same technology as used by ISDB-T for coding and modulating DTV signals, which means that the signals are transmitted with the BST technique and OFDM. The Japanese model was chosen because it was considered, at the time that the field tests were performed, to be the most advanced of the three available standards (ISDB, DVB, and ATSC). In the field and laboratory tests performed in Brazil, the modulation and coding scheme of ISDB-T presented the best performance (Bedicks *et al.*, 2006). There were also commercial interests, because the DVD standard was backed by the big European telecommunications companies, and the broadcasters in Brazil were afraid to lose their grip on the transport and transmission business (Alencar, 2005b).

The BST-OFDM scheme allows flexibility and mobility, making it possible for television signals to be received by fixed and mobile receivers. It allows high-quality digital modulation and supports HDTV. The subdivision of the digital channel allows simultaneous transmission of multiple services. The following subsection presents a brief description of the modulation scheme.

9.5.1   BST-OFDM

OFDM is a digital multicarrier modulation scheme that multiplexes a high number of subcarriers in the same channel bandwidth, while maintaining orthogonality among them. Each subcarrier is modulated at a low symbol rate with conventional modulation schemes. In comparison to single-carrier modulation schemes, OFDM has the advantage of being able to suppress or attenuate some types of channel distortion. For example, multipath interference is minimized by inserting a guard interval in the time domain.

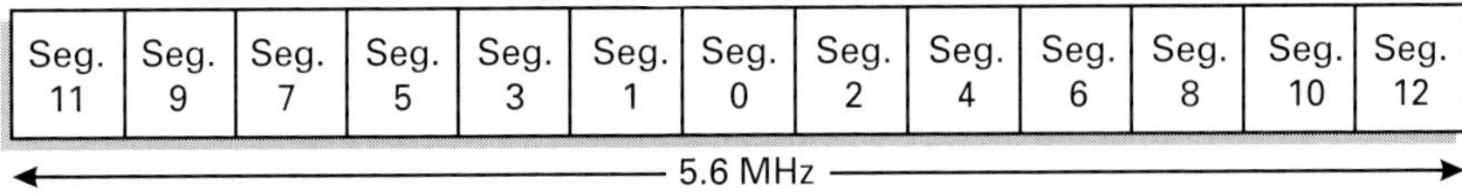

| Seg. 11 | Seg. 9 | Seg. 7 | Seg. 5 | Seg. 3 | Seg. 1 | Seg. 0 | Seg. 2 | Seg. 4 | Seg. 6 | Seg. 8 | Seg. 10 | Seg. 12 |

$$\longleftarrow\text{—— 5.6 MHz ——}\longrightarrow$$

Figure 9.12   Arrangement of segments in the channel bandwidth

As the name suggests, BST-OFDM is a segmented version of OFDM. The BST-OFDM scheme uses a set of basic frequency blocks called *segments*, which have a common carrier structure. The assignment of segments in the frequency bandwidth is illustrated in Figure 9.12. Each segment has a bandwidth corresponding to 1/14 of the terrestrial television channel bandwidth. Thirteen segments are active within one terrestrial television channel. In Brazil, where the channel bandwidth is 6 MHz, the useful bandwidth is $6\,\text{MHz} \times 13/14 = 5.57\,\text{MHz}$. Each segment occupies $6/14\,\text{MHz} = 428.6$ kHz. The net bit rate for one segment ranges from 280.85 to 1787.28 kbits/s and the data throughput ranges from 3.65 to 23.23 Mbit/s. Although BST-OFDM was developed and tested with 6 MHz channels, it can be scaled to any channel bandwidth with corresponding variations in data capacity.

The main characteristic of BST-OFDM is the flexibility of setting individual modulation parameters for each group of one or more segments, which are known as layers. This modulation scheme is known as *hierarchical modulation* and supports up to three layers of segments (layers A, B, and C). Depending on the transmission requirements, one, two, or three layers can be used. If the signal is transmitted using only one layer, the layer transmitted is A. If the signal is transmitted using two layers, the center layer is A and the outer layer is B. If the three layers are to be transmitted, the center layer is A, the middle layer is B, and the outer layer is C. Partial reception of services of the transmission channel is possible using narrow-band receivers that have a bandwidth as low as one OFDM segment. This is known as 'one-segment' and corresponds to the reception of only segment numbered zero.

For each of the three layers, the transmission parameters can be independently selected for choosing the modulation scheme (DQPSK, QPSK, 16-QAM, 64-QAM), the coding rate of the inner error correcting code ($\frac{1}{2}, \frac{2}{3}, \frac{3}{4}, \frac{5}{6}, \frac{7}{8}$), and the length of time interleaving. The error correcting schemes used are concatenated codes: Reed–Solomon (204, 188) code for the outer code and a convolutional code for the inner code. The resultant bit rate depends on the selected modulation scheme, inner coding rate, and guard-interval ratio.

Each segment can meet different service requirements and a number of segments can be combined to provide wideband services, such as HDTV. It is possible, for example, to send an audio service on one segment, a data service on another segment, and a television service on another segment – all within the same channel bandwidth. Furthermore, each segment may be modulated with different parameters so that, for example, the audio and data services are optimized for mobile reception, while the television service is optimized for stationary reception in a high-multipath environment.

BST-OFDM has three transmission modes (modes 1, 2, and 3) with different subcarrier intervals to allow for a wide range of transmitting frequencies. In mode 1, one segment consists of 108 subcarriers, while modes 2 and 3 have 216 and 432 carriers, respectively. The system has four choices of guard-interval length to enable better design of a single frequency network (SFN). There are also three different spacings among carriers to guarantee an adequate distance between SFN stations and sufficient robustness to Doppler effect during mobile reception.

9.6   Middleware

The greatest technological innovation introduced by DTV is *interactivity*. While the user experience with analog television is passive, the DTV technology allows interaction between users and the broadcasting companies. In practice, this means that the user is able to participate in polls, play games, surf the Web and send/receive emails.

Since the application data are transmitted along with the traditional television content, DTV receivers must be capable of separating and processing the different information formats and interpreting and executing instructions.

One of the biggest challenges for digital television systems is to guarantee the interpretation and execution of instructions in a wide variety of heterogeneous receivers, with different resources and capacities, and from different manufacturers. It is also necessary to allow software updates or upgrades, as needed. The subsystem that implements the mechanisms that provide the semantics for applications is known as *middleware*. The term middleware is used to describe integration software that connects other software together, acting as a mediator between independent programs.

The type of middleware used by a DTV standard generally defines the types of services available at the receiver. The choice of middleware affects the interactivity resources and the implementation complexity. In the case of ISDTV, the choice of technologies to be used in the middleware subsystem was considered a big step towards establishing a DTV standard that was fitted to the country's requirements. As a consequence, a significant number of the technological advances developed by Brazilian researchers were in the area of middleware.

The middleware adopted by ISDTV is called *Ginga*, and was jointly developed by researchers from the Catholic University of Rio de Janeiro and from the Federal University of Paraíba. Ginga middleware meets the requirements of the ITU J.200, ITU J.201, and ITU J.202 recommendations (ITU-T, 2001, 2003, 2004). Ginga is also compatible with the Globally Executable MHP (GEM) standard. GEM is a unified specification for DTV middleware, which was proposed by the DVB group and later adopted in the ISDB (ARIB, 2004a) and ATSC standards (ATSC, 2005). In the next subsection a brief introduction to Ginga is given.

9.6.1   Ginga middleware

The Ginga standard specifies a set of common functionalities (Ginga-Core) that supports the Ginga application environments. The Ginga-Core is composed of the common content decoders and procedures that can be used to prepare the data to be transported through

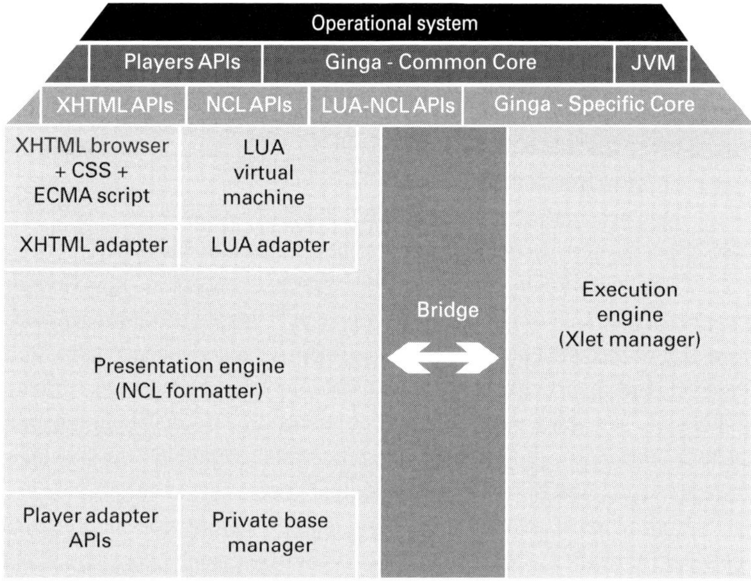

Figure 9.13   Ginga middleware architecture

the interactivity channel. The Ginga-Core also supports the conceptual display model of ISDTV. Ginga specifications for the architecture and applications are designed to work on terrestrial broadcasting DTV receivers, but may also be used for other systems such as satellite and cable DTV systems. Figure 9.13 shows the architecture of the Ginga middleware.

In applications for DTV, two types of languages coexist: declarative and procedural. The most appropriate language depends on the application objective. Declarative middleware gives support to applications developed in declarative languages, which do not require that the programmer specifies each step to be executed. It is only necessary to provide the set of tasks to be performed. It is up to the compiler to implement these tasks. Declarative languages simplify the development stage by not requiring that the professional to know advanced programming tools and, consequently, reducing development costs.

Procedural middleware gives support to applications developed in procedural languages. Procedural languages give a greater control over the program, but require knowledge of more advanced implementation features. To develop applications in this platform the programmer needs to specify the program control and execution flow. Procedural middleware is presented in the form of a Java Virtual Machine and a set of APIs. Java is the development language that is supported by most DTV middleware.

The Ginga applications can be divided into declarative (Ginga-NCL (Soares *et al.*, 2007)), procedural (Ginga-J (Soares *et al.*, 2007)), and hybrid. An application is hybrid when it contains both declarative and procedural content types. For example, declarative applications often make use of script content, which is procedural in nature, or reference and embedded JavaTV Xlet. Also, a procedural application may reference declarative content, such as graphic content, or construct and initiate the presentation of declarative content.

Ginga-NCL is the declarative application environment of the Ginga middleware that has as its core a nested context language, (NCL). NCL is a declarative language developed at the Catholic University of Rio de Janeiro (Silva *et al.*, 2004), which focuses on how media objects are structured and related in time and space. NCL does not restrict or prescribe the media content object types and it can include XHTML-based media objects, as defined in other common DTV standards. A main component of Ginga-NCL is the declarative content decoding engine (NCL formatter). Other important modules are the XHTML-based user agent, which includes a stylesheet (CSS), an ECMAScript interpreter, and the LUA engine (responsible for interpreting LUA scripts).

Ginga-J is the procedural application environment of the Ginga middleware. An important component of Ginga-J is the procedural content execution engine, composed by a Java Virtual Machine. Common content decoders can be used for both procedural and declarative applications for decoding and presenting common content types, such as PNG, JPEG, MPEG, H.264 and other types of formats.

9.7   Interactivity channel

The interactivity channel is responsible for all exchanges of information between the interactive applications, running on users' receivers, and the application servers, sited at broadcasting television stations and content/service providers which are distributed all over the Internet. A simplified diagram of the interactivity channel is presented in Figure 9.14, which illustrates the communication flow between the users' receivers

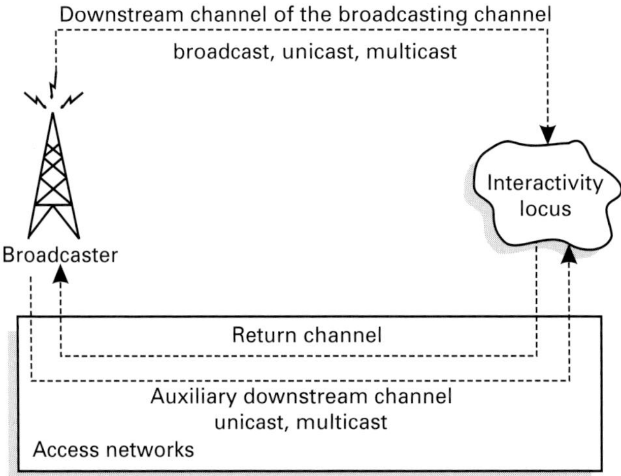

Figure 9.14   Simplified diagram of the interactivity channel

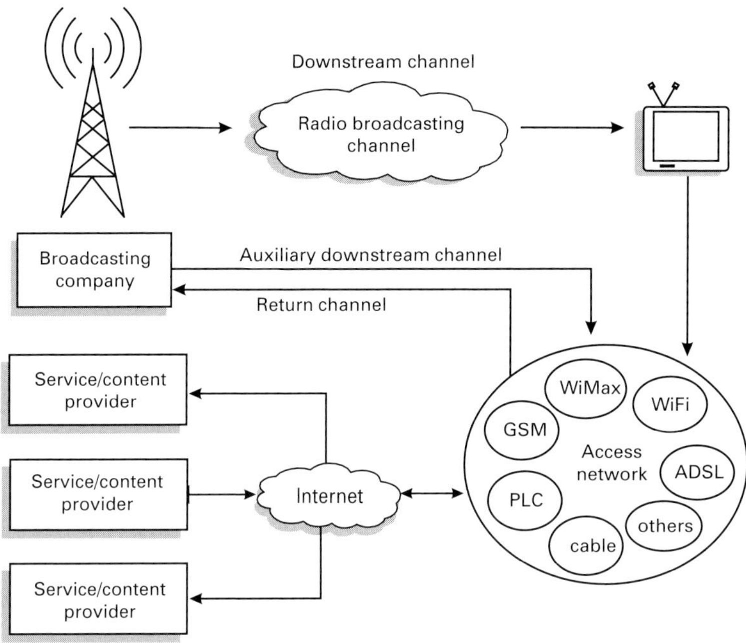

Figure 9.15   Interactivity channel subsystem

and the broadcasting television stations. The two main components of the interactivity channel are shown in this figure: the return channel and the downstream channel.

A more detailed diagram is presented in Figure 9.15, which shows the basic architecture of the interactivity channel. It can be seen in this figure that there are two types of communication routes. The first establishes the bidirectional data stream between the multiple communication networks and application servers. The second establishes the bidirectional data stream between the application server and the broadcasting companies/ programmers. Multiple access technologies can be adopted for the interactivity channel, which gives the system a very interesting feature, with capillarity, coverage, number of users and transmission rate.

The components of the interactivity channel are the following:

- Downstream channel: this is used by broadcasting companies and content providers to deliver data to end-users. The downstream channel is composed of (1) the broadcasting channel and (2) the communications platform adopted for the return channel, which can act as an extra downstream channel.
- Return channel: this is employed by users to request or send information to broadcasting companies or content providers distributed over the Internet. The return channel can be built out of any access network technology, such as Ethernet, Wi-Fi, GPRS, and so forth.
- Gateway: this connects the communication networks with the broadcasting companies/programmers. It is an access server used by the users to request information

through the communication networks. Before being transmitted, the requested information must be adequately processed, adapted, and forwarded to their final destination. The gateway is also the locality where security measures of the network are implemented.

- Application server: The information or requests coming from the users are processed and/or interpreted in the application server in order to generate an adequate response.
- Modem: The connection of the communication network with the receiver happens through a modem appropriate for the communication network being used. The input/output (I/O) interface of the modem is the communication gate that is used for sending data back or sending requests from the user to the broadcasting company/programmer. It also allows customized data, which is sent through the complementary download channel, to be received.
- Broadcasting adaptation unity: This is responsible for the adaptation of the information or requests to be transmitted in the TS of the broadcasting channel. Depending on the type of interactivity, it may be necessary to use a carousel generator or some type of processing.
- Access networks (communication networks): It is through the access networks that data from the return channel and the downstream channel are transmitted between modems and gateways. The access network chosen depends on geographic and demographic factors, and on the types of communication networks available. The access technologies are not definitive; new technologies may be added later.

The ISDTV standard has not specified any particular technology for the return channel. Consequently, manufacturers are free to build televisions and set-top boxes tailored to any access network platform currently available. But, unlike in some countries where the return channel can be implemented over already installed cable networks, such an option is not always readily available in Brazil. Therefore, its broad adoption can become very costly, especially in some of the remote areas. In order for Brazil to accomplish its goals of promoting digital inclusion through educational and government services, as well as large-scale (free) Internet access, a far-reaching access technology is needed.

A number of technologies have been studied by the Institute of Advanced Studies in Communications (Iecom) for the return channel, including DVB-RCT, CDMA-450 MHz-EVDO, Wi-Fi, WiMAX, and power line communications (PLC). The analysis showed that the best alternative for ISDTV is the WiMAX technology. In particular, the WiMAX-700 technology, which is a new WiMAX profile (Meloni, 2007). The profile operates in the 400–900 MHz primary frequency band (UHF band) and, optionally, from 54 MHz to 400 MHz as a secondary band (VHF band). It is argued that WiMAX-700 presents several advantages over current WiMAX profiles: better indoor penetration, higher propagation range (up to 65 km), and lower operational costs. A brief description of the WiMAX technology is given in the next subsection.

9.7.1   WiMAX

The WiMAX IEEE802.16-2004 standard specifies wireless communication systems that can be used for several types of services. One of the main applications of WiMAX is to provide Internet access, as an alternative to ADSL. WiMAX was developed targeting fixed, nomad, portable, and mobile receivers. WiMAX has three modes of operation: single carrier, OFDM 256, or OFDMA 2K. The most commonly used mode is OFDM 256.

With a very robust modulation scheme, WiMAX delivers high throughput rates with long range and great spectral efficiency, as well as good tolerance to signal reflection. The data transmission speed varies from 1 Mbits/s to 75 Mbits/s, depending on the propagation conditions. The typical radius of a WiMAX cell is around 6–9 km.

A dynamic adaptative modulation allows a base station station to set its throughput and range accordingly. For example, suppose that a base station station using a modulation scheme of higher order, such as 64-QAM, cannot establish a robust link with a very distant user. In this case, the modulation is reduced to a 16-QAM or QPSK, which increases the signal range by reducing the throughput. To accommodate the design of a WiMAX cell the standard supports several bandwidths. For example, if an operator has available a bandwidth of 20 MHz, it can be divided into two sectors of 10 MHz or four sectors of 5 MHz each. The operator can also increase the number of users keeping the signal range and throughput, or can reuse the spectrum in two or more sectors isolating the station antenna from the base station.

The standard also supports technologies that allow the expansion of the coverage, including the technology of smart antennas as well as mesh technology. WiMAX provides premium service levels for business clients, as well as a higher service volume with a QoS equivalent to services offered by ADSL and cable modem, all in the same base station. Characteristics of privacy and cryptography, including authentication procedures, are predicted in the standard, which allows safer transmissions.

10   Digital terrestrial television multimedia broadcasting (DTMB)

with Marcelo Menezes de Carvalho and Mylène Christine Queiroz de Farias

10.1   Introduction

China has the world's largest population of electronics consumers. There are currently more than 400 million television sets in China, which represents around 30% of all television sets worldwide. Like other developing countries, China decided to develop its own standard in order to avoid having to pay licensing fees. Both the US and the European digital television systems, for example, depend on patented technology. Although paying to use that technology may not represent much in more developed countries, it is certainly an important issue for China. Adapting the digital television technology to the local reality is also an important reason for developing a national standard.

Efforts to develop China's digital television standard started in 1994. In 1995, the government created the high-definition television (HDTV) Technical Executive Experts Group (TEEG), with members from the research community. The first HDTV prototype was developed three years later. It included a high-definition encoder and decoder, as well as a multiplexer, a modulator, a demodulator, and a demultiplexer. This system was used to broadcast a live program celebrating the 50th National Anniversary Day in 1999. In 2001, the government started to accept proposals for a digital terrestrial television broadcasting standard. In order to evaluate the suitability of the proposals for the Chinese market, laboratory and field tests were performed, along with intellectual property analysis. Three of the submitted proposals were chosen:

- Advanced Digital Television Broadcasting – Terrestrial (ADTB-T): a single-carrier approach from HDTV TEEG.
- Digital Multimedia Broadcasting – Terrestrial (DMB-T): a multicarrier approach from Tsinghua University.
- Terrestrial Interactive Multiservice Infrastructure (TiMi): a multicarrier approach, proposed by the Academy of Broadcasting Science.

In 2004, the Chinese Academy of Engineering led a working group to merge these three leading proposals. Finally, in August 2006, the working group agreed a merged standard (Chinese National Standard Organization, 2006).

The official name of the Chinese digital television standard is "Framing structure, channel coding, and modulation for digital television terrestrial broadcasting system (GB 20600-2006)." But, it has become known as Digital Terrestrial Television

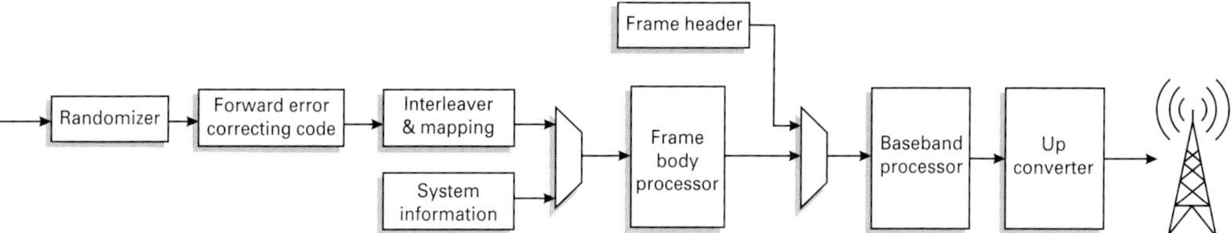

Figure 10.1   The transmission system diagram

Multimedia Broadcasting (DTMB). The Chinese digital television standard supports many combinations of modes, including single-carrier and multicarrier modulation, three frame header options, three forward error correcting (FEC) code rates, five constellation mappings, two interleaving depths, and some other features that result in hundreds of operational modes to support various multiprogram standard and high-definition digital television services.

The main characteristics of the Chinese digital television standard are:

(1) Time-domain pseudorandom sequences acting as frame headers;
(2) The use of low-density parity-check codes (LDPCs);
(3) System information protected with spread spectrum technology.

This chapter describes the main components of the transmission system of DTMB. Figure 10.1 shows a diagram of the main components of the Chinese transmission system. The following sections describe each component.

10.2   Scrambler

The input to the transmission system is a bit stream called the transport stream (TS) that typically carries a combination of audio, video, and data information that is compressed and multiplexed at the MPEG layer. Depending on its content, the TS may contain long sequences with a certain periodicity. Such periodicity leads to spectral concentration, which can reduce the system performance for selective fading channels and cause synchronism problems at the receiver. To avoid such problems, the TS is scrambled before performing channel coding. The DTMB scrambler multiplies the bits of the TS with a pseudorandom sequence of $2^{15} - 1$ bits generated by the polynomial

$$G(x) = 1 + x^{14} + x^{15},$$

with an initial state 100101010000000. The scrambler is reset at the beginning of each *signal frame* (to be defined later).

10.3   Forward error correction (FEC) code

To improve resilience to channel errors, the scrambled TS passes through an FEC code subsystem. The Chinese digital television standard has adopted the concatenation of outer BCH (762, 752) and inner LDPC codes. BCH (762, 752) is derived from BCH (1023, 1013) by adding 261 zeros in front of every 752 information bits before BCH encoding, and these bits are then removed from the output of the BCH encoder.

10.3.1   Low-density parity-check (LDPC) codes

One of the main novelties of the Chinese digital television standard is the adoption of *LDPC codes* as part of the FEC code subsystem. LDPC codes were invented by Gallager in 1961 (Gallager, 1962), and were largely ignored until turbo codes were introduced in 1993 (Berrou *et al.*, 1993). After turbo codes were introduced, LDPC codes were reinvented by Mackay and Neal (MacKay and Neal, 1996), and by Wiberg (Wiberg *et al.*, 1995). It did not take too long to realize that these new codes were actually reinventions of Gallager's original work.

LDPC codes have the distinction of being the only class of codes for which theorems giving guarantees on asymptotic performance for large block lengths are available, with an analytical framework for evaluating how far the performance is from Shannon capacity (Madhow, 2008). As a practical consequence, it is possible to design LDPC codes that, for large block lengths, outperform turbo codes constructed from serial or parallel concatenation of convolutional codes. In fact, provided that the codeword length is long, LDPC codes can achieve a performance close to the Shannon limit. Compared to turbo codes, LDPC codes tend to have relatively high encoding complexity but low decoding complexity (the opposite happens with turbo codes). In addition, LDPC decoding may be implemented with parallel architectures and relatively low complexity algorithms.

Fundamentally, LDPC codes are linear block codes with a particular structure for the parity-check matrix $\mathbf{H}$, whose number of "ones" is much lower than the number of "zeros," i.e. the fraction (or density) of non-zero entries in $\mathbf{H}$ is small (and hence the name of the code). More specifically, a (n_c, n_r) regular binary LDPC code has a parity-check matrix $\mathbf{H}$ with n_c ones in each column and n_r ones in each row, in which n_c and n_r are chosen as part of the codeword design and are small relative to the codeword length.

As with any linear block code, LDPC codes can be compactly specified by either the parity-check matrix $\mathbf{H}$ or the associated generator matrix $\mathbf{G}$. In the case of the Chinese digital television standard, the generator matrix of the LDPC code is given by

$$
\mathbf{G}_{kc} = \begin{bmatrix}
\mathbf{G}_{0,0} & \mathbf{G}_{0,1} & \cdots & \mathbf{G}_{0,c-1} & \mathbf{I} & \mathbf{0} & \cdots & \mathbf{0} \\
\mathbf{G}_{1,0} & \mathbf{G}_{1,1} & \cdots & \mathbf{G}_{1,c-1} & \mathbf{0} & \mathbf{I} & \cdots & \mathbf{0} \\
\vdots & \vdots & \mathbf{G}_{i,j} & \vdots & \vdots & \vdots & \ddots & \vdots \\
\mathbf{G}_{k-1,0} & \mathbf{G}_{k-1,1} & \cdots & \mathbf{G}_{k-1,c-1} & \mathbf{0} & \mathbf{0} & \cdots & \mathbf{I}
\end{bmatrix}, \tag{10.1}
$$

Table 10.1. LDPC parameters and associated performance (BER $< 3 \times 10^{-6}$)

Code rate	Block length (bits)	Information bits	k	c	E_b/N_0 (dB)	Performance away from Shannon limit (dB)
≈ 0.4	7 488	3 008	24	35	2.1	2.3
≈ 0.6	7 488	4 512	36	23	2.3	1.6
≈ 0.8	7 488	6 016	48	11	3.3	1.2

Table 10.2. System payload data rates in megabytes per second for the signal frame of 4200 symbols

FEC Rate	0.4	0.6	0.8
4-QAM-NR	–	–	5 414
4-QAM	5 414	8 122	10 829
16-QAM	10 829	16 243	21 658
32-QAM	–	–	27 072
64-QAM	16 243	24 365	32 486

in which $\mathbf{I}$ is a $b \times b$ identity matrix with $b = 127$, $\mathbf{0}$ is a $b \times b$ zero matrix, and $\mathbf{G}_{i,j}$ is a $b \times b$ circulant matrix with $0 \leq i \leq k - 1$ and $0 \leq j \leq c - 1$. The parameters k and c are used to define the LDPC code and, for the Chinese digital television standard, three LDPC coding rates have been defined: LDPC (7493, 3048), LDPC (7493, 4572), LDPC (7493, 6096). The encoded block length obtained from the generator matrix in (10.1) is 7493, but the first five check bits are then punctured to reduce the block length to 7488. The LDPC parameters, coding rates, and associated performance are shown in Table 10.1.

10.4 Signal constellation and mapping

The output binary sequence of the FEC code is converted into an M-ary quadrature amplitude modulation (m-QAM) symbol stream, with the first encoded input bit of each symbol as the least significant bit (LSB). The signal constellations that are supported are 64-QAM, 32-QAM, 16-QAM, 4-QAM, and 4-QAM-NR (Nordstrom–Robinson). 4-QAM-NR is a combination of Nordstrom–Robinson coding (block coding) and 4-QAM modulation. During symbol mapping, power normalization is enforced to keep the average power the same for all mappings. Tables 10.2, 10.3, and 10.4 contain the system payload data rates under different combinations of frame header length, code rate, and constellation. Blank cells indicate the cases that are not supported by the standard.

Table 10.3. System payload data rates in megabytes per second for the signal frame of 4375 symbols

FEC Rate	0.4	0.6	0.8
4-QAM-NR	–	–	5 198
4-QAM	5 198	7 797	10 396
16-QAM	10 396	15 593	20 791
32-QAM	–	–	25 989
64-QAM	15 593	23 390	31 187

Table 10.4. System payload data rates in megabytes per second for the signal frame of 4725 symbols

FEC Rate	0.4	0.6	0.8
4-QAM-NR	–	–	4 813
4-QAM	4 813	7 219	9 626
16-QAM	9 626	14 438	19 251
32-QAM	–	–	24 064
64-QAM	14 438	21 658	28 877

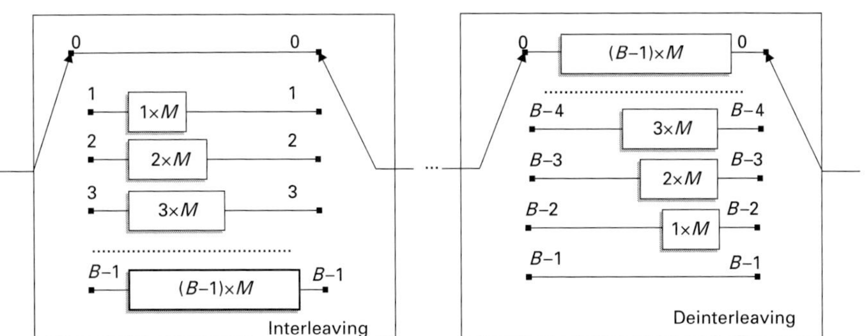

Figure 10.2   The structure of the convolutional interleaver

10.5   Interleaving

A convolutional interleaver is used across many OFDM signal frames, as shown in Figure 10.2. The number of interleaving branches is $B = 52$, and there are two modes for the interleaving depth M: $M = 240$ and $M = 720$, respectively. The overall delay of the time interleaving/deinterleaving is $B \times (B - 1) \times M$, and the corresponding delay of these modes is 170 and 510 OFDM signal frames, respectively. Frequency interleaving is also employed within each OFDM frame body and only in the multicarrier mode.

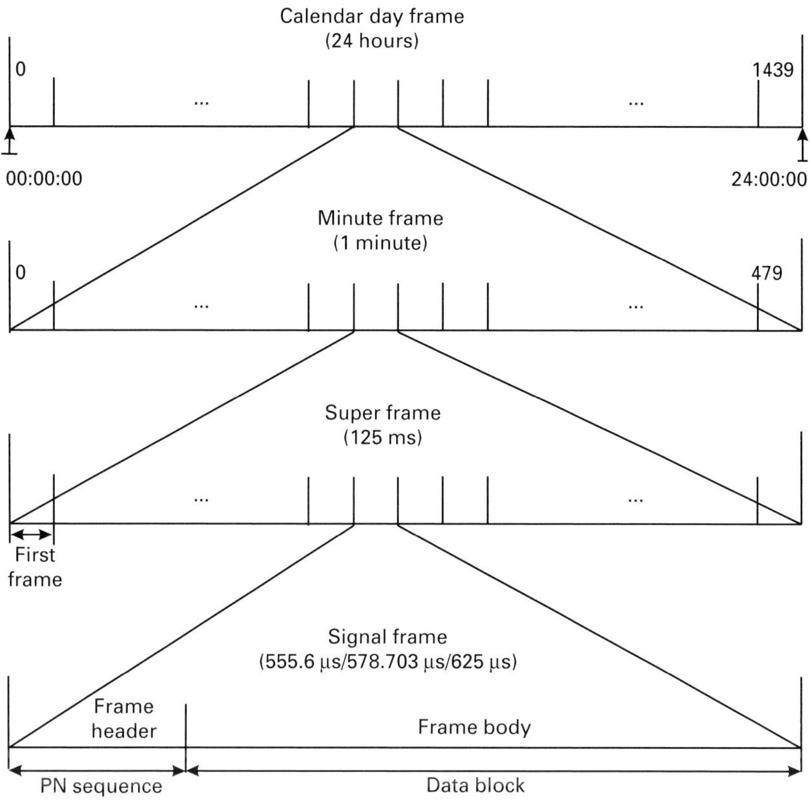

Figure 10.3 Frame structure of DTMB

10.6 Frame structure

The frame structure follows a hierarchy of decreasing frame lengths, as depicted in Figure 10.3. At the top level of the hierarchy is the calendar day frame, which lasts exactly 24 hours (starting from 00:00:00 a.m. each day). Next, there is the minute frame, which lasts exactly 1 minute. Each minute frame contains super frames, which have a fixed duration of 125 ms. Finally, at the bottom of the hierarchy, there is the signal frame, which is the fundamental unit of the frame structure. The signal frame consists of a *frame header* (FH) and a *frame body* (FB), as can be seen in Figure 10.4. Both FH and FB have the same baseband symbol rate of 7.56 Msymbol/s.

10.6.1 Frame header

The frame header is a pseudonoise (PN) sequence. Three types of frame headers are defined, based on their lengths: 420, 595, or 945 symbols. The three types of frame headers are conventionally abbreviated as PN420, PN595, and PN945, respectively.

Table 10.5. The frame structures

| preamble (82 symbols) | PN255, *M*-sequence of 255 | post-amble (83 symbols) |
| preamble (217 symbols) | PN511, *M*-sequence of 511 | post-amble (217 symbols) |

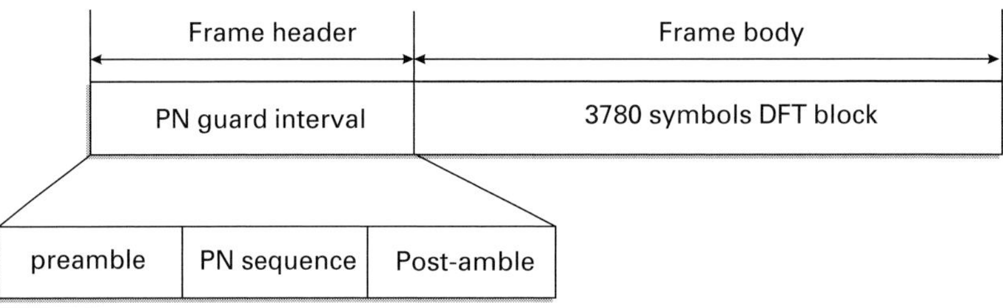

Figure 10.4   Signal frame format for TDS-OFDM

Two different methods are used to construct the frame headers. One method is used to construct both the PN420 and PN945 frame headers; see Table 10.5. PN420 and PN945 are made up of complete *M*-sequences of lengths 255 and 511, respectively, with their respective cyclical extensions acting as preambles and post-ambles. For these frame headers, the average power is twice the average power of the frame body. The frame header of this structure is rotated in each super frame, which means that each signal frame within the same super frame has a unique address (frame header) and can be identified exclusively. As an alternative, the frame header may be fixed if the frame addressing is not needed. The PN595 frame header is made up of the first 595 symbols of an *M*-sequence of length 1 023, without cyclical extensions. The PN595 frame header has the same average power as that of the frame body.

The frame headers are the key to receiver design. They can be used as training sequences for equalizers and serve as the guard interval for the frame body. Regardless of the length, cyclical characteristics, or power level, the frame headers are always modulated in the time domain (single-carrier) with BPSK mapping. This is a fundamental characteristic of the Chinese standard that allows a number of operations, such as frame synchronization, carrier recovery, channel estimation, and equalization, to be done using time-domain signal processing techniques, in addition to allowing single-carrier (C1) and multiple-carrier (C3780) modes to share the same architecture.

10.6.2   Frame body

The frame body consists of system information and coded data. Each frame body is made of 3780 symbols and, therefore, it lasts exactly 500 μs (3780 $\times$ 1/7.56 μs). There is 6-bit physical layer system information (TPS bits) in the frame body to provide the receiver with information such as demodulation/decoding information, including constellation mapping, LDPC rate, interleaving mode, carrier mode information, as well as frame header information. The spread spectrum technique is used to protect these bits, resulting

Table 10.6. The frame structure possibilities

Frame header	Frame body
420 symbols (55.6 μs)	3780 symbols (500 μs)
595 symbols (78.7 μs)	3780 symbols (500 μs)
945 symbols (125 μs)	3780 symbols (500 μs)

in 36 TPS symbols mapped according to BPSK modulation. The first four symbols indicate the mode of operation, i.e. they denote the number of carriers to be used. One is the C1 mode, which indicates that a single-carrier modulation is being used. The other option is the C3780 mode, which indicates that a multiple-carrier modulation with 3780 subcarriers is to be used. The last 32 symbols indicate information such as constellation mapping, LDPC code rate, and interleaving mode. The lengths of the frame headers may vary, and three signal frame structures are defined, as shown in Table 10.6.

10.7 Time-domain synchronous orthogonal frequency division multiplexing (TDS-OFDM)

Another innovation of the Chinese digital television standard is the adoption of TDS-OFDM. The time-domain approach employed in the Chinese standard is based on the idea of using a known, short PN sequence as the OFDM preamble (Tufvesson *et al.*, 1999, Muck *et al.*, 2003). As opposed to the cyclic prefix that is used in DVB-T, for instance, PN sequences are inserted as guard intervals and act as training symbols for synchronization and estimation purposes, so that no additional pilot symbol is needed. The use of PN sequences brings the benefits of fast channel acquisition, since this can be done directly in the time domain (in about 5% of the time required by DVB-T, for instance). DVB-T sends a large number of training symbols (more than 10% of the data symbols) in order to facilitate the fine symbol synchronization and channel estimation.

10.7.1 System model for TDS-OFDM

The ith information sequence to be transmitted, represented by $S_i[n]$, $n = 0, \ldots, N-1$, is first modulated by an N-point IFFT:

$$s_i[k] = \frac{1}{\sqrt{N}} \sum_{n=0}^{N-1} S_i[n] e^{j2\pi kn/N}, \quad 0 \le k \le N-1. \tag{10.2}$$

Following that, a predefined PN sequence, represented by $c_i[k]$, $k = 0, \ldots, M-1$, is inserted before each inverse fast Fourier transform (IFFT) output $\{s_i[k]\}_{k=0}^{N-1}$.

The PN sequences that TDS-OFDM uses are defined as a set of shifted m-sequences. Such sequences satisfy the following orthogonality condition to provide the unique frame address:

$$c_i[k] * c_j[k] = \delta(i,j), \tag{10.3}$$

with $*$ indicating the convolution operation, and $\delta(i,j)$ representing a modified delta function. The signal frame vectors are constructed after the insertion of the PN sequence. Thus, each OFDM frame can be decomposed into two non-overlapping parts in the time domain: the PN sequence $\{c_i[k]\}_{k=0}^{M-1}$, and the information sequence $\{s_i[k]\}_{k=0}^{N-1}$. Assuming the channel is modeled as a quasi-static lth order FIR filter with impulse response $\{h_i[k]\}_{k=0}^{L-1}$, the TDS-OFDM system is designed in such a way that the duration of the PN sequence exceeds the channel memory, i.e. $M \geq L$.

At the receiver, the ith received OFDM frame $r_i[k]$, $k = 1, \ldots, M + N + L - 2$, consists of two overlapping parts (ignoring the effects of noise, for the sake of clarity): the sequence

$$y_i[k] = c_i * h_i = \sum_{l=0}^{L-1} c_i[k-l]h_i[l], \quad 0 \leq k \leq M + L - 1, \tag{10.4}$$

which represents the linear convolution between the PN sequence $c_i[k]$ and the channel impulse response $h_i[k]$, and the sequence

$$x_i[k] = s_i * h_i = \sum_{l=0}^{L-1} s_i[k-l]h_i[l], \quad 0 \leq k \leq N + L - 1, \tag{10.5}$$

which indicates the linear convolution between the information sequence $s_i[k]$ and the channel impulse response $h_i[k]$. If we take into consideration the effect of additive white Gaussian noise (AWGN), represented by the sequence $n_i[k]$, the received signal frame $r_i[k]$ is given by

$$r_i[k] = u_i[k] + n_i[k], \quad 0 \leq k < M + N + L - 1, \tag{10.6}$$

with

$$u_i[k] = \begin{cases} x_{i-1}[k+N] + y_i[k], & \text{if } 0 \leq k < L - 1 \\ y_i[k], & \text{if } L - 1 \leq k < M \\ x_i[k-M] + y_i[k], & \text{if } M \leq k < M + L - 1 \\ x_i[k-M], & \text{if } M + L - 1 \leq k < N + M \\ x_i[k-M] + y_{i+1}[k-N-M], & \text{if } N + M \leq k < N + M + L - 1. \end{cases} \tag{10.7}$$

If the sequence $y_i[k]$ is subtracted from the received signal $r_i[k]$, the remaining signal $x_i[k]$ is equivalent to a zero padded OFDM (ZP-OFDM) signal (Giannakis, 1997), (Scaglione *et al.*, 1999a, 1999b) (Muquet *et al.*, 2002). Hence, all the well-established methods related to ZP-OFDM can be applied.

Unlike with traditional cyclic prefix OFDM (CP-OFDM), in each block of a ZP-OFDM transmission, zero symbols are appended after the IFFT-precoded information

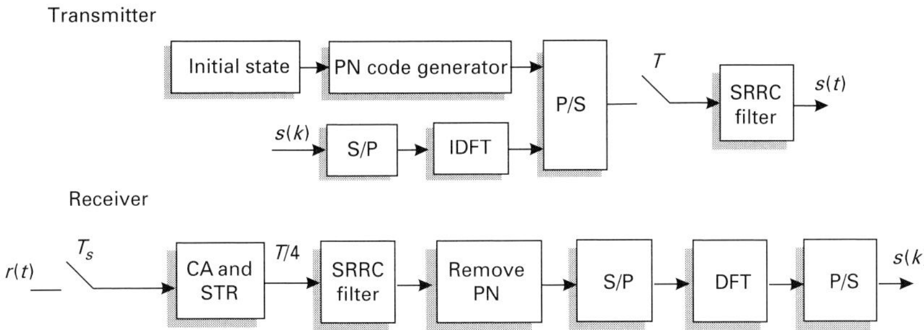

Figure 10.5   TDS-OFDM transmitter and receiver

symbols. If the number of zero symbols equals the cyclic prefix length, then ZP-OFDM and CP-OFDM transmissions have the same spectral efficiency. Unlike CP-OFDM and without bandwidth-consuming channel coding, ZP-OFDM guarantees symbol recovery and ensures FIR equalization of FIR channels regardless of the channel zero locations. The price that is paid is somewhat increased receiver complexity: instead of the single FFT required by CP-OFDM, an FIR filtering is required. The solution of relying on a larger fast Fourier transform (FFT) demodulator has the merit of guaranteeing symbol recovery irrespective of channel null locations in the absence of noise when the channel is known (coherent modulations are assumed). The baseband equivalent of the TDS-OFDM system is shown in Figure 10.5.

10.8   Post-baseband processing

A square-root raised cosine (SRRC) filter with a roll-off factor of 5% is used as a shaping filter to limit the bandwidth of the transmitted signal to 8 MHz. The frequency response of the filter is given by

$$H(f) = \begin{cases} 1, & \text{if} \quad |f| \le (1-\alpha)/2T_s \\ \left\{ \frac{1}{2} + \frac{1}{2} \cos\left[\frac{(2T_s|f| - 1 + \alpha)\pi}{2\alpha} \right] \right\}^{\frac{1}{2}}, & \text{if} \quad (1-\alpha)/2T_s < |f| \le (1+\alpha)/2T_s \\ 0, & \text{if} \quad |f| > (1+\alpha)/2T_s, \end{cases}$$

$$(10.8)$$

with $T_s = 1/7.56 \,\mu$s. In addition, in C1 mode, dual pilots may be optionally inserted at $\pm$ 0.5 the symbol rate, with an average power -16 dB lower than the total average power.

10.9   Receiver design

Although receiver signal processing technology is not defined as part of the standard, the time-domain modulated frame header and the number of carriers (C1, C3780) are the most significant options in the standard which have direct impacts on receiver performance and complexity. Two types of channel estimation and compensation methods were developed. One is an all-time domain processing approach, which employs a time-domain, code enhanced, data-directed adaptive decision feedback equalizer with the LMS algorithm. The other is a hybrid time- and frequency-domain processing approach, which implements a channel estimator in the time domain with a known frame header, and compensates the frame body in the frequency domain (Zheng *et al.*, 2004, Wang *et al.*, 2003, 2005, Song *et al.*, 2005, Yang *et al.*, 2002) .

Appendix A
Evolution of television standards

A.1   The beginning of image broadcasting

The history of television owes a lot to engineers, mathematicians, and physicists, who have introduced an important vehicle for information transmission to civilization. Since the beginning of the nineteenth century, scientists have been investigating long-distance transmission of images, and it was with Alexander Bain's invention, in 1892, that the telegraphic transmission of an image (facsimile) was obtained.

In 1817, the Swedish chemist Jacob Berzelius discovered selenium, but it was not until 56 years later, in 1873, that the Englishman Willoughby Smith proved that this substance possessed the property of turning light into electricity. This property was used by Father Roberto Landell de Moura in his inventions (Alencar *et al.*, 2004).

Julius Elster and Hanz Geitel invented the photoelectric cell in 1892. In Russia, Boris Rosing made similar discoveries. The system made use of moving mirrors in addition to the cathode ray tube.

In 1920, broadcasts were made with the help of John Logie Baird, from the United Kingdom, who used a mechanical system based on Paul Nipkow's invention. Nipkow's disk consisted of a device, invented in 1884 by Nipkow, that sent moving images using an electric conductor. To accomplish this, he used selenium's property to vary electrical conductivity proportionately to the intensity of light.

In 1924, Baird made the first distance broadcasts of object outlines, and, in the following year, of people's faces. In 1926, Baird gave the first demonstration to the scientific community at the Royal Institution in London, and then signed a contract with the BBC for experimental broadcasts. The definition standard had 30 lines and was mechanical.

Also in the 1920s, the engineers of Bell Telephone developed a system that synchronized sound with movie images, which marked the beginning of sound movies. The audio signal was recorded electrically on a 16-inch disk. Its commercial exploration came with the formation of Vitaphone Corp., which produced the first sound film, "Don Juan," on August 6, 1926. The soundtrack of the film was just music. In 1927, Warner Brothers, owner of Vitaphone Corp., in association with Walter J. Rich, produced the first truly spoken film, "The Jazz Singer" (Nebeker, 2001).

In 1923, the Russian Vladimir Kosma Zworykin patented the iconoscope, an invention that used tubes of cathode rays. In 1927, Philo Farnsworth patented a cathode ray image dissecting system, although with an unsatisfactory level of resolution. Zworykin was invited to join RCA by David Sarnoff to lead the team that would produce the first television tube, called the orticon, which started to be mass produced in 1945.

In March 1935, the first official broadcasts in Germany took place in which the signal was picked up in 22 public rooms. This was followed in France in November of that year with broadcasts from the Eiffel Tower. One of the first significant broadcasts was the 1936 Olympic Games in Berlin.

In London, the BBC began regular television broadcasting in 1936 using a 405-line system. The following year three electronic cameras were used to broadcast the coronation of King George VI to 50 thousand viewers.

Television started operating in the Soviet Union in 1938, and in the USA in 1939. Also in 1939, there was the first closed circuit television broadcast in Brazil. It happened during the Samples International Fair in Rio de Janeiro, by then the capital of the country, using German equipment.

During World War II, Germany was the only European country to keep television on the air. Paris started broadcasting again in October 1944, Moscow in December 1945, and the BBC in June 1946, when it broadcast the victory parade.

In 1950, France had a television network that used a definition of 819 lines, United Kingdom 405 lines, the USSR 625 lines, and the USA and Japan 525 lines. In September of the same year, Tupi Network was inaugurated in São Paulo. "TV Tupi" belonged to journalist Assis Chateaubriand, owner of the "Diários Associados," and was based on the American system.

A.2   The beginning of color broadcast

The American network NBC began regular color broadcasting in the USA in 1954. However, in 1929, Hebert Eugene Ives made the first color image broadcast, through wire, from New York, using 50-line definition. Peter Goldmark improved the invention and gave demonstrations using 343 lines in 1940.

In the USA, a special committee was created to introduce color into the monochromatic system. This committee was called the National Television System Committee (NTSC). The Federal Communication Commission approved color television broadcasting standards, compatible with the existing 525-line monochrome system, on December 17, 1953. However, it was not until 1964 that a significant number of households were equipped with color receivers (Benson and Fink, 1991).

The system developed was based on the black and white standard that worked with luminance levels (Y), in addition to chrominance (C). The luminance signal was modulated in amplitude, with a partial suppression of one of the sidebands, and became known as vestigial sideband (VSB). To insert color, the chrominance signal was modulated in quadrature (QUAM). The principle of capturing and receiving color images lies in the decomposition of the white light into three primary colors, which are red (R), green (G), and blue (B), in the following proportions: 30% R, 59% G, and 11% B.

Table A.1. Color formation on television

Color	R (%)	G (%)	B (%)
Red	100	0	0
Green	0	100	0
Blue	0	0	100
Yellow	100	100	0
Magenta	100	0	100
Cyan	0	100	100
White	100	100	100
Black	0	0	0
Yellowish	100	80	0
Purple	100	0	30
Violet	0	70	100
Light blue	80	80	100
Light yellow	100	100	30
Pink	90	70	70
Brown	30	20	10

In the reception, the process is inverted. The image is composed through the addition of colors in the pixel, i.e., the dots on the screen of the television. Table A.1 shows, as percentages, how the colors are formed from the three primary colors (RGB).

In 1967, a variation of the American system was introduced in Germany. It was intended to solve some problems of the NTSC in relation to the stability of the colors. This system was called phase alternated line, or PAL. In the same year, Séquentielle Couleur avec Memoire (SECAM), which was incompatible with the French black and white system, started operating in France.

A.3 The PAL standard

At the beginning of television development, when the vertical synchronism frequency was chosen, the electronic oscillators did not offer enough stability at this low frequency, and the electric distribution network frequency was used to synchronize the image vertically.

In the USA, the frequency used was 60 Hz, and the television broadcasting system generated 60 image fields per second to avoid the scintillation effect. The image was composed of 525 lines per frame, with two fields, and 30 frames per second to give the idea of movement. These features of the North American television broadcasting system are part of the norms established by the Radio Manufacturers Association (RMA) and were adopted by other countries with the same network frequency, such as Brazil.

In countries where the electric power is generated with a frequency of 50 Hz, namely Germany, Argentina, and others, mainly in Europe, the image is formed with 625 lines, of which 576 are visible, per frame at 25 frames per second to give the idea of movement.

Table A.2. Technical information about the PAL standard

System	PAL-B, G, H	PAL-I	PAL-D	PAL-N	PAL-M
Lines/field	625/50	625/50	625/50	625/50	525/60
Hor. frequency	15 625 kHz	15 625 kHz	15 625 kHz	15 625 kHz	15 750 kHz
Vert. frequency	50 Hz	50 hertz	50 Hz	50 Hz	60 Hz
Color carrier	4.433 618 MHz	4.433 618 MHz	4.433 618 MHz	3.582 056 MHz	3.575 611 MHz
Video bandwidth	5.0 MHz	5.5 MHz	6.0 MHz	4.2 MHz	4.2 MHz
Sound Carrier	5.5 MHz	5.5 MHz	5.5 MHz	4.5 MHz	4.5 MHz

Developed in the USA, NTSC was the first color television system, with its first broadcasts dating back to 1954. It had a number of limitations such as, for instance, requiring shading control and presenting problems with color stability. As an enhancement of NTSC, the phase alternating line (PAL) color signal encoding system was proposed at the end of the 1960s. PAL was created in Germany with the aim of eliminating several problems in NTSC related to color reproduction. Its color reproduction was more precise than in NTSC, and the system was adopted in various countries around the world, except those already committed with the NTSC system.

Table A.2 shows the technical information about the PAL systems. The PAL-M standard has a resolution of 525 lines, uses a repetition rate of 29.97 frames per second (M standard) and a horizontal frequency of 15 750 kHz, which is close to that of the NTSC standard. The vertical frequency of the PAL-M and the NTSC systems is approximately 60 Hz, unlike in Europe, where the frequency of the PAL-B, PAL-G, and SECAM systems is 50 Hz. The frequency of the secondary carrier of color is 3 575 611 MHz, and the bandwidth is 6 MHz. The PAL-N system is used in Argentina, Paraguay, and Uruguay, and employs a vertical synchronism frequency equal to that of the electric power network of 50 Hz.

PAL-M is the color television system that has been used in Brazil since the first official broadcast. It uses the PAL color encoding system in a subcarrier, in the "M" standard of image formation. It was the solution found at the time of the adoption of the color system by the committee coordinated by Euclides Quandt de Oliveira, so that color broadcasts could be received by the black and white devices without the need for adapters. The German enterprise Telefunken developed the standard for Brazil and conceded the royalties from the patents.

The following use the PAL standard: Afghanistan, Albania, Algeria, Angola, Argentina**, Australia, Austria, Azores, Bahrain, Bangladesh, Belgium, Botswana, Brazil*, Brunel, Cameroon, Canary Islands, Cyprus, Denmark, Dubai, Ethiopia, Faeroe Islands, Finland, Germany, Ghana, Gibraltar, Greenland, Guinea, Holland, Hong Kong, Iceland, India, Indonesia, Ireland, Israel, Italy, Jordan, Kenya, Kuwait, Laos, Liberia, Madeira, Malaysia, Malta, Mozambique, Nepal, New Guinea, New Zealand, Nigeria, North Korea, Norway, Oman, Pakistan, Paraguay**, Poland, Portugal, Qatar, Romania, Singapore, Somalia, South Africa, SW Africa, Spain, Sri Lanka, Sudan, Swaziland,

Table A.3. Technical information about the SECAM standard

System	SECAM B, G, H	SECAM D, K, K1, L
Lines/field	625/50	625/50
Hor. frequency	15 625 kHz	15 625 kHz
Vert. frequency	50 Hz	50 Hz
Video bandwidth	5.0 MHz	6.0 MHz
Sound carrier	5.5 MHz	6.5 MHz

Sweden, Switzerland, Tanzania, Thailand, Turkey, Uganda, United Arab Emirates, United Kingdom, Uruguay, Yemen, the former Yugoslavia, Zambia, Zimbabwe. (Note: PAL-N = **, PAL-M = *.)

A.4 The SECAM standard

The Séquentielle Couleur Avec Mémoire (SECAM) standard, from the French "sequential color with memory," is an analog color system, created in France by Henri de France. The development of SECAM started in 1956. The technology was ready at the end of the 1950s, but it was early for a large-scale introduction. SECAM did not use the 819-line standard used on French television at that time. France changed the standard to 625 lines, which happened at the beginning of the 1960s with the introduction of a second network.

SECAM was inaugurated in France on October 1, 1967, on the "La Seconde Chaîne" (the second network), today called France 2. At that time, a color television cost 5000 francs. Color television was not initially very popular and only 1500 people watched the inaugural program. One year later, only 200 000 sets had been sold, when there had been an expectation of sales reaching up to 1 million units. SECAM was later adopted by the old French and Belgian colonies, by the Eastern European countries, by the USSR, and by the Middle Eastern countries. However, with the changes in the USSR, and following a period in which multistandard television sets had become common, several Eastern European countries decided to change to PAL.

There are three varieties of the standard: French SECAM (SECAM-L), used in France and its old colonies; SECAM B/G, used in the Middle East, for some time in Greece and in the old East Germany; SECAM D/K, used in the Independent States Community (ISC) and in Eastern Europe.

France, also introduced the SECAM standard in its possessions. Nevertheless, the SECAM standard used in the overseas French territories, as well as the African countries that had been French colonies, was slightly different from the SECAM used in France. The version standardized in France was SECAM-L. Overseas French territories and many African countries adopted the SECAM-K standard. Table A.3 presents technical information about the SECAM standard.

The SECAM standard is used in: Armenia, Azerbaijan, Belarus, Benin, Bosnia, Bulgaria, Burundi, Chad, Congo, Croatia, the former Czechoslavakia, the former East

Table A.4. Technical information about the NTSC standard

System	NTSC M
Lines/field	525/60
Horizontal frequency	15 734 kHz
Vertical frequency	60 hertz
Color carrier	3.579545 MHz
Video bandwidth	4.2 MHz
Audio carrier	4.5 MHz

Germany, Egypt, Estonia, France, French Guyana, Gabon, Greece, Guadeloupe, Guyana Republic, Hungary, Iran, Iraq, Ivory Coast, Latvia, Lebanon, Libya, Liechtenstein, Lithuania, Luxemburg, Madagascar, Mali, Martinique, Mauritania, Mauritius, Monaco, Mongolia, Morocco, New Caledonia, Niger, Russia, Saint-Pierre, Saudi Arabia*, Senegal, Syria, Tahiti, Togo, Tunisia, Ukraine, Zaire. (Note: MESECAM=*.)

A.5   The NTSC standard

NTSC is the analog television system used in the USA and in many other countries, including most countries in the Americas and some in parts of East Asia. It was named after the National Television Systems Committee, the sector's representative organization, responsible for the creation of this standard.

NTSC uses a frame consisting of 486 visible horizontal lines in the active area, out of a total of 525 scan lines. The remainder of the lines, called the vertical blanking interval, are used for synchronization and vertical retrace, and can contain other data such as closed captioning and vertical interval timecode. The frame rate is 29.97 frames/s. The frame is interlaced, which means that it is composed of two fields, with a field rate of 59.94 fields per second. The bandwidth occupied by a composite NTSC signal is 6 MHz.

The NTSC signal uses amplitude modulation vestigial sideband (AM-VSB) for the transmission of the video information, and frequency modulation for the audio signal, with a subcarrier 4.5 MHz higher than the video carrier. The amplifier non-linear distortion can cause the 3.58 MHz color carrier to beat with the sound carrier to produce a dot pattern on the screen. In order to minimize the resulting effect, the original 60 Hz field rate is adjusted down by a factor of 1.001%, to approximately 59.94 fields/s. Table A.4 presents technical information about the NTSC standard.

The following use the NTSC standard: Antigua, Bahamas, Barbados, Barbuda, Belize, Bermuda, Bolivia, Burma, Cambodia, Canada, Cayman Islands, Chile, Colombia, Costa Rica, Cuba, Dutch Antilles, Ecuador, El Salvador, Guam, Guatemala, Haiti, Honduras, Jamaica, Japan, Mexico, Midway Islands, Nicaragua, North Mariana Island, Panama, Peru, Philippines, Puerto Rico, Saint Kitts, Saint Lucia, Saint Vincent, Saipan, Samoa, South Korea, Surinam, Taiwan, Tobago, Trinidad, USA, Venezuela, Virgin Islands.

Appendix B
Signal analysis

B.1 Introduction

The objective of this appendix is to provide the reader with the necessary mathematical basis for understanding the characteristics of signals which are related to their analysis and transmission through communication channels. Concepts and equations involving the Fourier transform, which constitute powerful tools for spectral analysis, are presented (Alencar, 1999).

B.2 Fourier transform

The direct Fourier transform is a mapping from the time to the frequency domain,

$$F(\omega) = \int_{-\infty}^{\infty} f(t) e^{-j\omega t} \, dt;$$ (B.1)

it is sometimes denoted in the literature as $F(\omega) = \mathcal{F}[f(t)]$.

The inverse Fourier transform can be defined in a similar manner, as

$$f(t) = \frac{1}{2\pi} \int_{-\infty}^{\infty} F(\omega) e^{j\omega t} \, d\omega.$$ (B.2)

A Fourier transform pair is often denoted as $f(t) \longleftrightarrow F(\omega)$.

Fourier transforms of known signals are presented in the following (Haykin, 1988).

Pulse function

The pulse function is used to aid in the simulation of digital signals, and is defined by the expression $p_T(t) = A[u(t + T/2) - u(t - T/2)]$, or

$$p_T(t) = \begin{cases} A & \text{if } |t| \leq T/2 \\ 0 & \text{if } |t| > T/2 \end{cases},$$ (B.3)

in which $u(t)$ denotes the *unit step* function, defined as

$$u(t) = \begin{cases} 1 & \text{if } t \geq 0 \\ 0 & \text{if } t < 0 \end{cases}.$$ (B.4)

The pulse function is illustrated in Figure B.1. The Fourier transform of the pulse function can be calculated as

$$F(\omega) = \int_{-T/2}^{T/2} A e^{-j\omega t} dt \tag{B.5}$$

$$= \frac{A}{j\omega}(e^{j\omega T/2} - e^{-j\omega T/2}),$$

which can be put in the form

$$F(\omega) = AT\left(\frac{\sin(\omega T/2)}{\omega T/2}\right),$$

and finally

$$F(\omega) = AT\,\mathrm{Sa}\left(\frac{\omega T}{2}\right), \tag{B.6}$$

in which $\mathrm{Sa}\,(x) = \sin(x)/x$ is the *sampling* function. This function converges to 1, as x goes to zero. The sampling function, whose magnitude is illustrated in Figure B.2, is relevant in communication theory.

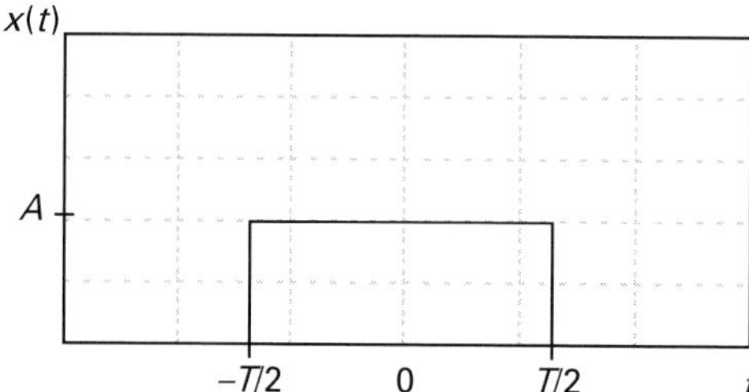

Figure B.1   Pulse function

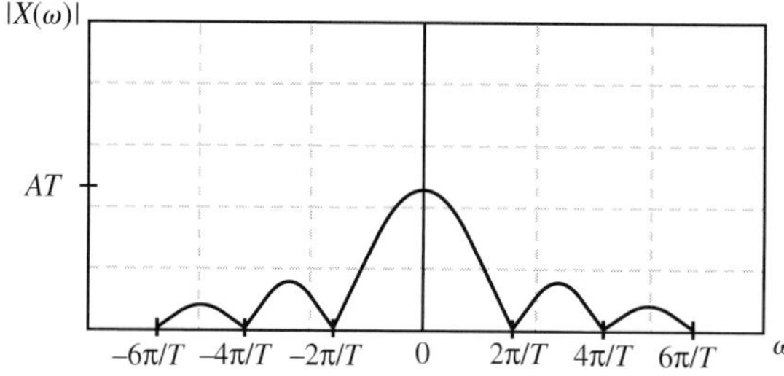

Figure B.2   Plot of the magnitude of the Fourier transform of the pulse function

Impulse function

If $f(t) = \delta(t)$ in (B.1), one concludes that

$$F(\omega) = \int_{-\infty}^{\infty} \delta(t)e^{-j\omega t}\,dt. \qquad (B.7)$$

The impulse filtering property implies that $F(\omega) = 1$. Therefore, the impulse function represents a continuum of equal amplitude spectral components, as shown in Figure B.3.

The constant function

If $f(t)$ is constant, then its Fourier transform, in principle, would not exist since this function does not satisfy the absolute integrability criterion. The Fourier transform of $f(t)$ is expected to be finite, i.e.

$$|F(\omega)| \le \int_{-\infty}^{\infty} |f(t)||e^{-j\omega t}|\,dt < \infty, \qquad (B.8)$$

since $|e^{-j\omega t}| = 1$, then

$$\int_{-\infty}^{\infty} |f(t)|\,dt < \infty. \qquad (B.9)$$

This is a sufficiency condition, but not a necessary condition for the existence of the Fourier transform, since there exist functions which, although they do not satisfy the condition of absolute integrability, in the limit have a Fourier transform (Carlson, 1975).

This important observation leads to the computation of Fourier transforms of several functions. Considering the constant function, it can be approximated by a pulse function with amplitude A and width τ, and then making τ approach very large values,

$$\mathcal{F}[A] = \lim_{\tau \to \infty} A\tau \operatorname{Sa}\left(\frac{\omega\tau}{2}\right) \qquad (B.10)$$

$$= 2\pi A \lim_{\tau \to \infty} \frac{\tau}{2\pi} \operatorname{Sa}\left(\frac{\omega\tau}{2}\right)$$

$$= 2\pi A\delta(\omega). \qquad (B.11)$$

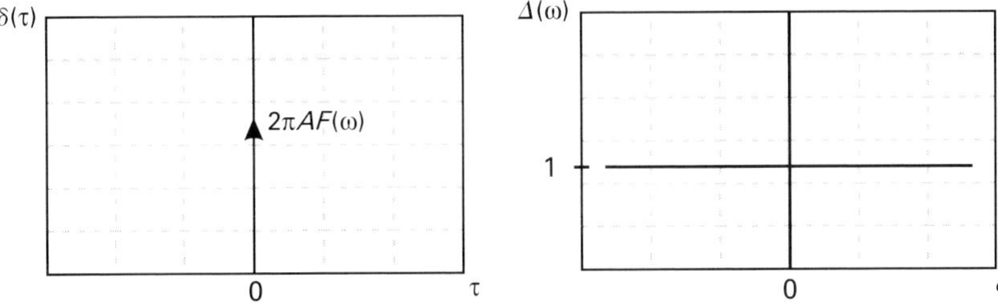

Figure B.3   Impulse function and its Fourier transform

This result is intuitive, since a constant function in time represents a DC level and, as is to be expected, contains no spectral component except for the one at $\omega = 0$, as shown in Figure B.4.

The sine and the cosine Fourier transforms

The sine and the cosine are periodic functions that do not satisfy the condition of absolute integrability, but their Fourier transforms exist in the limit when τ goes to infinity. A plot of the sine function is shown in Figure B.5.

Assuming the function to exist only in the interval $(-\tau/2, \tau/2)$ and to be zero outside this interval, and considering the limit of the expression when τ goes to infinity

$$
\begin{aligned}
\mathcal{F}(\sin \omega_0 t) &= \lim_{\tau \to \infty} \int_{-\tau/2}^{\tau/2} \sin \omega_0 t \; e^{-j\omega t} dt \\
&= \lim_{\tau \to \infty} \int_{-\tau/2}^{\tau/2} \frac{e^{-j(\omega-\omega_0)t}}{2j} - \frac{e^{-j(\omega+\omega_0)t}}{2j} dt \\
&= \lim_{\tau \to \infty} \left[\frac{j\tau \sin (\omega + \omega_0)\tau/2}{2(\omega + \omega_0)\tau/2} - \frac{j\tau \sin (\omega - \omega_0)\tau/2}{2(\omega - \omega_0)\tau/2} \right] \\
&= \lim_{\tau \to \infty} \left\{ j\frac{\tau}{2}\mathrm{Sa}\left[\frac{(\omega + \omega_0)}{2} \right] - j\frac{\tau}{2}\mathrm{Sa}\left[\frac{\tau(\omega + \omega_0)}{2} \right] \right\}.
\end{aligned}
\tag{B.12}
$$

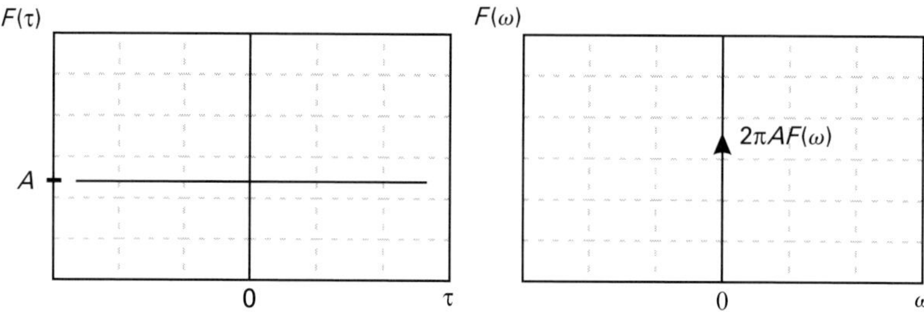

Figure B.4   Constant function and its transform

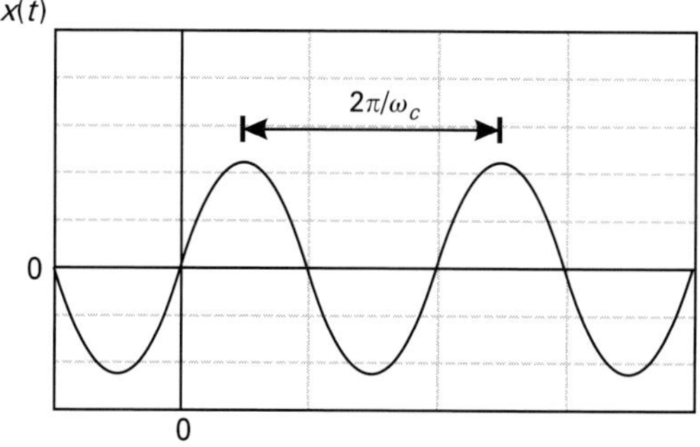

Figure B.5   The sine function

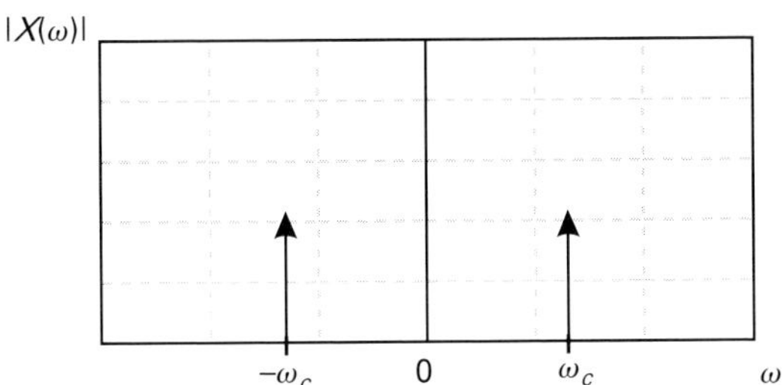

Figure B.6 Plot of the magnitude of the Fourier transform of the sine function

Therefore,

$$\mathcal{F}(\sin\omega_0 t) = \mathrm{j}\pi\,[\delta(\omega + \omega_0) - \delta(\omega - \omega_0)].$$

A plot of the magnitude of the Fourier transform of the sine function is shown in Figure B.6.

Applying a similar line of thought, one obtains

$$\mathcal{F}(\cos\omega_0 t) = \pi\,[\delta(\omega - \omega_0) + \delta(\omega + \omega_0)]. \tag{B.13}$$

Bilateral exponential signal

If $f(t) = \mathrm{e}^{-a|t|}$, it follows from (B.1) that

$$F(\omega) = \int_{-\infty}^{\infty} \mathrm{e}^{-a|t|}\mathrm{e}^{-\mathrm{j}\omega t}\,\mathrm{d}t \tag{B.14}$$

$$= \int_{-\infty}^{0} \mathrm{e}^{at}\mathrm{e}^{-\mathrm{j}\omega t}\,\mathrm{d}t + \int_{0}^{\infty} \mathrm{e}^{-at}\mathrm{e}^{-\mathrm{j}\omega t}\,\mathrm{d}t$$

$$= \frac{1}{a - \mathrm{j}\omega} + \frac{1}{a + \mathrm{j}\omega},$$

$$= \frac{2a}{a^2 + \omega^2}. \tag{B.15}$$

The Fourier transform of the complex exponential

Using Euler's identity, $\mathrm{e}^{\mathrm{j}\omega_0 t} = \cos\omega_0 t + \mathrm{j}\sin\omega_0 t$, it follows that

$$\mathcal{F}[\mathrm{e}^{\mathrm{j}\omega_0 t}] = \mathcal{F}[\cos\omega_0 t + \mathrm{j}\sin\omega_0 t]. \tag{B.16}$$

Substituting in (B.16) the Fourier transforms of the sine and of the cosine functions, respectively, it follows that

$$\mathcal{F}[\mathrm{e}^{\mathrm{j}\omega_0 t}] = 2\pi\,\delta(\omega - \omega_0). \tag{B.17}$$

The transform of a periodic function

Consider next the exponential Fourier series representation of a periodic function $f_T(t)$ of period T:

$$f_T(t) = \sum_{n=-\infty}^{\infty} F_n e^{jn\omega_0 t}. \tag{B.18}$$

Applying the Fourier transform to both sides in (B.18), one obtains

$$\mathcal{F}[f_T(t)] = \mathcal{F}\left[\sum_{n=-\infty}^{\infty} F_n e^{jn\omega_0 t} \right] \tag{B.19}$$

$$= \sum_{n=-\infty}^{\infty} F_n \mathcal{F}[e^{jn\omega_0 t}]. \tag{B.20}$$

Now, applying in (B.20) the result from (B.17), it follows that

$$F(\omega) = 2\pi \sum_{n=-\infty}^{\infty} F_n \delta(\omega - n\omega_0). \tag{B.21}$$

B.3   Properties of the Fourier transform

Linearity

Linearity is an important property in the study of communication systems. A system is defined to be a linear system if it satisfies the properties of homogeneity and additivity.

(1) Homogeneity: if the application of the signal $x(t)$ at the system input produces $y(t)$ at the system output, then the application of the input $\alpha x(t)$, in which α is a constant, produces $\alpha y(t)$ at the output.
(2) Additivity: if the application of the signals $x_1(t)$ and $x_2(t)$ at the system input produces respectively $y_1(t)$ and $y_2(t)$ at the system output, then the application of the input $x_1(t) + x_2(t)$ produces $y_1(t) + y_2(t)$ at the output.

Application of the tests for homogeneity and additivity indicates that the process that generates the signal $s(t) = A\cos(\omega_c t + \Delta m(t) + \theta)$ from an input signal $m(t)$ is non-linear. By applying the same test to the signal $r(t) = m(t)\cos(\omega_c t + \theta)$ it is immediate to show that the process generating $r(t)$ is linear.

The Fourier transform is a linear operator, i.e. if a function can be written as a linear combination of well-behaved functions, the corresponding Fourier transform is given by a linear combination of the corresponding Fourier transforms of each one of the functions involved in the linear combination (Gagliardi, 1988).

If $f(t) \longleftrightarrow F(\omega)$ and $g(t) \longleftrightarrow G(\omega)$ it then follows that

$$\alpha f(t) + \beta g(t) \longleftrightarrow \alpha F(\omega) + \beta G(\omega). \tag{B.22}$$

Proof: Let $h(t) = \alpha f(t) + \beta g(t)$, then it follows that

$$H(\omega) = \int_{-\infty}^{\infty} h(t)e^{-j\omega t}\,dt$$

$$= \alpha \int_{-\infty}^{\infty} f(t)e^{-j\omega t}\,dt + \beta \int_{-\infty}^{\infty} g(t)e^{-j\omega t}\,dt,$$

and finally

$$H(\omega) = \alpha F(\omega) + \beta G(\omega). \tag{B.23}$$

Time scaling

$$\mathcal{F}[f(at)] = \int_{-\infty}^{\infty} f(at)e^{-j\omega t}\,dt. \tag{B.24}$$

Consider, for the moment, the restriction $a > 0$ in (B.24). By letting $u = at$, it follows that $dt = (1/a)du$. Replacing u for at in (B.24), one obtains

$$\mathcal{F}[f(at)] = \int_{-\infty}^{\infty} \frac{f(u)}{a} e^{-j\frac{\omega}{a}u}\,du,$$

which simplifies to

$$\mathcal{F}[f(at)] = \frac{1}{a}F\left(\frac{\omega}{a}\right).$$

Consider now the case $a < 0$. By a similar procedure, it follows that

$$\mathcal{F}[f(at)] = -\frac{1}{a}F\left(\frac{\omega}{a}\right).$$

Therefore, finally

$$\mathcal{F}[f(at)] = \frac{1}{|a|}F\left(\frac{\omega}{a}\right). \tag{B.25}$$

This result indicates that if a signal is compressed in the time domain by a factor a, then its Fourier transform expands in the frequency domain by that same factor.

Symmetry property

This interesting property can be observed in even functions. The symmetry property states that if

$$f(t) \longleftrightarrow F(\omega), \tag{B.26}$$

then it follows that

$$F(t) \longleftrightarrow 2\pi f(-\omega). \tag{B.27}$$

Proof: By definition,

$$f(t) = \frac{1}{2\pi} \int_{-\infty}^{+\infty} F(\omega) e^{j\omega t} \, d\omega,$$

which after multiplication of both sides by 2π becomes

$$2\pi f(t) = \int_{-\infty}^{+\infty} F(\omega) e^{j\omega t} \, d\omega.$$

By letting $u = -t$ it follows that

$$2\pi f(-u) = \int_{-\infty}^{+\infty} F(\omega) e^{-j\omega u} \, d\omega,$$

and now by making $t = \omega$, one obtains

$$2\pi f(-u) = \int_{-\infty}^{+\infty} F(t) e^{-jtu} \, dt.$$

Finally, by letting $u = \omega$ it follows that

$$2\pi f(-\omega) = \int_{-\infty}^{+\infty} F(t) e^{-j\omega t} \, dt. \tag{B.28}$$

Time domain shift

Consider that $f(t) \longleftrightarrow F(\omega)$, therefore $f(t-t_0) \longleftrightarrow F(\omega) e^{-j\omega t_0}$. Let $g(t) = f(t-t_0)$. In this case it follows that

$$G(\omega) = \mathcal{F}[g(t)] = \int_{-\infty}^{\infty} f(t-t_0) e^{-j\omega t} \, dt. \tag{B.29}$$

By making $\tau = t - t_0$, it follows that

$$G(\omega) = \int_{-\infty}^{\infty} f(\tau) e^{-j\omega(\tau+t_0)} \, d\tau \tag{B.30}$$

$$= \int_{-\infty}^{\infty} f(\tau) e^{-j\omega\tau} e^{-j\omega t_0} \, d\tau, \tag{B.31}$$

and finally

$$G(\omega) = e^{-j\omega t_0} F(\omega). \tag{B.32}$$

This last result implies that if a function is shifted in time its amplitude spectrum remains the same. However, the corresponding phase spectrum experiences a rotation proportional to ωt_0.

Frequency domain shift

Given that $f(t) \longleftrightarrow F(\omega)$, it then follows that $f(t)e^{j\omega_0 t} \longleftrightarrow F(\omega - \omega_0)$.

$$\mathcal{F}[f(t)e^{j\omega_0 t}] = \int_{-\infty}^{\infty} f(t)e^{j\omega_0 t}e^{-j\omega t}\,\mathrm{d}t \tag{B.33}$$

$$= \int_{-\infty}^{\infty} f(t)e^{-j(\omega - \omega_0)t}\,\mathrm{d}t,$$

$$= F(\omega - \omega_0). \tag{B.34}$$

Differentiation in the time domain

Consider a function and its Fourier transform,

$$f(t) \longleftrightarrow F(\omega). \tag{B.35}$$

Differentiating the function in the time domain,

$$\frac{\mathrm{d}f(t)}{\mathrm{d}t} \longleftrightarrow j\omega F(\omega). \tag{B.36}$$

Proof: Consider the expression for the inverse Fourier transform

$$f(t) = \frac{1}{2\pi} \int_{-\infty}^{\infty} F(\omega)e^{j\omega t}\,\mathrm{d}\omega. \tag{B.37}$$

The derivative of the function in time gives

$$\frac{\mathrm{d}f(t)}{\mathrm{d}t} = \frac{1}{2\pi}\frac{\mathrm{d}}{\mathrm{d}t} \int_{-\infty}^{\infty} F(\omega)e^{j\omega t}\,\mathrm{d}\omega$$

$$= \frac{1}{2\pi} \int_{-\infty}^{\infty} \frac{\mathrm{d}}{\mathrm{d}t}F(\omega)e^{j\omega t}\,\mathrm{d}\omega$$

$$= \frac{1}{2\pi} \int_{-\infty}^{\infty} j\omega F(\omega)e^{j\omega t}\,\mathrm{d}\omega,$$

and then

$$\frac{\mathrm{d}f(t)}{\mathrm{d}t} \longleftrightarrow j\omega F(\omega). \tag{B.38}$$

In general, one obtains

$$\frac{\mathrm{d}^n f(t)}{\mathrm{d}t} \longleftrightarrow (j\omega)^n f(\omega). \tag{B.39}$$

Integration in the time domain

Consider that $f(t)$ is a signal with zero average value, i.e. let $\int_{-\infty}^{\infty} f(t)dt = 0$. By defining

$$g(t) = \int_{-\infty}^{t} f(\tau)d\tau, \tag{B.40}$$

it follows that

$$\frac{dg(t)}{dt} = f(t),$$

and since

$$g(t) \longleftrightarrow G(\omega), \tag{B.41}$$

then

$$f(t) \longleftrightarrow j\omega G(\omega),$$

and

$$G(\omega) = \frac{F(\omega)}{j\omega}. \tag{B.42}$$

Therefore, for a signal with zero average value one obtains

$$f(t) \longleftrightarrow F(\omega)$$

$$\int_{-\infty}^{t} f(\tau)d\tau \longleftrightarrow \frac{F(\omega)}{j\omega}. \tag{B.43}$$

For the case when $f(t)$ has a non-zero average value, the formula can be generalized to

$$\int_{-\infty}^{t} f(\tau)d\tau \longleftrightarrow \frac{F(\omega)}{j\omega} + \pi\delta(\omega)F(0). \tag{B.44}$$

The convolution theorem

The convolution operation is a powerful tool for obtaining the response of linear systems to input signals. The theorem is used to analyze the frequency contents of a signal. The use of the convolution theorem is of fundamental importance in digital communication.

The convolution between two time functions $f(t)$ and $g(t)$ is defined by the following integral:

$$\int_{-\infty}^{\infty} f(\tau)g(t - \tau)d\tau, \tag{B.45}$$

which is often written as $f(t) * g(t)$.

Let $h(t) = f(t) * g(t)$ and let $h(t) \longleftrightarrow H(\omega)$. It follows that

$$H(\omega) = \int_{-\infty}^{\infty} h(t)e^{-j\omega t}\,dt = \int_{-\infty}^{\infty}\int_{-\infty}^{\infty} f(\tau)g(t-\tau)e^{-j\omega t}\,dt\,d\tau. \tag{B.46}$$

$$= \int_{-\infty}^{\infty} f(\tau)\int_{-\infty}^{\infty} g(t-\tau)e^{-j\omega t}\,dt\,d\tau, \tag{B.47}$$

$$= \int_{-\infty}^{\infty} f(\tau)G(\omega)e^{-j\omega \tau}\,d\tau \tag{B.48}$$

and finally,

$$H(\omega) = F(\omega)G(\omega). \tag{B.49}$$

The convolution of two functions in time is equivalent in the frequency domain to the product of their respective Fourier transforms. For the case in which $h(t) = f(t) \cdot g(t)$, proceeding in a similar manner one obtains

$$H(\omega) = \frac{1}{2\pi}[F(\omega) * G(\omega)]. \tag{B.50}$$

The product of two time functions has a Fourier transform given by the convolution of their respective Fourier transforms. The convolution operation is often used when computing the response of a linear circuit, given its impulse response and an input signal.

The application of the unit impulse $x(t) = \delta(t)$ as the input to this circuit causes an output $y(t) = h(t) * x(t)$. In the frequency domain, by the convolution theorem it follows that $Y(\omega) = H(\omega)X(\omega) = H(\omega)$, i.e. the Fourier transform of the impulse response of a linear system is the system transfer function.

Using the frequency convolution theorem it can be shown that

$$\cos(\omega_c t)u(t) \longleftrightarrow \frac{\pi}{2}[\delta(\omega + \omega_c) + \delta(\omega - \omega_c)] + j\frac{\omega}{\omega_c^2 - \omega^2}.$$

B.4 The Nyquist theorem

A band-limited signal $f(t)$, which has no frequency components above $\omega_M = 2\pi f_M$, can be reconstructed from its samples, collected at uniform time intervals $T_s = 1/f_s$, i.e. at a sampling rate f_s, in which $f_s \geq 2f_M$.

By a band-limited signal $f(t) \longleftrightarrow F(\omega)$ it is understood that there is a frequency ω_M above which $F(\omega) = 0$, i.e. that $F(\omega) = 0$ for $|\omega| > \omega_M$.

Harry Nyquist concluded, in a seminal paper, that all the information about $f(t)$, as illustrated in Figure B.7, is contained in the samples of this signal, collected at regular time intervals T_s. In this manner the signal can be completely recovered from its samples. For a band-limited signal $f(t)$, i.e. such that $F(\omega) = 0$ for $|\omega| > \omega_M$, it follows that

$$f(t) * \frac{\sin(at)}{\pi t} = f(t), \text{ if } a > \omega_M,$$

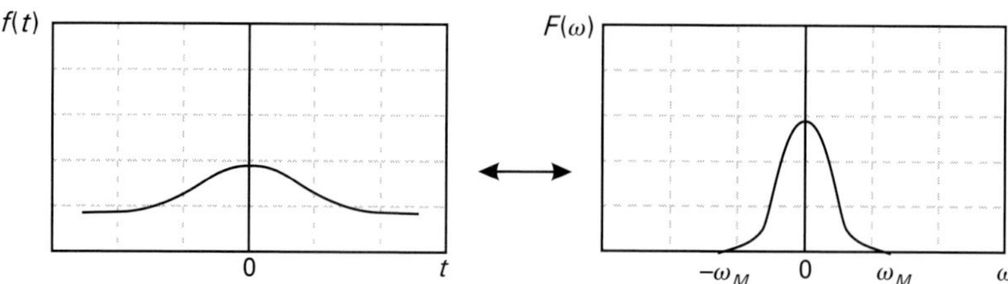

Figure B.7   Band-limited signal $f(t)$ and its spectrum

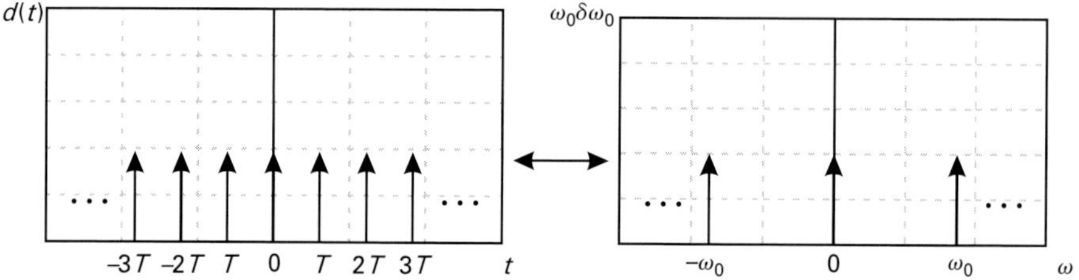

Figure B.8   Impulse train used for sampling

because in the frequency domain this corresponds to the product of $F(\omega)$ with a pulse function of width greater than $2\omega_M$.

The function $f(t)$ is sampled once every T_s seconds or, equivalently, sampled with a sampling frequency f_s, in which $f_s = 1/T_s \geq 2f_M$.

Consider the signal $f_s(t) = f(t)\delta_T(t)$, in which

$$\delta_T(t) = \sum_{n=-\infty}^{\infty} \delta(t - nT) \longleftrightarrow \omega_0\delta_{\omega_0} = \omega_0 \sum_{n=-\infty}^{\infty} \delta(\omega - n\omega_0). \tag{B.51}$$

The signal $\delta_T(t)$ is illustrated in Figure B.8. The signal $f_s(t)$ represents $f(t)$ sampled at uniform time intervals of T_s seconds. From the frequency convolution theorem it follows that the Fourier transform of the product of two functions in the time domain is given by the convolution of their respective Fourier transforms. It now follows that

$$f_s(t) \longleftrightarrow \frac{1}{2\pi}[F(\omega) * \omega_0\delta_{\omega_0}(\omega)] \tag{B.52}$$

and thus

$$f_s(t) \longleftrightarrow \frac{1}{T}[F(\omega) * \delta_{\omega_0}(\omega)] = \frac{1}{T} \sum_{n=-\infty}^{\infty} F(\omega - n\omega_0). \tag{B.53}$$

Figure B.9 shows that if the sampling frequency ω_s is less than $2\omega_M$, there is an overlap of spectral components. This causes a loss of information because the original signal

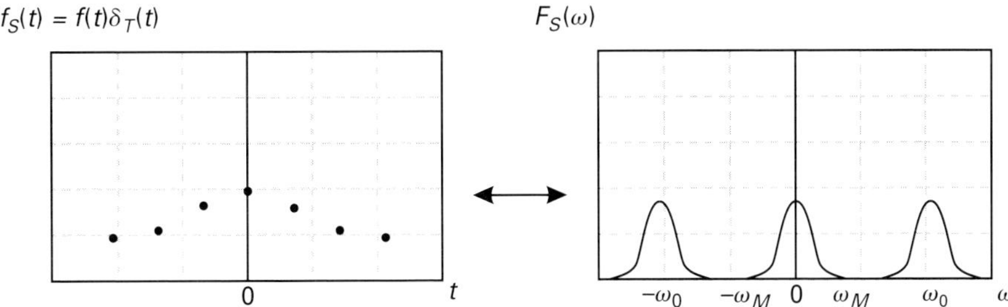

Figure B.9   Sampled signal and its spectrum

can no longer be fully recovered from its samples. As ω_s becomes smaller than $2\omega_M$, the sampling rate diminishes causing a partial loss of information.

The minimum sampling frequency that allows perfect recovery of the signal is $\omega_s = 2\omega_M$, and is known as the Nyquist sampling rate. In order to recover the original spectrum $F(\omega)$, it is enough to pass the sampled signal through a low-pass filter with cut-off frequency ω_M.

If the sampling frequency ω_S is lower than $2\pi B$, there is spectral overlap and, as a consequence, information loss. As long as ω_S is lower than $2\pi B$, the sampling rate decreases, leading to partial loss of information. Therefore, the sampling frequency for a baseband signal to be recovered without loss is $\omega_S = 2\pi B$; this is known as the Nyquist sampling frequency.

A sampling rate lower than the Nyquist frequency implies that the signal will not be completely recovered, since there will be spectral superposition, leading to distortion in the highest frequencies. This phenomenon is known as aliasing. On the other hand, increasing the sampling frequency to a value higher than the Nyquist frequency leads to spectral separation higher than the minimum necessary to recover the signal.

B.5   Fourier transform in two dimensions

The two-dimensional Fourier transform is used in the analysis of images, wave propagation, digital television and optical communications (Hsu, 1973, Bracewell, 1965, 2004). Its usual representation is

$$X(\omega, \phi) = \int_{-\infty}^{\infty} \int_{-\infty}^{\infty} x(t, \tau)e^{-j\omega t - j\phi \tau} \, dt d\tau. \tag{B.54}$$

The inverse transform in two dimensions is given by the formula

$$x(t, \tau) = \frac{1}{(4\pi)^2} \int_{-\infty}^{\infty} \int_{-\infty}^{\infty} X(\omega, \phi)e^{j\omega t + j\phi \tau} \, d\omega d\phi. \tag{B.55}$$

B.6   Discrete Fourier transform (DFT)

Consider a signal sampled in time:

$$x_A(t) = x(t) \sum_{n=-\infty}^{\infty} \delta(t - nT) = \sum_{n=-\infty}^{\infty} x_n \delta(t - nT). \tag{B.56}$$

Substituting (B.56) into the formula for the Fourier transform, one obtains

$$X_A(\omega) = \int_{-\infty}^{\infty} \sum_{n=-\infty}^{\infty} x_n \delta(t - nT) e^{-j\omega t} \, dt. \tag{B.57}$$

Using the impulse function properties, it is possible to simplify the result to

$$X_A(\omega) = \sum_{n=-\infty}^{\infty} x_n e^{-j\omega nT}. \tag{B.58}$$

This formula represents the DFT, which is widely applied in signal processing, including audio and video. It can be expressed as

$$X(\omega) = \sum_{n=-\infty}^{\infty} x(n) e^{-j\omega n}, \tag{B.59}$$

and the anti-transform is

$$x(n) = \frac{1}{2\pi} \int_{-\pi}^{\pi} X(\omega) e^{j\omega n} \, d\omega. \tag{B.60}$$

The properties of the continuous Fourier transform prevail in the discrete domain. For finite duration signals, it is common to use the following representation for the DFT, also known as the analysis equation (Oppenheim and Schafer, 1989):

$$X(k) = \sum_{n=0}^{N-1} x(n) e^{-j(2\pi k/N)n}, \tag{B.61}$$

The inverse transform, known as the synthesis equation, is given by

$$x(n) = \frac{1}{N} \sum_{n=0}^{N-1} X(k) e^{j(2\pi k/N)n}. \tag{B.62}$$

Periodicity properties of the complex exponential are exploited to compute the transform rapidly and efficiently. The algorithm is known as the fast Fourier transform (FFT), and has applications in various areas, including digital television and orthogonal frequency division multiplexing (OFDM).

B.7   Discrete cosine transform (DCT)

The DCT is used in digital television to change the image representation domain. This process does not introduce any loss in video quality and permits a more explicit representation of the image components, which facilitates compression.

The DCT is defined in different ways. The following definition is adopted by the JPEG and MPEG international standards (Jurgen, 1997), for an image with dimension $N \times M$:

$$X(u, v) = \frac{2c(u)c(v)}{\sqrt{NM}} \sum_{j=0}^{N-1} \sum_{k=0}^{M-1} x(j, k) \cos\left[\frac{(2j + 1)\pi u}{2N}\right] \cos\left[\frac{(2k + 1)\pi v}{2M}\right], \quad \text{(B.63)}$$

in which $c(u) = 1/\sqrt{2}$, for $u = 0$, $c(u) = 1$, for $u = 1, 2, \ldots, N - 1$ and $c(v)$ is defined in a similar manner.

The inverse discrete transform is defined as

$$x(j, k) = \frac{2}{\sqrt{NM}} \sum_{u=0}^{N-1} \sum_{v=0}^{M-1} c(u)c(v)X(u, v) \cos\left[\frac{(2j + 1)\pi u}{2N}\right] \cos\left[\frac{(2k + 1)\pi v}{2M}\right].$$

$$\text{(B.64)}$$

For fast computation of the DCT, for instance in the MPEG-2 standard, it is commonplace to use a basis functions table, with 8×8 blocks, for the transformation of blocks of pixels of the same dimension, as shown in Figure B.10.

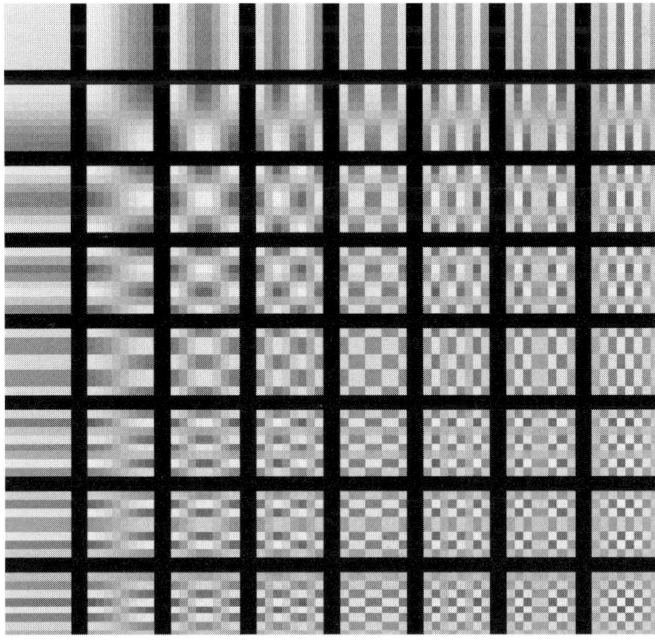

Figure B.10   Basis functions for DCT

B.8   The Hilbert transform

The Hilbert transform of a signal $f(t)$ is denoted as $\hat{f}(t)$, and is defined as (Gagliardi, 1978)

$$\hat{f}(t) = \frac{1}{\pi} \int_{-\infty}^{\infty} \frac{f(\tau)}{t - \tau} \, d\tau. \tag{B.65}$$

The Hilbert transform is a linear operation and its inverse transform is given by

$$f(t) = -\frac{1}{\pi} \int_{-\infty}^{\infty} \frac{\hat{f}(t)}{t - \tau} \, d\tau. \tag{B.66}$$

The functions $f(t)$ and $\hat{f}(t)$ form a pair of Hilbert transforms. This transform shifts all frequency components of an input signal by $\pm 90°$. The positive frequency components are shifted by $-90°$, and the negative frequency components are shifted by $+90°$. The spectral amplitudes are not affected.

From the definition, $\hat{f}(t)$ can be interpreted as the convolution of $f(t)$ and $1/\pi t$. As already known, the Fourier transform of the convolution of two signals is given by the product of the transforms of each signal. Therefore,

$$G(\omega) = \mathcal{F}\left[f(t) * \frac{1}{\pi t}\right] = \mathcal{F}[f(t)] \cdot \mathcal{F}\left[\frac{1}{\pi t}\right]. \tag{B.67}$$

Fourier transform of $1/\pi t$

To obtain the Fourier transform of $1/\pi t$, one can resort to the following:

- Consider the signal function in Figure B.11.
- The derivative of this function is an impulse centered at the origin, whose Fourier transform is the constant 2.
- Using the Fourier integral property,

$$u(t) - u(-t) \leftrightarrow \frac{2}{j\omega}.$$

- Finally, using the symmetry property, it follows that

$$\frac{1}{\pi t} \longleftrightarrow j(u(-\omega) - u(\omega)).$$

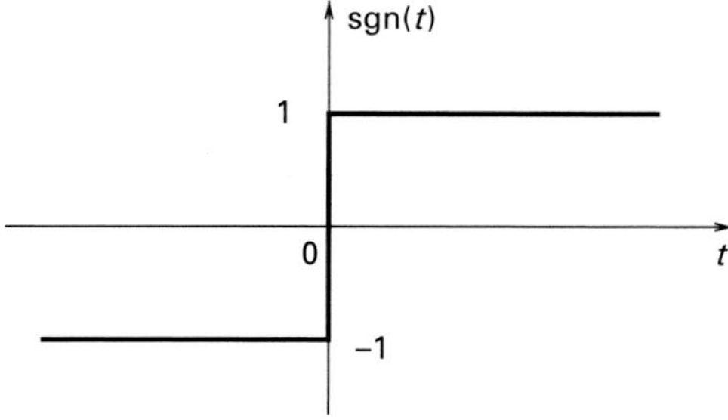

Figure B.11   Signal function

Therefore,

$$G(\omega) = j[u(-\omega) - u(\omega)] \cdot F(\omega). \tag{B.68}$$

Properties of the Hilbert transform

The Hilbert transform has the following properties

- Signal $f(t)$ and its transform $\hat{f}(t)$ have the same power spectral density.
- Signal $f(t)$ and its transform $\hat{f}(t)$ have the same power autocorrelation.
- Signal $f(t)$ and its transform $\hat{f}(t)$ are orthogonal.
- If $\hat{f}(t)$ is the Hilbert transform of $f(t)$, then the Hilbert transform of $\hat{f}(t)$ is $-f(t)$.

B.9 Useful Fourier transforms

This section contains a set of Fourier transforms, and properties, that are useful to solve problems (Hsu, 1973, Spiegel, 1976, Lathi, 1989, Gradshteyn and Ryzhik, 1990, Oberhettinger, 1990).

$f(t)$	$F(\omega)$
$f(at)$	$\dfrac{1}{\|a\|}F(\dfrac{\omega}{a})$
$f(-t)$	$F(-\omega)$
$f^*(t)$	$F^*(-\omega)$
$f(t-\tau)$	$F(\omega)e^{-j\omega\tau}$
$f(t)e^{j\omega_0 t}$	$F(\omega-\omega_0)$
$f(t)\cos\omega_0 t$	$\dfrac{1}{2}F(\omega-\omega_0) + \dfrac{1}{2}F(\omega+\omega_0)$
$f(t)\sin\omega_0 t$	$\dfrac{1}{2j}F(\omega-\omega_o) - \dfrac{1}{2j}F(\omega+\omega_0)$
$F(t)$	$2\pi f(-\omega)$
$f'(t)$	$j\omega F(\omega)$
$f^{(n)}(t)$	$(j\omega)^n F(\omega)$
$\int_{-\infty}^{t} f(x)dx$	$\dfrac{1}{j\omega}F(\omega) + \pi F(0)\delta(\omega)$
$-jtf(t)$	$F'(\omega)$
$(-jt)^n f(t)$	$F^{(n)}(\omega)$
$f(t)*g(t) = \int_{-\infty}^{\infty} f(\tau)g(t-\tau)dx$	$F(\omega)G(\omega)$
$\delta(t)$	1

(**cont.**)

$f(t)$	$F(\omega)$
$\delta(t-\tau)$	$\mathrm{e}^{-\mathrm{j}\omega\tau}$
$\delta'(t)$	$\mathrm{j}\omega$
$\delta^{(n)}(t)$	$(\mathrm{j}\omega)^n$
$f(t)g(t)$	$\dfrac{1}{2\pi}F(\omega)*G(\omega)=\dfrac{1}{2\pi}\int_{-\infty}^{\infty}F(\phi)G(\omega-\phi)\mathrm{d}\phi$
$\mathrm{e}^{-at}u(t)$	$\dfrac{1}{a+\mathrm{j}\omega}$
$\mathrm{e}^{-a\lvert t\rvert}$	$\dfrac{2a}{a^2+\omega^2}$
e^{-at^2}	$\sqrt{\dfrac{\pi}{a}}\,\mathrm{e}^{-\omega^2/(4a)}$
$t\mathrm{e}^{-at^2}$	$\mathrm{j}\sqrt{\dfrac{\pi}{4a^3}}\,\omega\mathrm{e}^{-\omega^2/(4a)}$
$p_T(t)=\begin{cases}0 & \text{for }\lvert t\rvert>T/2\\[4pt] A & \text{for }\lvert t\rvert\le T/2\end{cases}$	$AT\dfrac{\sin(\omega T/2)}{(\omega T/2)}$
$\dfrac{\sin at}{\pi t}$	$p_{2a}(\omega)$
$t\mathrm{e}^{-at}u(t)$	$\dfrac{1}{(a+\mathrm{j}\omega)^2}$
$\dfrac{t^{n-1}}{(n-1)!}\mathrm{e}^{-at}u(t)$	$\dfrac{1}{(a+\mathrm{j}\omega)^n}$
$\mathrm{e}^{-at}\sin bt\,u(t)$	$\dfrac{b}{(a+\mathrm{j}\omega)^2+b^2}$
$\mathrm{e}^{-at}\cos bt\,u(t)$	$\dfrac{a+\mathrm{j}\omega}{(a+\mathrm{j}\omega)^2+b^2}$
$\dfrac{1}{a^2+t^2}$	$\dfrac{\pi}{a}\mathrm{e}^{-a\lvert\omega\rvert}$
$\dfrac{t}{a^2+t^2}$	$\mathrm{j}\pi\mathrm{e}^{-a\lvert\omega\rvert}[u(-\omega)-u(\omega)]$
$\dfrac{\cos bt}{a^2+t^2}$	$\dfrac{\pi}{2a}[\mathrm{e}^{-a\lvert\omega-b\rvert}+\mathrm{e}^{-a\lvert\omega+b\rvert}]$
$\dfrac{\sin bt}{a^2+t^2}$	$\dfrac{\pi}{2a\mathrm{j}}[\mathrm{e}^{-a\lvert\omega-b\rvert}-\mathrm{e}^{-a\lvert\omega+b\rvert}]$
$\sin bt^2$	$\dfrac{\pi}{2b}\left[\cos\dfrac{\omega^2}{4b}-\sin\dfrac{\omega^2}{4b}\right]$

$f(t)$	$F(\omega)$
$\cos bt^2$	$\dfrac{\pi}{2b}\left[\cos\dfrac{\omega^2}{4b} + \sin\dfrac{\omega^2}{4b}\right]$
$\operatorname{sech} bt$	$\dfrac{\pi}{b}\operatorname{sech}\dfrac{\pi\omega}{2b}$
$\ln\left[\dfrac{x^2+a^2}{x^2+b^2}\right]$	$\dfrac{2\mathrm{e}^{-b\omega} - 2\mathrm{e}^{-a\omega}}{\pi\omega}$
$f_P(t) = \dfrac{1}{2}[f(t)+f(-t)]$	$\mathrm{Re}\,(\omega)$
$f_I(t) = \dfrac{1}{2}[f(t)-f(-t)]$	$\mathrm{jIm}\,(\omega)$
$f(t) = f_P(t)+f_I(t)$	$F(\omega) = \mathrm{Re}\,(\omega)+\mathrm{jIm}(\omega)$
$\mathrm{e}^{\mathrm{j}\omega_0 t}$	$2\pi\,\delta(\omega-\omega_0)$
$\cos\omega_0 t$	$\pi[\delta(\omega-\omega_0)+\delta(\omega+\omega_0)]$
$\sin\omega_0 t$	$-\mathrm{j}\pi[\delta(\omega-\omega_0)-\delta(\omega+\omega_0)]$
$\sin\omega_0 t\,u(t)$	$\dfrac{\omega_0}{\omega_0^2-\omega^2} + \dfrac{\pi}{2\mathrm{j}}[\delta(\omega-\omega_0)-\delta(\omega+\omega_0)]$
$\cos\omega_0 t\,u(t)$	$\dfrac{\mathrm{j}\omega}{\omega_0^2-\omega^2} + \dfrac{\pi}{2}[\delta(\omega-\omega_0)+\delta(\omega+\omega_0)]$
$u(t)$	$\pi\,\delta(\omega) + \dfrac{1}{\mathrm{j}\omega}$
$u(t-\tau)$	$\pi\,\delta(\omega) + \dfrac{1}{\mathrm{j}\omega}\mathrm{e}^{-\mathrm{j}\omega\tau}$
$tu(t)$	$\mathrm{j}\pi\,\delta'(\omega) - \dfrac{1}{\omega^2}$
1	$2\pi\,\delta(\omega)$
t	$2\pi\,\mathrm{j}\delta'(\omega)$
t^n	$2\pi\,\mathrm{j}^n\delta^{(n)}(\omega)$
$|t|$	$\dfrac{-2}{\omega^2}$
$\dfrac{1}{t}$	$\pi\mathrm{j} - 2\pi\,\mathrm{j}u(\omega)$
$\dfrac{1}{t^n}$	$\dfrac{(-\mathrm{j}\omega)^{n-1}}{(n-1)!}[\pi\mathrm{j} - 2\pi\,\mathrm{j}u(\omega)]$
$u(t)-u(-t)$	$\dfrac{2}{\mathrm{j}\omega}$

(cont.)

$f(t)$	$F(\omega)$
$\dfrac{1}{e^{2t}-1}$	$\dfrac{-\mathrm{j}\pi}{2}\coth\dfrac{\pi\omega}{2}+\dfrac{\mathrm{j}}{\omega}$
$\delta_T(t)=\sum_{n=-\infty}^{\infty}\delta(t-nT)$	$\omega_0\delta_{\omega_0}(\omega)=\omega_0\sum_{n=-\infty}^{\infty}\delta(\omega-n\omega_0)$
$\cos\left(\dfrac{t^2}{4a}-\dfrac{1}{4}\pi\right)$	$2\sqrt{\pi a}\cos(a\omega^2)$
$\sin\left(\dfrac{t^2}{4a}+\dfrac{1}{4}\pi\right)$	$2\sqrt{\pi a}\sin(a\omega^2)$
$\dfrac{\Gamma(1-s)\sin\left(\dfrac{1}{2}s\pi\right)}{\lvert t\rvert^{1-s}}$	$\pi\lvert\omega\rvert^{-s}\quad 0<\mathrm{Re}\,(s)<1$
$\dfrac{1}{\lvert t\rvert}$	$\dfrac{\sqrt{2\pi}}{\lvert\omega\rvert}$
$\dfrac{\sqrt{\sqrt{\alpha^2+t^2}+\alpha}}{\sqrt{\alpha^2+t^2}}$	$\sqrt{\dfrac{2\pi}{\lvert\omega\rvert}}\,e^{-\alpha\lvert\omega\rvert}$
$\dfrac{\cos\left(\dfrac{1}{2}\alpha\right)\cosh\left(\dfrac{1}{2}t\right)}{\cosh(t)+\cos(\alpha)}$	$\dfrac{\pi\cosh(\alpha\omega)}{\cosh(\pi\omega)}\quad-\pi<\alpha<\pi$
$\dfrac{\sin(\alpha)}{\cosh(t)+\cos(\alpha)}$	$\dfrac{\pi\sin,(\alpha\omega)}{\sin,(\pi\omega)}\quad-\pi<\alpha<\pi$
$J_0(\alpha t)$	$\dfrac{2}{\sqrt{\alpha^2-\omega^2}}\quad\lvert\omega\rvert<\alpha$
	$0\qquad\qquad\lvert\omega\rvert>\alpha$
$J_0(\alpha\sqrt{b^2-t^2})\quad\lvert t\rvert<b$	$2\dfrac{\sin[b(\alpha^2+\omega^2)^{\frac{1}{2}}]}{\sqrt{\alpha^2+\omega^2}}$
$0\qquad\qquad\lvert t\rvert>b$	
$j^n J_{n+\frac{1}{2}}(t)$	$\sqrt{2}P_n(\omega)\lvert\omega\rvert<1$
	$0\qquad\lvert\omega\rvert>1$
$J_0(\alpha\sqrt{t^2+b^2})$	$\dfrac{2\cos(b\sqrt{\alpha^2-\omega^2})}{\sqrt{\alpha^2-\omega^2}}\quad\lvert\omega\rvert<\alpha$
	$0\qquad\qquad\lvert\omega\rvert>\alpha$

$f(t)$	$F(\omega)$
$J_0(\alpha\sqrt{t^2 - b^2})$	$\dfrac{2\cosh(b\sqrt{\alpha^2 - \omega^2})}{\sqrt{\alpha^2 - \omega^2}} \quad \lvert\omega\rvert < \alpha$
	$0 \qquad\qquad \lvert\omega\rvert > a$
$\dfrac{1}{(\alpha - \mathrm{j}t)^\nu}$	$2\pi\omega^{\nu-1}\mathrm{e}^{-\alpha\omega}/\Gamma(\nu), \quad \omega > 0$
$\mathrm{Re}\,\alpha > 0, \quad \mathrm{Re}\,\nu > 0$	$0, \omega < 0$
$\dfrac{1}{(\alpha + \mathrm{j}t)^\nu}$	$0, \omega > 0$
$\mathrm{Re}\,\nu > 0, \quad \mathrm{Re}\,\alpha > 0$	$-2\pi(-\omega)^{\nu-1}\mathrm{e}^{\alpha\omega}/\Gamma(\nu), \quad \omega < 0$
$\dfrac{1}{(t^2 + \alpha^2)(\mathrm{j}t)^\nu}$	
$\lvert\nu\rvert < 1, \quad \mathrm{Re}\,\alpha > 0$	$\pi\alpha^{-\nu-1}\mathrm{e}^{-\lvert\omega\rvert\alpha}$
$\arg(\mathrm{j}t) = 1/2\pi, t > 0$	
$\arg(\mathrm{j}t) = -1/2\pi, t < 0$	
$\dfrac{1}{(t^2 + \alpha^2)(\beta + \mathrm{j}t)^\nu} \quad \mathrm{Re}\,\nu > -1$	$\dfrac{\pi\mathrm{e}^{-\alpha\omega}}{\alpha(\alpha + \beta)^\nu}, \quad \omega > 0$
$\mathrm{Re}\,\alpha > 0, \quad \mathrm{Re}\,\beta > 0$	
$\dfrac{1}{(x^2 + \alpha^2)(\beta - \mathrm{j}t)^\nu}$	$\dfrac{\pi(\beta - \alpha)^\nu\mathrm{e}^{\alpha\omega}}{\alpha}, \quad \omega > 0$
$\mathrm{Re}\,\nu > -1, \quad \mathrm{Re}\,\alpha > 0$	
$\mathrm{Re}\,\beta > 0, \quad \alpha \neq \beta$	
$\dfrac{1}{(\alpha - \mathrm{e}^{-t})\mathrm{e}^{\lambda t}}$	$\pi\alpha^{\lambda-1+\mathrm{j}\omega}\cot(\pi\lambda + \mathrm{j}\pi\omega)$
$0 < \mathrm{Re}\,\lambda < 1, \quad \alpha > 0$	
$\dfrac{1}{(\alpha + \mathrm{e}^{-t})\mathrm{e}^{\lambda t}}$	$\pi\alpha^{\lambda-1+\mathrm{j}\omega}\mathrm{cosec}(\pi\lambda + \mathrm{j}\pi\omega)$
$0 < \mathrm{Re}\,\lambda < 1, \quad -\pi < \arg\alpha < \pi$	
$\dfrac{t}{(\alpha + \mathrm{e}^{-t})\mathrm{e}^{\lambda t}}$	$\pi\alpha^{\lambda-1+\mathrm{j}\omega}\mathrm{cosec}(\pi\lambda + \mathrm{j}\pi\omega)$
$0 < \mathrm{Re}\,\lambda < 1, \quad -\pi < \arg\alpha < \pi$	$\times[\log\alpha - \pi\cot(\pi\lambda + \mathrm{j}\pi\omega)]$

(cont.)

$f(t)$	$F(\omega)$
$\dfrac{t^2}{(1+e^{-t})e^{\lambda t}}$	$\pi^3\mathrm{cosec}^3(\pi\lambda+j\omega\pi)[2-\sin(\pi\lambda+j\omega\pi)]$
$0<\mathrm{Re}\,\lambda<1$	
$\dfrac{1}{(\alpha+e^{-t})(\beta+e^{-t})e^{\lambda t}}$	$\pi(\beta-\alpha)^{-1}(\alpha^{\lambda-1+j\omega}-\beta^{\lambda-1+j\omega})$
$0<\mathrm{Re}\,\lambda<2,\quad \beta\neq\alpha$	$\times\,\mathrm{cosec}\,(\pi\lambda+j\omega\pi)$
$|\arg\alpha|<\pi,\quad |\arg\beta|<\pi$	
$\dfrac{t}{(\alpha+e^{-t})(\beta+e^{-t})e^{\lambda t}}$	$\dfrac{\pi(\alpha^{\lambda-1+j\omega}\log\alpha-\beta^{\lambda-1+j\omega}\log\beta)}{(\alpha-\beta)\sin(\lambda\pi+j\omega\pi)}$
$0<\mathrm{Re}\,\lambda<2,\quad \alpha\neq\beta$	$+\dfrac{\pi^2(\alpha^{\lambda-1+j\omega}-\beta^{\lambda-1+j\omega})\cos(\lambda\pi+j\omega\pi)}{(\beta-\alpha)\sin^2(\lambda\pi+j\omega\pi)}$
$|\arg\alpha|<\pi,\quad |\arg\beta|<\pi$	
$\dfrac{1}{(1+e^{-t})^n e^{\lambda t}}$	$\pi\,\mathrm{cosec}\,(\pi\lambda+j\omega\pi)\prod_{j=1}^{n-1}(j-\lambda-j\omega)/(n-1)!$
$n=1,2,3,\ldots,\quad 0<\mathrm{Re}\,\alpha<n$	
$e^{-\lambda t}\log|1-e^{-t}|$	$\pi(\lambda+j\omega)^{-1}\cot(\pi\lambda+j\omega\pi)$
$-1<\mathrm{Re}\,\lambda<0$	
$e^{-\lambda t}\log(1+e^{-t})$	$\pi(\lambda+j\omega)^{-1}\mathrm{cosec}\,(\pi\lambda+j\omega\pi)$
$-1<\mathrm{Re}\,\lambda<0$	
$e^{-\lambda t}\log\left(\dfrac{|1+e^{-t}|}{|1-e^{-t}|}\right)$	$\pi(\lambda+j\omega)^{-1}\tan(\tfrac{1}{2}\pi\lambda+\tfrac{1}{2}j\omega\pi)$
$|\mathrm{Re}\,\lambda|<1$	
$\dfrac{1}{(\sin t+\sin\alpha)}\quad \alpha>0$	$-\pi j e^{j\alpha\omega}\mathrm{sech}\,\alpha\,\mathrm{cosech}\,(\pi\omega)\times[\cosh(\pi\omega)-e^{-2j\alpha\omega}]$
$\dfrac{1}{[\Gamma(\nu-t)\Gamma(\mu+t)]}$	$\left[2\cos\left(\tfrac{1}{2}\omega\right)\right]^{\mu+\nu-2}e^{\frac{1}{2}j\omega(\mu-\nu)}$
	$\times[\Gamma(\mu+\nu-1)]^{-1},\ |\omega|<\pi$
	$0,\ |\omega|>\pi$

B.10   Two-dimensional Fourier transform

This section presents some Fourier transforms in two dimensions, along with useful properties (Bracewell, 1965, Hsu, 1973, Spiegel, 1976, Lathi, 1989, Gradshteyn and Ryzhik, 1990, Oberhettinger, 1990).

$f(t, \tau)$	$F(\omega, \phi)$
$\alpha f(t, \tau) + \beta g(t, \tau)$	$\alpha F(\omega, \phi) + \beta G(\omega, \phi)$
$f(at, b\tau)$	$\dfrac{1}{\lvert ab \rvert} F\left(\dfrac{\omega}{a}, \dfrac{\phi}{b}\right)$
$f(-t, -\tau)$	$F(-\omega, -\phi)$
$f^*(t, \tau)$	$F^*(-\omega, -\phi)$
$f(t - \sigma, \tau - \zeta)$	$F(\omega, \phi)\mathrm{e}^{-\mathrm{j}\omega\sigma - \mathrm{j}\phi\zeta}$
$f(t, \tau)\mathrm{e}^{\mathrm{j}\omega_0 t}$	$F(\omega - \omega_0, \phi)$
$f(t, \tau)\cos \omega_0 t$	$\dfrac{1}{2}F(\omega - \omega_0, \phi) + \dfrac{1}{2}F(\omega + \omega_0, \phi)$
$f(t, \tau)\sin \omega_0 t$	$\dfrac{1}{2\mathrm{j}}F(\omega - \omega_o, \phi) - \dfrac{1}{2\mathrm{j}}F(\omega + \omega_0, \phi)$
$F(t, \tau)$	$(2\pi)^2 f(-\omega, -\phi)$
$\dfrac{\partial}{\partial t}f(t, \tau)$	$\mathrm{j}\omega F(\omega, \phi)$
$\dfrac{\partial}{\partial \tau}f(t, \tau)$	$\mathrm{j}\phi F(\omega, \phi)$
$\dfrac{\partial^2}{\partial t\, \partial \tau}f(t, \tau)$	$-\omega\phi F(\omega, \phi)$
$\dfrac{\partial^n}{\partial t^n}f(t, \tau)$	$(\mathrm{j}\omega)^n F(\omega, \phi)$
$\dfrac{\partial^n}{\partial \tau^n}f(t, \tau)$	$(\mathrm{j}\phi)^n F(\omega, \phi)$
$\int_{-\infty}^{t} f(t, \tau)\mathrm{d}t$	$\dfrac{1}{\mathrm{j}\omega}F(\omega, \phi) + \pi F(0, \phi)\delta(\omega, \phi)$
$\int_{-\infty}^{\infty} \int_{-\infty}^{\infty} f(t, \tau)\mathrm{d}t\mathrm{d}\tau$	$F(0, 0)$
$\int_{-\infty}^{\infty} \int_{-\infty}^{\infty} \lvert f(t, \tau)\rvert^2 \mathrm{d}t\mathrm{d}\tau$	$\int_{-\infty}^{\infty} \int_{-\infty}^{\infty} \lvert F(\omega, \phi)\rvert^2 \mathrm{d}\omega\mathrm{d}\phi$
$\int_{-\infty}^{\infty} \int_{-\infty}^{\infty} f(t, \tau)g^*(t, \tau)\mathrm{d}t\mathrm{d}\tau$	$\int_{-\infty}^{\infty} \int_{-\infty}^{\infty} F(\omega, \phi)G^*(\omega, \phi)\mathrm{d}\omega\mathrm{d}\phi$
$f(t, \tau) * g(t, \tau)$	$F(\omega, \phi)G(\omega, \phi)$
$f(t, \tau) * f^*(-t, -\tau)$	$\lvert F(\omega, \phi)\rvert^2$

B.11   Use of the radiofrequency spectrum

The standardization organizations and regulatory agencies are responsible for establishing regulations and enforcing them. Table B.1 presents some regulatory agencies, institutes, and consortia.

Tables B.2–B.5 present the allocation of services in part of the radiofrequency spectrum, including some of the power levels standardized by the Federal Communications Commission (FCC) in the USA (Stremler, 1982).

Table B.1. Standardization organizations

ANSI	American National Standards Institute
CEPT	Council of European PTTs
EIA	Electronics Industry Association
ETSI	European Telecommunication Standardization Institute
FCC	Federal Communications Commission
IEEE	Institute of Electrical and Electronics Engineers
ITU	International Telecommunication Union
JTC	Joint Technical Committee
RACE	Research and Development into Advanced Communications Technology for Europe
RCR	Research Center for Radio
TIA	Telecommunications Industry Association

Table B.2. Frequency bands

VLF (very low frequencies)	3 Hz up to 30 kHz
LF (low frequencies)	30 kHz up to 300 kHz
MF (medium frequencies)	300 kHz up to 3 MHz
HF (high frequencies)	3 MHz up to 30 MHz
VHF (very high frequencies)	30 MHz up to 300 MHz
UHF (ultrahigh frequencies)	300 MHz up to 3 GHz
SHF (superhigh frequencies)	3 GHz up to 30 GHz
EHF (extra-high frequencies)	30 GHz up to 300 GHz

Table B.3(a). Permitted power levels: radio

AM Band	Transmission class	Average power, kW
	Local	0.1–1.0
	Regional	0.5–5.0
	Line of sight	0.25–50

FM band: 0.25, 1, 3, 5, 10, 25, 50, 100 kW, it depends on service class (size of community) and coverage area.

Table B.3(b). Permitted power levels: television

Television (video)	Channels	Efective iradiated power (average), kW
	VHF (2–6)	100
	VHF (7–13)	316
	UHF	5 000

Television (audio): 10% (-10dB) to 20% (-7dB) of the video carrier power.

Table B.4. Frequency allocation for VHF television

Channel number	Frequency band (MHz)	Video carrier (MHz)
1		Not used
2	54–60	55.25
3	60–66	61.25
4	66–72	67.25
5	76–82	77.25
6	82–88	83.25
		(Commercial FM, band 88–108)
7	174–180	175.25
8	180–186	181.25
9	186–192	187.25
10	192–198	193.25
11	198–204	199.25
12	204–210	205.25
13	210–216	211.25

Table B.5. Frequency allocation for UHF television

Channel number	Frequency band (MHz)	Video carrier (MHz)	Channel number	Frequency band (MHz)	Video carrier (MHz)
14	470–476	471.25	42	638–644	639.25
15	476–482	477.25	43	644–650	645.25
16	482–488	483.25	44	650–656	651.25
17	488–494	489.25	45	656–662	657.25
18	494–500	495.25	46	662–668	663.25
19	500–506	501.25	47	668–674	669.25
20	506–512	507.25	48	674–680	675.25
21	512–518	513.25	49	680–686	681.25
22	518–524	519.25	50	686–692	687.25
23	524–530	525.25	51	692–698	693.25
24	530–536	531.25	52	698–704	699.25
25	536–542	537.25	53	704–710	705.25
26	542–548	543.25	54	710–716	711.25
27	548–554	549.25	55	716–722	717.25
28	554–560	555.25	56	722–728	723.25
29	560–566	561.25	57	728–734	729.25
30	566–572	567.25	58	734–740	735.25
31	572–578	573.25	59	740–746	741.25
32	578–584	579.25	60	746–752	747.25
33	584–590	585.25	61	752–758	753.25
34	590–596	591.25	62	758–764	759.25
35	596–602	597.25	63	764–770	765.25
36	602–608	603.25	64	770–776	771.25
37	608–614	609.25	65	776–782	777.25
38	614–620	615.25	66	782–788	783.25
39	620–626	621.25	67	788–794	789.25
40	626–632	627.25	68	794–800	795.25
41	632–638	633.25	69	800–806	801.25

Appendix C
Random signals and noise

A random signal, or a stochastic process, is an extension of the concept of a random variable, involving a sample space, a set of signals, and the associated probability density functions. Figure C.1 illustrates a random signal and its associated probability density function.

A stochastic process $X(t)$ defines a random variable for each point on the time axis, and it is said to be stationary if the probability densities associated with the process are time-independent.

C.1 The autocorrelation function

The autocorrelation function is an important joint moment of the random process $X(t)$, and is defined as

$$R_X(t, \sigma) = E[X(t)X(\sigma)], \tag{C.1}$$

in which $E[\cdot]$ represents the expected value of the process.

The random process is called wide-sense stationary if its autocorrelation depends only on the interval of time separating $X(t)$ and $X(\sigma)$, i.e. it depends only on $\tau = \sigma - t$. Equation (C.1) in this case can be written as

$$R_X(\tau) = E[X(t)X(t + \tau)]. \tag{C.2}$$

In general, the statistical mean of a time signal is a function of time. Thus, the mean value

$$E[X(t)] = m_X(t),$$

the power

$$E[X^2(t)] = P_X(t),$$

and the autocorrelation

$$R_X(\tau, t) = E[X(t)X(t + \tau)]$$

are, in general, time-dependent.

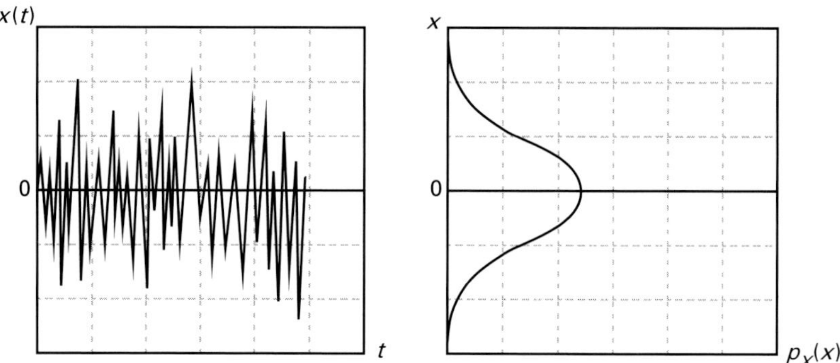

Figure C.1   Example of a random process and its corresponding probability density function

However, there exists a set of signals the mean values of which are time-independent. These signals are called stationary signals. A signal is stationary whenever its probability density function is time-independent, i.e. whenever $p_X(x, t) = p_X(x)$.

C.1.1   Properties of the autocorrelation function

The autocorrelation function has the following properties:

(1) total power: $R_X(0) = E[X^2(t)] = P_X$;
(2) average power or DC level: $R_X(\infty) = E^2[X(t)]$;
(3) autocovariance: $C_X(\tau) = R_X(\tau) - E^2[X(t)]$;
(4) variance: $V[X(t)] = E[(X(t) - E[X(t)])^2] = E[X^2(t)] - E^2[X(t)]$
     or $P_{AC}(0) = R_X(0) - R_X(\infty)$;
(5) maximum at the origin: $R_X(0) \geq |R_X(\tau)|$;
(6) symmetry: $R_X(\tau) = R_X(-\tau)$;
(7) mean value of the signal: $E[X(t)] = \sqrt{R_X(\infty)}$.

C.2   The power spectral density

The autocorrelation function can be used to compute the power spectral density of a signal. It suffices to apply the Wiener–Khintchin theorem to obtain

$$S_X(\omega) = \int_{-\infty}^{+\infty} R_X(\tau) e^{-j\omega\tau} \, d\tau. \tag{C.3}$$

The function $S_X(\omega)$ is called the power spectral density of the random process. The inverse Fourier transform gives the autocorrelation function:

$$R_X(\tau) = \frac{1}{2\pi} \int_{-\infty}^{+\infty} S_X(\omega) e^{j\omega\tau} \, d\omega. \tag{C.4}$$

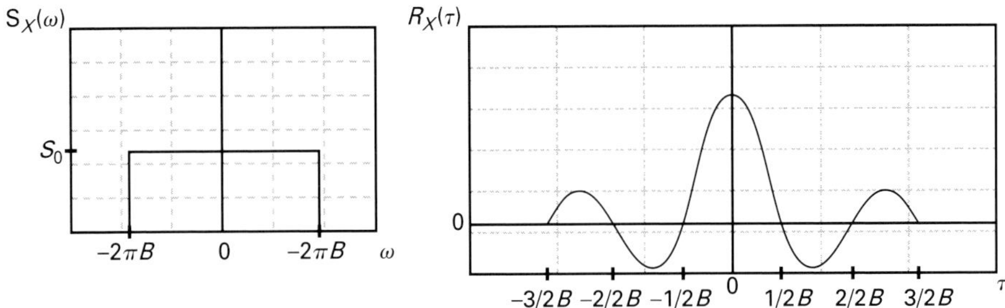

Figure C.2   Power spectral density and the autocorrelation function for a band-limited signal

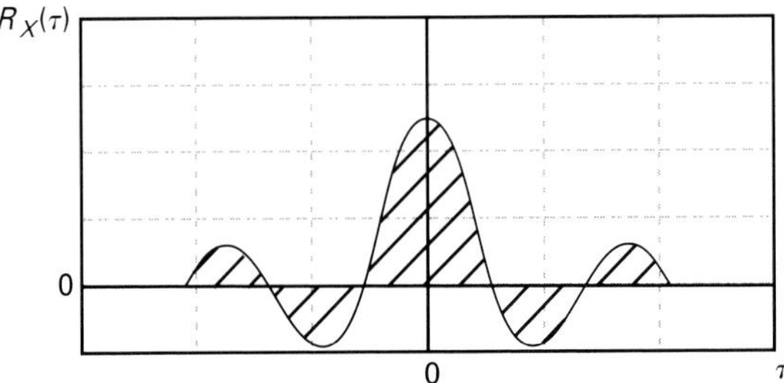

Figure C.3   Area under the curve of the autocorrelation function

Figure C.2 illustrates the power spectral density and the autocorrelation function for a band-limited signal.

C.2.1   Properties of the power spectral density

A few properties of the power spectral density are listed in the following.

- The area under the curve of the power spectral density is equal to the total power of the random process:

$$P_X = \frac{1}{2\pi} \int_{-\infty}^{+\infty} S_X(\omega)\,d\omega. \tag{C.5}$$

- $S_X(0) = \int_{-\infty}^{\infty} R_X(\tau)\,d\tau.$
  Figure C.3 illustrates the fact that the area under the curve of the autocorrelation function is the value of the power spectral density at the origin.

- If $R_X(\tau)$ is real and even, then

$$S_X(\omega) = \int_{-\infty}^{\infty} R_X(\tau)[\cos \omega\tau - \text{jsin}\, \omega\tau]\, d\tau,$$

$$= \int_{-\infty}^{\infty} R_X(\tau) \cos \omega\tau\, d\tau, \tag{C.6}$$

i.e., $S_X(\omega)$ is real and even.
- $S_X(\omega) \geq 0$, since the density reflects a power measure.

The cross-correlation between two random processes $X(t)$ and $Y(t)$ is defined as

$$R_{XY}(\tau) = E[X(t)Y(t+\tau)], \tag{C.7}$$

which leads to the definition of the cross-power spectral density $S_{XY}(\omega)$.

$$S_{XY}(\omega) = \int_{-\infty}^{+\infty} R_{XY}(\tau)e^{-j\omega\tau}\, d\tau. \tag{C.8}$$

If two stationary processes, $X(t)$ and $Y(t)$, are added to form a new process $Z(t) = X(t) + Y(t)$, then the autocorrelation function of the new process is given by

$$R_Z(\tau) = E[Z(t) \cdot Z(t+\tau)]$$

$$= E[(x(t) + y(t))(x(t+\tau) + y(t+\tau))],$$

which implies

$$R_Z(\tau) = E[x(t)x(t+\tau) + y(t)y(t+\tau) + x(t)y(t+\tau) + x(t+\tau)y(t)].$$

By applying properties of the expected value to the above expression it follows that

$$R_Z(\tau) = R_X(\tau) + R_Y(\tau) + R_{XY}(\tau) + R_{YX}(\tau). \tag{C.9}$$

If the processes $X(t)$ and $Y(t)$ are uncorrelated, then $R_{XY}(\tau) = R_{YX}(\tau) = 0$. Thus, $R_Z(\tau)$ can be written as

$$R_Z(\tau) = R_X(\tau) + R_Y(\tau), \tag{C.10}$$

and the associated power can be written as

$$P_Z = R_Z(0) = P_X + P_Y.$$

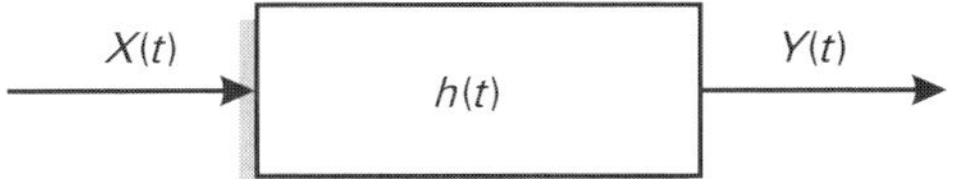

Figure C.4   A linear system fed with a random input signal

The corresponding power spectral density is given by

$$S_Z(\omega) = S_X(\omega) + S_Y(\omega).$$

C.3   Linear systems

The theory of stochastic processes can be useful when analyzing linear systems, and provides more general and more interesting solutions than those resulting from classical analysis. This section deals with the response of linear systems to a random input $X(t)$.

For a linear system, as illustrated in Figure C.4, the Fourier transform of its impulse response $h(t)$ is given by

$$H(\omega) = \int_{-\infty}^{\infty} h(t) e^{-j\omega t} \, dt. \tag{C.11}$$

The linear system response $Y(t)$ is obtained by means of the convolution of the input signal with the impulse response as follows:

$$Y(t) = X(t) * h(t) \Rightarrow Y(t) = \int_{-\infty}^{\infty} X(t - \alpha) \, h(\alpha) \, d\alpha$$

$$= \int_{-\infty}^{\infty} X(\alpha) \, h(t - \alpha) \, d\alpha.$$

C.3.1   Expected value of output signal

The mean value of the random signal at the output of a linear system is calculated as follows:

$$E[Y(t)] = E\left[\int_{-\infty}^{\infty} X(t - \alpha) \, h(\alpha) \, d\alpha \right] = \int_{-\infty}^{\infty} E[X(t - \alpha)] \, h(\alpha) \, d\alpha.$$

Considering the random signal $X(t)$ to be strict-sense stationary, it follows that $E[X(t - \alpha)] = E[X(t)] = m_X$, and thus

$$E[Y(t)] = m_X \int_{-\infty}^{\infty} h(\alpha) \, d\alpha = m_X H(0),$$

in which $H(0) = \int_{-\infty}^{\infty} h(\alpha) \, d\alpha$ follows from (C.11) computed at $\omega = 0$. Therefore, the mean value of the output signal depends only on the mean value of the input signal and on the value assumed by the transfer function at $\omega = 0$.

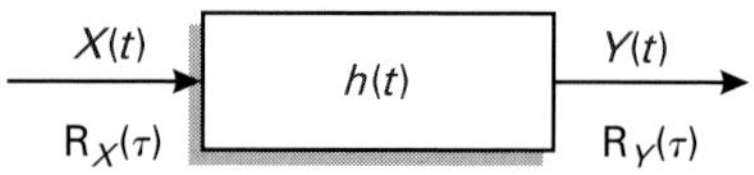

Figure C.5   A linear system and input–output relationships

C.3.2   The response of linear systems to random signals

The relationship between the input and the output of a linear system was shown earlier to be given by

$$Y(t) = \int_{-\infty}^{\infty} X(\rho)\, h(t-\rho)\, \mathrm{d}\rho = \int_{-\infty}^{\infty} X(t-\rho)\, h(\rho)\, \mathrm{d}\rho = X(t) * h(t).$$

The output autocorrelation function can be calculated directly from its definition as

$$R_Y(\tau) = E[Y(t)Y(t+\tau)]$$

$$= E\left[\int_{-\infty}^{\infty} X(t-\rho)\, h(\rho)\, \mathrm{d}\rho \cdot \int_{-\infty}^{\infty} X(t+\tau-\sigma)\, h(\sigma)\, \mathrm{d}\sigma\right]$$

$$= \int_{-\infty}^{\infty}\int_{-\infty}^{\infty} E[X(t-\rho)X(t+\tau-\sigma)] \cdot h(\rho) \cdot h(\sigma)\, \mathrm{d}\rho\, \mathrm{d}\sigma$$

$$= \int_{=\infty}^{\infty}\int_{-\infty}^{\infty} R_{XX}(\tau+\rho-\sigma)\, h(\rho)\, h(\sigma)\, \mathrm{d}\rho\, \mathrm{d}\sigma.$$

In general the output power spectral density can be computed by applying the Wiener–Khintchin theorem $S_Y(\omega) = \mathcal{F}[R_Y(\tau)]$:

$$S_Y(\omega) = \int_{-\infty}^{\infty}\int_{-\infty}^{\infty}\int_{-\infty}^{\infty} R_X(\tau+\rho-\sigma)\, h(\rho)\, h(\sigma) \cdot \mathrm{e}^{-\mathrm{j}\omega\tau}\, \mathrm{d}\rho\, \mathrm{d}\sigma\, \mathrm{d}\tau.$$

Integrating with regard to τ, it follows that

$$S_Y(\omega) = \int_{-\infty}^{\infty}\int_{-\infty}^{\infty} S_X(\omega)\mathrm{e}^{\mathrm{j}\omega(\rho-\sigma)}\, h(\rho)\, h(\sigma)\, \mathrm{d}\rho\, \mathrm{d}\sigma.$$

Finally, removing $S_X(\omega)$ from the double integral and then separating the two variables in this double integral it follows that

$$S_Y(\omega) = S_X(\omega) \int_{-\infty}^{\infty} h(\rho)\mathrm{e}^{\mathrm{j}\omega\rho}\, \mathrm{d}\rho \int_{-\infty}^{\infty} h(\sigma)\mathrm{e}^{-\mathrm{j}\omega\sigma}\, \mathrm{d}\sigma$$

$$= S_X(\omega) \cdot H(-\omega) \cdot H(\omega).$$

Therefore, $S_Y(\omega) = S_X(\omega) \cdot |H(\omega)|^2$ will result whenever the system impulse response is a real function.

Consider again white noise with autocorrelation function $R_X(\tau) = \delta(\tau)$ applied to a linear system. The white noise power spectral density is calculated as follows:

$$S_X(\omega) = \int_{-\infty}^{\infty} R_X(\tau)e^{-j\omega\tau}\,d\tau = \int_{-\infty}^{\infty} \delta(\tau)e^{-j\omega\tau}\,d\tau = 1,$$

from which it follows that

$$S_Y(\omega) = |H(\omega)|^2.$$

Other relationships can be derived between the different correlation functions, as illustrated in Figure C.6. The correlation measures between input and output (input–output cross-correlation) and correlation between output and input can also be calculated from the input signal autocorrelation.

The correlation between the input and the output can be calculated with the formula

$$R_{XY}(\tau) = E[X(t)Y(t+\tau)],$$

and in an analogous manner, the correlation between the output and the input can be calculated from

$$R_{YX}(\tau) = E[Y(t)X(t+\tau)].$$

For a linear system, the correlation between the output and the input is given by

$$R_{YX}(\tau) = E\left[\int_{-\infty}^{\infty} X(t-\rho)h(\rho)\,d\rho \cdot X(t+\tau)\right].$$

Exchanging the order of the expected value and integral computations, due to their linearity, it follows that

$$R_{YX}(\tau) = \int_{-\infty}^{\infty} E[X(t-\rho)X(t+\tau)]\,h(\rho)\,d\rho = \int_{-\infty}^{\infty} R_X(\tau+\rho)\,h(\rho)\,d\rho.$$

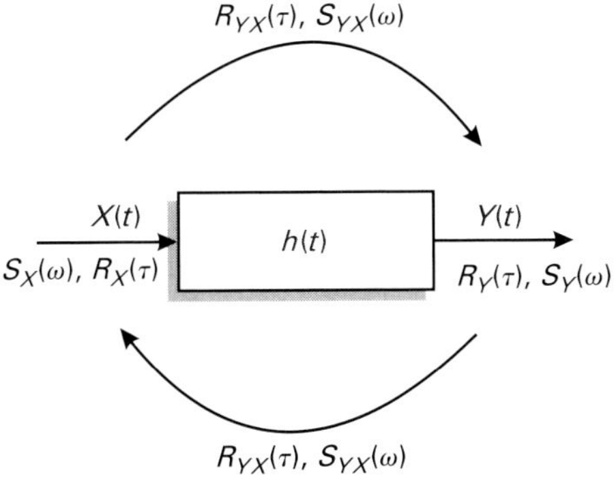

Figure C.6 A linear system and its correlation measures

In a similar manner, the correlation between the input and the output is calculated as

$$R_{XY}(\tau) = E\left[X(t) \cdot \int_{-\infty}^{\infty} X(t + \tau - \rho)\, h(\rho)\, d\rho \right]$$
$$= \int_{-\infty}^{\infty} E[X(t)X(t + \tau - \rho)]\, h(\rho)\, d\rho,$$

and finally,

$$R_{XY}(\tau) = \int_{-\infty}^{\infty} R_X(\tau - \rho)\, h(\rho)\, d\rho.$$

Therefore

$$R_{XY}(\tau) = R_X(\tau) * h(\tau)$$

and

$$R_{YX}(\tau) = R_X(\tau) * h(-\tau).$$

The resulting cross-power spectral densities between input–output and between output–input, respectively, are given by

$$S_{XY}(\tau) = S_X(\omega) \cdot H(\omega),$$
$$S_{YX}(\tau) = S_X(\omega) \cdot H^*(\omega).$$

C.4   Analysis of the digital random signal

Digital television relies on the transmission of discrete signals. A mathematical formulation for the digital signal, including the computation of the autocorrelation function and the power spectral density, is presented in the following.

The digital signal, which can be produced by the digitization of the audio or video signals, can be expressed mathematically as

$$m(t) = \sum_{k=-\infty}^{\infty} m_k p(t - kT_b), \tag{C.12}$$

in which m_k represents the kth randomly generated symbol from the discrete alphabet, $p(t)$ is the pulse function that shapes the transmitted signal, and T_b is the bit interval.

C.4.1   Autocorrelation of the digital signal

The autocorrelation function for signal $m(t)$, which can be non-stationary, is given by the formula

$$R_M(\tau, t) = E[m(t)m(t + \tau)]. \tag{C.13}$$

Substituting $m(t)$ into (C.13),

$$R_M(\tau, t) = E\left[\sum_{k=-\infty}^{\infty}\sum_{i=-\infty}^{\infty} m_k p(t - kT_b)m_j p(t + \tau - iT_b)\right]. \tag{C.14}$$

The expected value operator applies directly to the random signals, because of the linearity property, giving

$$R_M(\tau, t) = \sum_{k=-\infty}^{\infty}\sum_{i=-\infty}^{\infty} E\left[m_k m_j\right] p(t - kT_b)p(t + \tau - iT_b). \tag{C.15}$$

In order to eliminate the time dependency, the time average is taken in Equation C.15, producing

$$R_M(\tau) = \frac{1}{T_b}\int_0^{T_b} R_M(\tau, t)\mathrm{d}t, \tag{C.16}$$

or, equivalently,

$$R_M(\tau) = \frac{1}{T_b}\int_0^{T_b}\sum_{k=-\infty}^{\infty}\sum_{i=-\infty}^{\infty} E\left[m_k m_j\right] p(t - kT_b)p(t + \tau - iT_b)\mathrm{d}t. \tag{C.17}$$

The integral and summation operations can be exchanged, resulting in

$$R_M(\tau) = \frac{1}{T_b}\sum_{k=-\infty}^{\infty}\sum_{i=-\infty}^{\infty} E\left[m_k m_j\right]\int_0^{T_b} p(t - kT_b)p(t + \tau - iT_b)\mathrm{d}t. \tag{C.18}$$

The discrete autocorrelation is defined as

$$R(k - i) = E\left[m_k m_j\right], \tag{C.19}$$

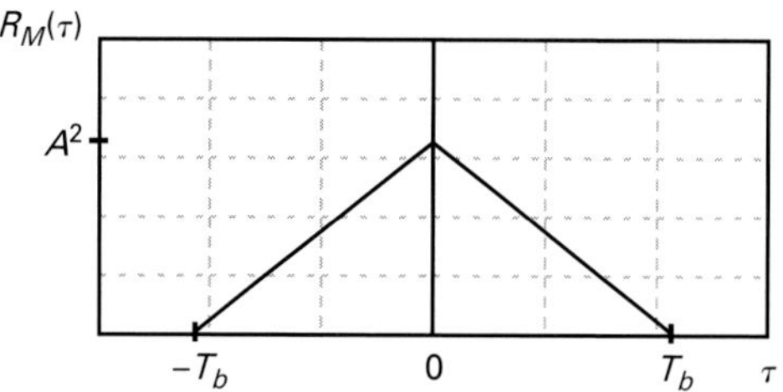

Figure C.7   Autocorrelation for the digital signal

and the signal autocorrelation can be written as

$$R_M(\tau) = \frac{1}{T_b} \sum_{k=-\infty}^{\infty} \sum_{i=-\infty}^{\infty} R(k-i) \int_0^{T_b} p(t - kT_b)p(t + \tau - iT_b)\mathrm{d}t. \tag{C.20}$$

For a rectangular pulse, with independent and equiprobable symbols, the autocorrelation function is given by

$$R_M(\tau) = A^2 \left[1 - \frac{|\tau|}{T_b}\right][u(\tau + T_b) - u(\tau - T_b)], \tag{C.21}$$

in which T_b is the bit interval and A represents the pulse amplitude.

This function has a triangular shape, as shown in Figure C.7. Its maximum occurs at the origin (signal power) and is equal to A^2. The autocorrelation decreases linearly with the time interval, and reaches zero at time T_b.

C.4.2   Spectrum of the digital signal

The Fourier transform of the autocorrelation function gives the power spectral density for the digital signal. It can be obtained from (C.20):

$$S_M(\omega) = \int_{-\infty}^{\infty} R_M(\tau)\mathrm{e}^{-\mathrm{j}\omega\tau}\,\mathrm{d}\tau, \tag{C.22}$$

Therefore,

$$S_M(\omega) = \frac{1}{T_b} \sum_{k=-\infty}^{\infty} \sum_{i=-\infty}^{\infty} R(k-i) \int_{-\infty}^{\infty} \int_0^{T_b} p(t - kT_b)p(t + \tau - iT_b)\mathrm{e}^{-\mathrm{j}\omega\tau}\,\mathrm{d}t\mathrm{d}\tau.$$

$$\tag{C.23}$$

One can compute the Fourier integral changing the order of integration of the shifted pulse,

$$S_M(\omega) = \frac{1}{T_b} \sum_{k=-\infty}^{\infty} \sum_{i=-\infty}^{\infty} R(k-i) \int_0^{T_b} p(t - kT_b)P(\omega)\mathrm{e}^{-\mathrm{j}\omega(kT_b - t)}\,\mathrm{d}t. \tag{C.24}$$

The term $P(\omega)\mathrm{e}^{-\mathrm{j}\omega kT_b}$ is independent of time and can be taken out of the integral, i.e.

$$S_M(\omega) = \frac{1}{T_b} \sum_{k=-\infty}^{\infty} \sum_{i=-\infty}^{\infty} R(k-i)P(\omega)\mathrm{e}^{-\mathrm{j}\omega kT_b} \int_0^{T_b} p(t - kT_b)\mathrm{e}^{\mathrm{j}\omega t}\,\mathrm{d}t. \qquad \text{(C.25)}$$

Computing the integral in (C.25) gives

$$S_M(\omega) = \frac{1}{T_b} \sum_{k=-\infty}^{\infty} \sum_{i=-\infty}^{\infty} R(k-i)P(\omega)P(-\omega)\mathrm{e}^{-\mathrm{j}\omega(k-i)T_b}. \qquad \text{(C.26)}$$

From the previous equation, the shape of the spectrum for the random digital signal depends on the pulse shape, defined by $P(\omega)$, and also on the way the symbols relate to each other, specified by the discrete autocorrelation function $R(k-i)$.

The signal design involves pulse shaping, and also depends on the correlation between the transmitted symbols, which can be obtained by signal processing.

For a real pulse, one can write $P(-\omega) = P^*(\omega)$, and the power spectral density can be written as

$$S_M(\omega) = \frac{|P(\omega)|^2}{T_b} \sum_{k=-\infty}^{\infty} \sum_{i=-\infty}^{\infty} R(k-i)\mathrm{e}^{-\mathrm{j}\omega(k-j)T_b}, \qquad \text{(C.27)}$$

which can be simplified to

$$S_M(\omega) = \frac{|P(\omega)|^2}{T_b} S(\omega). \qquad \text{(C.28)}$$

Letting $l = k - i$, the summations can be simplified and the power spectral density for the discrete sequence of symbols is given by

$$S(\omega) = \sum_{l=-\infty}^{\infty} R(l)\mathrm{e}^{-\mathrm{j}\omega lT_b}. \qquad \text{(C.29)}$$

For the example of (C.21), the corresponding power spectral density is

$$S_M(\omega) = A^2 T_b \frac{\sin^2(\omega T_b)}{(\omega T_b)^2}, \qquad \text{(C.30)}$$

which is the sample function squared, and shows that the random digital signal has a continuous spectrum that occupies a large portion of the spectrum. The function is sketched in Figure C.8. The first null is a usual measure of the bandwidth, and is given by $\omega_M = \pi/T_b$.

C.5 Noise

Figure C.9 illustrates the power spectral density and the corresponding autocorrelation function for white noise. It is noted that $S_X(\omega) = S_0$, which indicates a uniform

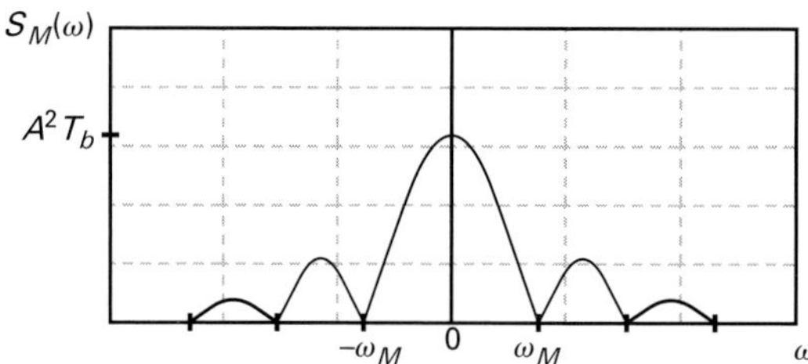

Figure C.8   Power spectral density for the random digital signal

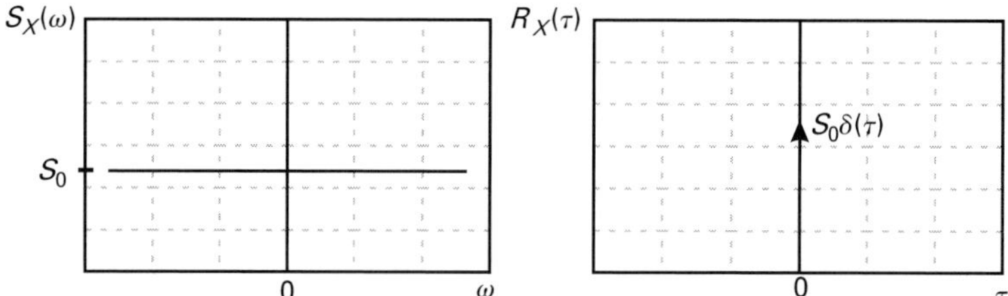

Figure C.9   Power spectral density and the autocorrelation function for white noise

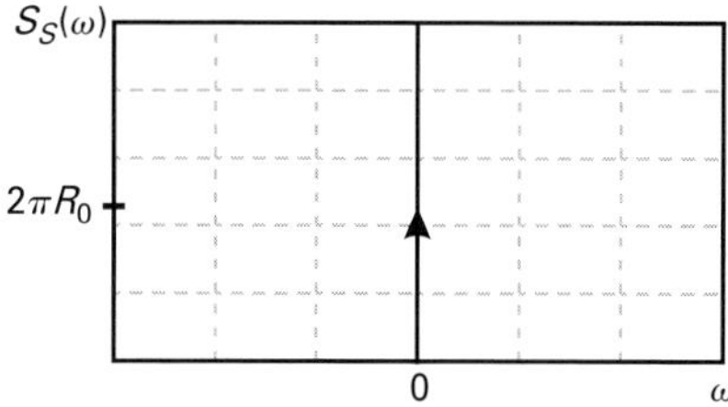

Figure C.10   Power spectral density for a constant signal

distribution for the power density along the spectrum, and $R_X(\tau) = S_0\delta(\tau)$, which shows white noise as the most uncorrelated, or random, of all signals.

The correlation is non-zero for the noise only at $\tau = 0$. On the other hand, Figure C.10 illustrates the power spectral density for a constant signal, with autocorrelation $R_S(\tau) = R_0$. This is, no doubt, the most predictable of all signals.

White noise, with autocorrelation $R_X(\tau) = \delta(\tau)$, can be used to test a linear system. The corresponding autocorrelation function of the output signal is given by

$$R_Y(\tau) = \int_{-\infty}^{\infty} \int_{-\infty}^{\infty} \delta(\tau + \rho - \sigma)\, h(\rho)\, h(\sigma)\, \mathrm{d}\rho\, \mathrm{d}\sigma$$

$$= \int_{-\infty}^{\infty} h(\sigma - \tau) \cdot h(\sigma)\, \mathrm{d}\sigma$$

$$= h(-\tau) * h(\tau).$$

Fourier transforming $R_Y(\tau)$ leads to the following result:

$$R_Y(\tau) = h(-t) * h(t) \iff S_Y(\omega) = H(-\omega) \cdot H(\omega),$$

and for $h(\tau)$ a real function of τ it follows that $H(-\omega) = H^*(\omega)$, and consequently

$$S_Y(\omega) = H(-\omega) \cdot H(\omega) = H^*(\omega) \cdot H(\omega) = |H(\omega)|^2.$$

Summarizing, the output power spectral density is $S_Y(\omega) = |H(\omega)|^2$ when white noise is the input to a linear system.

An interesting noise model can be obtained from the linear system shown in Figure C.11. It is a differentiator, used to detect frequency modulated signals. The output power spectral density for this circuit (or its frequency response) is

$$S_Y(\omega) = |j\omega|^2 \cdot S_X(\omega) = \omega^2 S_X(\omega).$$

It is thus shown that, for frequency modulated signals, the noise at the detector output follows a square law, i.e. the output power spectral density grows with the square of the frequency. In this manner, in a frequency division multiplexing of frequency modulated channels, the noise affects more intensely those channels occupying the higher-frequency region of the spectrum.

Figure C.12 shows, as an illustration of what has been discussed so far about square law noise, the spectrum of a low-pass flat noise (obtained by passing white noise through an ideal low-pass filter). This filtered white noise is applied to the differentiator circuit of the example, which in turn produces at the output the square law noise shown in Figure C.13. Pre-emphasis circuits are used in FM modulators to compensate for the effect of square noise.

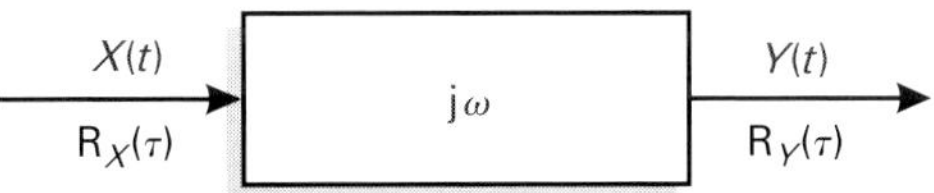

Figure C.11   Differentiator circuit

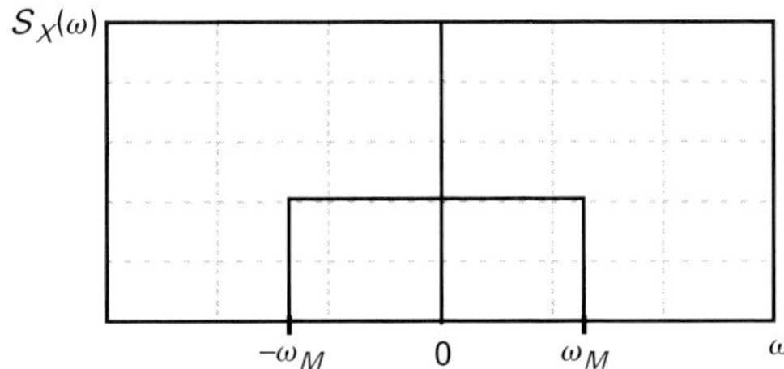

Figure C.12   Low-pass noise

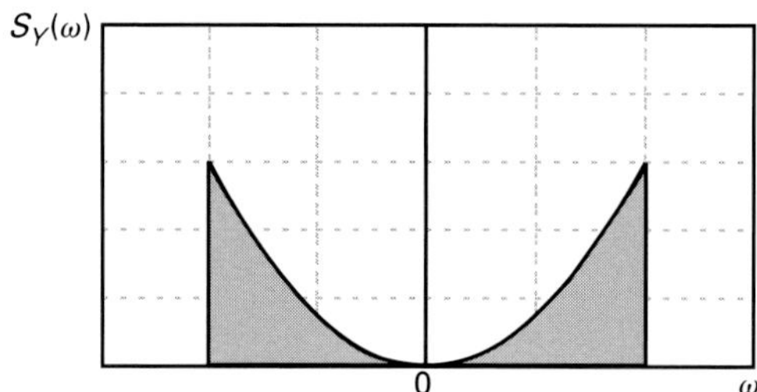

Figure C.13   Quadratic noise

C.5.1   Noise in modulated systems

For the performance analysis of modulated systems, as is the case with digital television, it is usual to define the quadrature representation of the noise $n(t)$, which is given as a function of the in-phase $n_I(t)$ and quadrature $n_Q(t)$ components

$$n(t) = n_I(t)\cos(\omega_c t + \phi) + n_Q(t)\sin(\omega_c t + \phi). \tag{C.31}$$

One can demonstrate that the in-phase and quadrature components have the same variance, which equals the noise $n(t)$ variance $\sigma_I^2 = \sigma_Q^2 = \sigma^2$, in which σ^2 represents the power of the noise $n(t)$.

The autocorrelation can be computed taking into account that the noise quadrature components are uncorrelated, $R_{N_I N_Q}(\tau) = 0$. Therefore,

$$R_N(\tau) = \frac{R_{N_I}(\tau)}{2}\cos\omega_c\tau + \frac{R_{N_Q}(\tau)}{2}\cos\omega_c\tau, \tag{C.32}$$

in which $R_{N_I}(\tau)$ and $R_{N_Q}(\tau)$ represent autocorrelations for the in-phase and quadrature noise components, respectively.

The noise power spectral density is similar to the one obtained for the quadrature amplitude modulation:

$$S_N(\omega) = \tfrac{1}{4}\left[S_{N_I}(\omega + \omega_c) + S_{N_I}(\omega - \omega_c) + S_{N_Q}(\omega + \omega_c) + S_{N_Q}(\omega - \omega_c)\right]. \tag{C.33}$$

Glossary

This glossary presents the usual acronyms and terminology used in communication and digital television standards (Alencar, 1999, 2002a, 2007b, Furiati, 1998, Design, 2001, Skycell, 2001, Mobile Word, 2001 and TIAB2B.com, 2001).

ADPCM   *Adaptive Differential Pulse Code Modulation.* ADPCM is a speech coding method that achieves bit rate reduction through the use of adaptive prediction and adaptive quantization.

ADSL   *Assymetric Digital Subscriber Line.*

ADVEF   *Advanced television Enhancement Forum.*

A/D   *Analog-to-Digital converter or conversion.*

AES   *Audio Engineering Society.*

AF   *Audio Frequency.* The band of frequencies (approximately 20 Hz–20 kHz) that, when transmitted as acoustic waves, can be heard by the normal human ear.

AGC   *Automatic Gain Control.*

AM   *Amplitude Modulation.*

AM-SC   *Amplitude Modulation Suppressed Carrier.*

AM-VSB   *AM Vestigial Sideband Modulation.* Classical method used to modulate carriers in traditional analog video.

AMPS   *Advanced Mobile Phone Service.* The first-generation analog cellular phone system that originated in the USA.

ANSI   *American National Standards Institute.* This is a US-based organization which develops standards and defines interfaces for telecommunications.

ARQ   *Automatic ReQuest for retransmission.* A type of communications link in which the receiver asks the transmitter to resend a block of data when errors are detected.

ASCII   *American Standard Code for Information Interchange.* ASCII data are standard seven-bit code with one parity bit. Otherwise referred to as text. ASCII data can be interchanged between almost every type of computer.

ASIC   *Application-Specific Integrated Circuit.*

ASK   *Amplitude Shift Keying.*

ATM   *Asynchronous Transfer Mode.*

ATSC	*Advanced Television System Committee.*
AWGN	*Additive White Gaussian Noise.*
BCH	*Bose–Chaudhury–Hocquenghem.*
BER	*Bit Error Rate.*
BPSK	*Binary Phase Shift Keying.*
BSS	*Broadcasting Satellite Service.*
BTA	*Broadcasting Technology Association* (Japanese industrial organisation).
BW	*Bandwidth.* The difference between the limiting frequencies within which the performance of a device, in respect to some characteristic, falls within specified limits.
C-QUAM	*Compatible QUAM.*
Cable modem	Modem used with optical fiber of coaxial cable.
CATV	*Cable TV.* A television distribution method in which signals from distant stations are received, amplified, and then transmitted via (coaxial or fiber) cable or microwave links to users.
CCI	*Cochannel Interference.*
CCITT	*Commité Consultatif International Telegraphique et Telephonique.* This is an international organization which develops standards and defines interfaces for telecommunications (now ITU-T).
CCS	*Cascading Style Sheets.*
CDMA	*Code Division Multiple Access.*
CELP	*Code-Excited Linear Prediction.* An analog-to-digital voice coding scheme.
CENELEC	*Comité Européen de Normalisation Électrotechnique.*
C/I	*Carrier-to-Interference Ratio.*
CNR	*Carrier-to-Noise Ratio.* The ratio of the level of the carrier to that of the noise in the intermediate frequency (IF) band before any non-linear process, such as amplitude limitation and detection, takes place.
Codec	*Coder–decoder.* Device that converts analog signals to digital signals and vice versa.
COFDM	*Coded Orthogonal Frequency Division Multiplex.*
CRC	*Cyclic Redundancy Check.* A method of detecting errors in the serial transmission of data. A CRC for a block of data is calculated before it is sent, and is then sent along with the data. A new CRC is calculated on the received data. If the new CRC does not match the one that has been sent with the data then an error has occurred.
D/A	*Digital-to-Analog converter.*
DASE	*DTV Application Software Environment.*

dB	Abbreviation for decibel(s). One tenth of the common logarithm of the ratio of relative powers, equal to 0.1 B (bel).
dBi	Unit used to express the gain related to the isotropic antenna.
dBm	Decibel related to 1 milliwatt
DCT	*Discrete Cosine Transform.*
DEMUX	*Demultiplexer.* The separation of two or more channels previously multiplexed, i.e. the reverse of multiplexing.
DFE	*Decision Feedback Equalizer.*
DOM	*Document Object Model.*
DPCM	*Differential Pulse Code Modulation.* Form of pulse code modulation for which efficiency is enhanced by transmitting the difference between the current signal strength and the previous pulse signal strength rather than the absolute values.
DQPSK	*Differential Quadrature Phase Shift Keying.*
DS-SS	*Direct Sequence Spread Spectrum.*
DSL	*Digital Subscriber Line.* A dedicated link using leased line or wireless for subscriber connection.
DTMF	*Dual Tone MultiFrequency.* The scheme used in "touch tone" telephones whereby each depressed key generates two audio tones.
DTV	*Digital TeleVision.*
DVB	*Digital Video Broadcasting.*
DVD	*Digital Video Disk.*
E_b/N_0	Energy-per-bit to noise density ratio.
EBU	*European Broadcasting Union.*
EDTV	*Enhanced-Definition Television.* The quality of EDTV is between SDTV and HDTV.
EPG	*Electronic Programming Guide.*
ETSI	*European Telecommunications Standards Institute.*
FCC	*Federal Communications Commission.* The US Government board of five presidential appointees that has the authority to regulate all non-federal government interstate telecommunications (including radio and television broadcasting), as well as all international communications that originate or terminate in the USA. Note: Similar authority for regulation of federal government telecommunications is vested in the National Telecommunications and Information Administration (NTIA)
FDMA	*Frequency Division Multiple Access.*
FEC	*Forward Error Correction.* System of error control for data transmission for which the receiving device has the capability to detect and correct any character or code block that contains fewer than a predetermined number of symbols in error.
FFT	*Fast Fourier Transform.*
FH-SS	*Frequency-Hopping Spread Spectrum.*
FM	*Frequency Modulation.*

FO	*Fiber Optics.* The branch of optical technology concerned with the transmission of light through fibers made of transparent materials, such as glasses and plastics.
FSK	*Frequency Shift Keying.*
FTP	*File Transfer Protocol.* File transfer protocol is the TCP/IP standard for remote file transfer.
FTTC	*Fiber To The Curb.*
FTTH	*Fiber-To-The-Home.*
FVC	*Forward Voice Channel.* The radio channel used for communication of voice and user data from the base station to a cellular device.
GEO	*Geostationary Earth Orbit.* Communications system with satellites in geosynchronous orbits $40,000$ km above the Earth, on the Equator plane.
GHz	Gigahertz. A unit of frequency denoting 10^9 Hz.
Ginga	The middleware used in the Brazilian digital television standard (ISDTV).
GMSK	*Gaussian Minimum Shift Keying.*
GSM	*Global System for Mobile Communication.* GSM originally stood for Groupe Speciale Mobile, but has been renamed to Global System for Mobile Communications, an international digital cellular standard.
HDSL	*High-data-rate Digital Subscriber Line.*
HDTV	*High-Definition Television.* Television that has approximately twice the horizontal and twice the vertical emitted resolution specified by the NTSC standard.
HEO	*Highly Elliptical Orbit.* This class of satellites covers orbits which have large eccentricities (are highly elliptical).
HF	*High Frequency.* From 3 to 30 MHz.
HFC	Hybrid network, including fiber and coaxial cable.
Hz	Hertz. Frequency unit equivalent to 1 cycle per second.
IEEE	*Institute of Electrical and Electronic Engineers.*
IEC	*International Electrotechnical Commission.*
IF	*Intermediate Frequency.* A frequency to which a carrier frequency is shifted as an intermediate phase in transmission or reception.
IFFT	*Inverse Fast Fourier Transform.*
ISB	*Independent Sideband.* Double-sideband transmission in which the information carried by each sideband is different.
ISDB	*Integrated Services Digital Broadcasting.*
ISDN	*Integrated Services Digital Network.* A digital phone service which provides two data channels, each with its own phone number
ISDTV	*International System for Digital Television.* HDTV standard developed in Brazil and based on the ISDB standard. Also called ISDB-Tb.

ISI	*Intersymbol Interference.* Digital communication system impairment in which adjacent symbols in a sequence are distorted by frequency response non-idealities, creating dispersion that interferes in the time domain with neighboring symbols.
ISO	*International Organization for Standardization.*
ITU	*International Telecommunications Union.*
ITU-R	*International Telecommunications Union, Radiocommunication Sector.*
ITU-T	*International Telecommunications Union, Telecommunication Standardization Sector.*

JPEG	*Joint Photographic Experts Group.*
JVM	*Java Virtual Machine.*

kbps	Kilobits per second. The use of Sh/s (shannon per second) is recommended.
kHz	Kilohertz.

LAN	*Local Area Network.* A computer network limited to the immediate area, usually the same building or floor of a building.
LED	*Light-Emitting Diode.* A semiconductor device that emits incoherent optical radiation when biased in the forward direction.
LEO	*Low Earth Orbit.* Mobile communications satellite between 700 and 2000 km above the Earth.
LF	*Low Frequency.* Frequency band between 30 kHz and 300 kHz.
LMDS	*Local Multipoint Distribution System.*
LSB	*Least Significant Bit.*

MAC	*Medium Access Control.*
Mbps	Megabits per second.
MCM	*MultiCarrier Modulation.* A technique of transmitting data by dividing it into several interleaved bit streams and using these to modulate several carriers.
MEO	*Medium Earth Orbit.* MEO satellites orbit about 10000 km above the earth.
MHP	*Multimedia Home Platform*
MHz	Megahertz.
ML	*Maximum Likelihood.*
Modem	*Modulator/demodulator.* A device that can encode digital signals from a computer into analog signals that can be transmitted over analog phone lines, and vice versa.
MPEG	*Motion Picture Experts Group.*
MPSK	*M-ary Phase Shift Keying.*
MSB	*Most Significant Bit.*
MSE	*Mean Square Error.*
MSK	*Minimum Shift Keying.*

MUX	*Multiplexer.* A device that combines multiple inputs into an aggregate signal to be transported via a single transmission channel.

NAFTA	*North American Free Trade Agreement.*
NBFM	*Narrow Band Frequency Modulation.*
NEXT	*Near-End Crosstalk.* Impairment typically associated with twisted-pair transmission, in which a local transmitter interferes with a local receiver.
NF	*Noise Figure.* This parameter indicates the amount of noise introduced in a device.
NRZ	*Non-Return-to-Zero.* Data encoding format in which each bit is represented for the entire duration of the bit period as a logic high or a logic low.
NTSC	*National Television System Committee.* The NTSC is the analog television system in use in the USA and in many other countries, including most countries in the Americas and some inparts of East Asia.

OFDM	*Orthogonal Frequency Division Multiplexing.* Multicarrier signaling technique designed to maximize throughput in channels with potentially poor frequency response.
OQPSK	*Offset Quadrature Phase Shift Keying.* QPSK system in which the two bits that compose a QPSK symbol are offset in time by a half-bit period for non-linear amplification.

PABX	*Private Automatic Branch eXchange.*
PAL	*Phase Alternated Line.* A television signal standard (625 lines, 50 Hz, 220 V primary power) used in the United Kingdom, much of the rest of western Europe, several South American countries, some Middle East and Asian countries, several African countries, Australia, New Zealand, and other Pacific island countries.
PAL-M	A modified version of the phase-alternated-line (PAL) television signal standard (525 lines, 50 Hz, 220 V primary power), used in Brazil.
PAM	*Pulse Amplitude Modulation.* Amplitude modulation of a carrier which uses pulses of varying amplitude to transmit information from source to destination.
PCM	*Pulse Code Modulation.* A common way of converting an analog signal to a digital signal. This is done by sampling the signal and coding the sample.
PCS	*Personal Communications Service.*
PDA	*Personal Digital Assistant.*
PID	*Packet Identification.*
PLL	*Phase-Locked Loop.*
PM	*Phase Modulation.*

PMD	*Principle of Majority Decision.*
PN	*Pseudorandom Noise.*
PNG	*Portable Networks Graphics.*
PRA	*Primary Rate Access.* An ISDN access method that uses maximum data rates of 2.048 Mbit/s in Europe, or 1.544 Mbit/s in the USA and Japan.
PSK	*Phase Shift Keying.*
PSTN	*Public Switched Telephone Network.* The traditional voice network infrastructure, including both local service and long distance service.
QAM	*Quadrature Amplitude Modulation.*
QCIF	*Quarter Common Interchange Format.*
QoS	*Quality of Service.* A parameter used to describe the attributes of a variety of network functions.
QPSK	*Quadrature Phase Shift Keying.*
QUAM	*Quadrature Amplitude Modulation.*
RACE	*Research and Development for Advanced Communications.* A European Community endeavor aimed at creating advanced communications networks.
RF	*Radio Frequency.*
RGB	*Red–Blue–Green.* Related to the use of three separate signals to carry the red, green, and blue components, respectively, of a color video image.
RLE	*Run-Length Encoding.*
RS	*Reed–Solomon (code).*
Rx	*Receiver.*
RZ	*Return-to-Zero.* Data encoding format in which each bit is represented for only a portion of the bit period as a logic high or a logic low, and what remains of the duration of the bit returns to logic zero.
SAN	*Small Area Network.* A network, generally limited to tens of meters, which uses specialized communications methods and is applied in such areas as process control and other specific real-time computer applications.
SBrT	*Brazilian Telecommunications Society.*
SBTVD	*Brazilian High-Definition Television Standard.* Also called ISDTV and ISDB-Tb
S-CDMA	*Synchronous Code Division Multiple Access.*
SDH	*Synchronous Digital Hierarchy.*
SDMA	*Space Division Multiple Access.*
SDTV	*Standard-Definition Television.*
SECAM	*Séquentielle Couleur avec Mémoire,* which means sequential color with memory; it is an analog color system, created in France by Henri de France.
SHF	*Super High Frequency.* Frequency band from 3 to 30 GHz.

SIT	*Satellite Interactive Terminal.*
SNR	*Signal-to-Noise Ratio.*
SONET	*Synchronous Optical Network.*
SQNR	*Signal-to-Quantization-Noise Ratio.*
SSB	*Single Sideband.*
STB	*Set-Top Box.* Equipment to receive digital television signals and convert them to be viewed on an analog TV set.
SWIFT	*Society for Worldwide Interbank Financial Telecommunications.* A network designed and operated to support the information transfer needs of banks and financial institutions.
TACS	*Total Access Communication System.*
TCM	*Trellis-Coded Modulation.*
TCP/IP	*Transmission Control Protocol/Internet Protocol.* This is the suite of protocols that defines the Internet. Originally designed for the UNIX operating system, TCP/IP software is now available for every major type of computer operating system.
TDM	*Time Division Multiplexing.*
TDMA	*Time Division Multiple Access.*
TIA	*Telecommunications Industry Association.* An organization concerned with various standards and aspects of the telecommunications industry.
TS	*Transport Stream.*
TWTA	*Traveling Wave Tube Amplifier.* An amplifier technology designed for generation of very high-power microwave signals, such as those used in satellite communication applications.
Tx	Transmitter.
UHF	*UltraHigh Frequency.* Frequency band from 300 MHz to 3 GHz.
ULF	*UltraLow Frequency.* Frequency band from 300 Hz to 3000 Hz.
VBR	*Variable Bit Rate.*
VCO	*Voltage-Controlled Oscillator.*
VDSL	*Very high-speed Digital Subscriber Line.* The next-generation twisted-pair technology (after ADSL) that targets higher data transmission rates than ADSL in exchange for shorter guaranteed distances.
VHF	*Very High Frequency.* Frequency band from 30 Hz to 300 Hz.
VLF	*Very Low Frequency.* Frequency band from 3 Hz to 30 Hz.
VOD	*Video On Demand.*
VSWR	*Voltage Standing Wave Ratio.* Measure of the RF interface quality between adjacent RF circuits that require adequate impedance matching for proper transfer of electrical energy at high frequencies. In a transmission line, the ratio of maximum to minimum voltage in a standing wave pattern.

WAN *Wide Area Network*. A network which covers a larger geographical area than a LAN and where telecommunications links are implemented.

WAP *Wireless Application Protocol.*
WBFM *Wide Band Frequency Modulation.*
WDM *Wavelength Division Multiplexing.*
WLL *Wireless Local Loop*. A wireless system meant to bypass a local landline telephone system.

X.25 A standard for packet transmission.

References

Alencar, M. S., Estimation of the power spectral density for Embratel's FDM-FM system (in Portuguese). In *Anais do Simpósio Brasileiro de Telecomunicações*, pp. 297–300, Campina Grande, Brazil (1988).

Alencar, M. S., Measurement of the probability density function of communication signals. In *Proceedings of the IEEE Instrumentation and Measurement Technology Conference – IMTC'89*, pp. 513–516, Washington, DC (1989).

Alencar, M. S., A model for evaluating the quantization noise power spectral density. In *Anais do IX Simpósio Brasileiro de Telecomunicações*, pp. 10.4.1–10.4.3, São Paulo, Brazil (1991a).

Alencar, M. S., Power spectrum density of quantization noise for uniform quantizers. In *Proceedings of the IASTED International Symposium on Computers, Electronics, Communication and Control*, pp. 274–275, Calgary, Canada (1991b).

Alencar, M. S., A frequency domain approach to the optimization of scalar quantizers. In *Proceedings of the IEEE International Symposium on Information Theory, San Antonio, USA*, p. 440 (1993a).

Alencar, M. S., A new bound on the estimation of the probability density function using spectral analysis. In *Proceedings of the IEEE International Symposium on Information Theory, San Antonio, USA*, p. 190 (1993b).

Alencar, M. S., Scalar quantization (in Portuguese). Thesis Approved for the Position of Chair Professor in the Areas of Electromagnetism, Microwaves and Telecommunications, Department of Electrical Engineering, Federal University of Paraíba (1995).

Alencar, M. S., Optimization of quantizers in the frequency domain. *Electronic Letters*, **34(2)**: 155–156 (1998).

Alencar, M. S., *Principles of Communications* (in Portuguese). João Pessoa, Brazil: University Press, UFPB (1999).

Alencar, M. S., *Communication Systems* (in Portuguese). São Paulo, Brazil: Érica Publisher Ltd., (2001).

Alencar, M. S., *Digital Telephony*, fourth edition (in Portuguese). Érica Publisher Ltd., São Paulo, Brazil: Érica Publisher Ltd., (2002a).

Alencar, M. S., On the digital television waves – Part III (in Portuguese). Article for an electronic newspaper over the Internet, *Jornal do Commercio Newspaper On Line*, Recife, Brazil (2002b).

Alencar, M. S., Digital television via cellular (in Portuguese). *Jornal do Commercio On Line*, Recife, Brazil (2005a).

Alencar, M. S., The secret history of digital television in Brazil (in Portuguese). *Jornal do Commercio On Line*, Recife, Brazil (2005b).

Alencar, M. S., What is the cost of digital television? (in Portuguese). *Jornal do Commercio On Line*, Recife, Brazil (2006).

Alencar, M. S., *Digital Cellular Telephony*, Second Edition (in Portuguese). São Paulo, Brazil: Érica Publisher Ltd. (2007a).

Alencar, M. S., *Digital Television* (in Portuguese). São Paulo, Brazil: Érica Pulisher Ltd., (2007b).

Alencar, M. S. and da Rocha Jr., V. C., *Communication Systems*. Boston, USA: Springer (2005).

Alencar, M. S. and Neto, B. G. A., Estimation of the probability density function by spectral analysis: a comparative study. In *Proceedings of the Treizième Colloque sur le Traitement du Signal et des Images – GRETSI*, pp. 377–380, Juan-Les-Pins, France (1991).

Alencar, M. S., Alencar, T. T., and Lopes, W. T. A., What father Landell de Moura used to do in his spare time. In *Proceedings of the 2004 IEEE Conference on the History of Electronics*, volume published on CD, Bletchley Park, England (2004).

ARIB, ISDB-T: Satellite television digital broadcasting transmission – STD-B20. Technical report (2004a).

ARIB, ISDB-T: Terrestrial television digital broadcasting transmission – STD-B31. Technical report (2004b).

ARIB, *ARIB STD-B23, Application Execution Engine Platform for Digital Broadcasting* (2004c).

ARIB Technical report, available at http://www.arib.or.jp, access February (2006).

Armstrong, E. H., A method of reducing disturbances in radio signaling by a system of frequency modulation. *Proceedings of the IRE*, **24**:689–674 (1936).

Asami, H. and Sasaki, M., Outline of ISDB systems. *IEEE Proceedings*, **94**:248–250 (2006).

ATSC, ATSC Standard A/52. Digital Audio Compression (AC-3) (1995a).

ATSC, ATSC Standard A/53. ATSC Digital Television Standard (1995b).

ATSC, ATSC Standard A/53B. ATSC Digital Television Standard, Revision B (1995c).

ATSC, ATSC Standard A/80. Modulation and Coding Requeriments for Digital TV (DTV) Applications over Satellite (1999).

ATSC, ATSC Standard A/90. Data Broadcast Standard (2000a).

ATSC, ATSC Standard A/90. Data Broadcast Standard (2000b).

ATSC, Advanced Common Application Platform (ACAP). Technical report, ATSC Forum (2005).

ATSC, ATSC Standard. Advanced Television Systems Committee, Inc., www.atsc.org (2008).

Banta, E. D., On the autocorrelation function of quantized signal plus noise. *IEEE Transactions on Information Theory*, **11(2)**:114–117 (1965).

Bedicks, J. G., Yamada, F., Sukys, F., *et al.*, *IEEE Transactions on Broadcasting*, **52(1)**: 38–44 (2006).

Bedrosian, E., Weighted PCM. *IRE Transactions on Information Theory*, **4(1)**:45–49 (1958).

Benedetto, S. and Biglieri, E., *Principles of Digital Transmission*. New York: Kluwer Academic/ Plenum Publishers (1999).

Bennett, W. R., Spectra of quantized signals. *The Bell System Technical Journal*, **27**:446–472 (1948).

Benson, K. B. and Fink, D. G., *HDTV – Advanced Television for the 1990s*. Intertext Publications. New York, USA: McGraw-Hill Book Company, Inc. (1991).

Berlekamp, E., *Algebraic Coding Theory*. New York, USA: McGraw-Hill (1968).

Berrou, C. and Glavieux, A., Near optimum error correcting coding and decoding: turbo-codes. *IEEE Transactions on Communications*, **44(10)**:1261–1271 (1996).

Berrou, C., Glavieux, A., and Thitimajshima, P., Near Shannon limit error-correcting coding and decoding: turbo codes. In *Proceedings of the 1993 IEEE International Conference on Communications (ICC'93)*, pp. 1064–1070 (1993).

Blachman, N. M. and McAlpine, G. A., The spectrum of a high-index FM waveform: Woodward's theorem revisited. *IEEE Transactions on Communications Technology*, **17(2)** (1969).

Blahut, R., *Theory and Practice of Error Control Codes*. Reading: Addison Wesley (1983).

Blum, R. S. and Kassam, S. A., Asymptotically optimum quantization with time invariant breakpoints for signal detection. *IEEE Transactions on Information Theory*, **37(2)**:402–407 (1991).

Bose, R. C. and Ray-Chaudhury, D. K., On a class of error correcting binary group codes. *Information and Control*, (**3**):68–79 (1960).

Bracewell, R., *The Fourier Transform and Its Applications*. New York, USA: McGraw-Hill Book Company (1965).

Bracewell, R., *Fourier Analysis and Imaging*. New York, USA: Kluwer Academic/Plenum Publishers (2004).

Cain, J. B., Clark, G. C., and Geist, J. M., Punctured convolutional codes of rate $(n-1)/n$ and simplified maximum likelihood decoding. *IEEE Transactions on Communications Technology*, **25(1)**:97–100 (1979).

Carlson, B. A., *Communication Systems*. Tokyo, Japan: McGraw-Hill (1975).

Carson, J. R., Notes on the theory of modulation. *Proceedings of the IRE* (1922).

Carvalho, F. B. S., Application of data transmission via the distribution network for the digital television return channel. Master's dissertation, Federal University of Campina Grande (2006).

Chinese National Standard Organization, *Framing Structure, Channel Coding and Modulation for Digital Television Terrestrial Broadcasting System* (in Chinese). Chinese National Standard GB 20600-2006 (2006).

Claasen, T. A. C. M. and Jongepier, A., Model for the power spectral density of quantization noise. *IEEE Transactions on Accoustics, Speech and Signal Processing*, **29**:914–917 (1981).

Clark, G. C. and Cain, J. B., *Error-Correction Coding for Digital Communications*. New York, USA: Plenum Press (1981).

CPqD, *World Panorama of Exploitation and Deployment Models*. Campinas, Brazil: CPqD Telecom and IT Solutions (2005).

CPqD, *Reference Model – The Brazilian Terrestrial Digital Television System*. Campinas, Brazil: CPqD Telecom and IT Solutions (2006).

Crinon, R. J., Bhat, D., Catapano, D., Thomas, G., Loo, J. T. V., and Bang, G., Data broadcasting and interactive television. *IEEE Proceedings*, **94**:102–118 (2006).

DASE, *DTV Application Software Environment Level 1 (DASE-1) - PART 1: Introduction, architecture, and common facilities* (2003).

Design, C. S., Communication systems design – reference library: Glossary. www.csdmag.com/glossary (2001).

Drury, G., Markarian, G., and Pickavance, K., *Coding and Modulation for Digital Television*. Berlin, Germany: Springer (2001).

DVB. Technical report, available at http://www.dvb.org, access February (2006).

Eisenberg, A., In pursuit of perfect TV color, with L.E.D.s and lasers. *New York Times*, www.nytimes.com (2007).

Elias, P., Error-free coding. *IRE Transactions on Inform. Theory*, **IT-4**:29–37 (1954).

ETSI, Digital Video Broadcasting (DVB); Framing structure, channel coding and modulation for 11/12 GHz satellite services (1997a).

ETSI, 300 744, Digital video broadcasting (DVB); framing structure, channel coding and modulation for digital terrestrial television (DVB-T). Dvb-tm document, ETSI (1997b).

ETSI, Digital Video Broadcasting (DVB); Framing structure, channel coding and modulation for cable systems (1998).

ETSI, DVB Project, Digital Video Broadcasting (DVB); interaction channel for digital terrestrial television (DVB-RCT) incorporating multiple access OFDM. DVB-TM document, ETSI (2001).

ETSI, Digital Video Broadcasting (DVB); Transmission System for Handheld Terminals (DVB-H) (2004).

ETSI, TS 102 427 V1.1.1 (2005-06): Digital Audio Broadcasting (DAB); DMB video service; user application specification. DVB-TM document (2005).

ETSI, Digital Video Broadcasting (DVB); Second generation framing structure, channel coding, and modulation systems for Broadcasting, Interactive Services, News Gathering, and other broadband satellite applications (2006a).

ETSI, TS 102 427 V1.1.1 (2005-06): Digital Audio Broadcasting (DAB); Data Broadcasting – MPEG-2 TS Streaming. DVB-TM document (2006b).

Faria, G., Henriksson, J. A., Stare, E., and Talmola, P., DVB-H digital broadcast services to handheld devices. *IEEE Proceedings*, **94**:194–209 (2006).

Farias, M. C. Q., Carvalho, M.M., and Alencar, M. S., Digital television broadcasting in Brazil. *IEEE MultiMedia* (2008).

Fernandes, J., Lemos, G., and Silveira, G., Introduction to digital interactive televison: architecture, protocols, standards and practice (in Portuguese). *Proceedings of the JAI-SBC* (2004).

Fischer, T. R. and Wang, M., Entropy-constrained trellis-coded quantization. *IEEE Transactions on Information Theory*, **38(2)**:415–426 (1992).

Flanagan, J. L. *et al.*, Speech coding. *IEEE Transactions on Communications*, **27(4)**:710–737 (1979).

Forney, J. G. D., *Concatenated Codes*. Cambridge, USA: MIT Press (1966).

Forney, J. G. D., Burst correcting codes for the classic bursty channel, *IEEE Transactions on Communications Technology*, **19(5)**:772–781 (1971).

Forney, J. G. D., The Viterbi algorithm. *IEEE Proceedings*, **68**:268–278 (1973).

Forum, M. I., MPEG-4 – The Media Standard. www.mpegif.org (2002).

Francis, M. and Green, R., Forward error correction in digital television broadcast systems. *WP270, Xilinx*, **1.0**, pp. 1–25 (2007).

Furiati, G., Services are Re-classified (in Portuguese). *Revista Nacional de Telecomunicações*, **(227)**:32–35 (1998).

Gagliardi, R., *Introduction to Communication Engineering*. New York, USA: John Wiley & Sons, (1978).

Gagliardi, R. M., *Introduction to Communications Engineering*. New York, USA: Wiley (1988).

Gallager, R., Low-density parity-check codes. *IRE Transactions on Information Theory*, **IT-8(1)**: 21–28 (1962).

Gallager, R. G., *Low-Density Parity Check Codes*. Cambridge, USA: MIT Press (1963).

Giannakis, G., Filterbanks for blind channel identification and equalization. *IEEE Signal Processing Letters*, **4(6)**:184–187 (1997).

Goblirsch, D. M. and Farvardin, N., Switched scalar quantizers for hidden Markov sources. *IEEE Transactions on Information Theory*, **38(5)**:1455–1473 (1992).

Graciosa, H. M. M., Tutorial on Digital TV in Brazil. Technical report, www.teleco.com. br/tutoriais/tutorialtvd2/default.asp (2006).

Gradshteyn, I. S. and Ryzhik, I. M., *Table of Integrals, Series, and Products*. San Diego, USA: Academic Press, Inc., (1990).

Gray, R. M., Quantization noise spectra. *IEEE Transactions on Information Theory*, **36(6)**: 1220–1244 (1990).

Gray, R. M., Dithered quantizers. *IEEE Transactions on Information Theory*, **39(3)**:805–812 (1993).

Hagenauer, J., *Advanced Methods for Satellite and Deep Space Communications. Lecture Notes in Control and Information Sciences, vol. 182*, Berlin, Germany: Springer-Verlag (1992).

Haskell, B. G., Puri, A., and Netravali, A. N., *Digital Video: An Introduction to MPEG-2*. Digital multimedia standards. New York, USA: Chapman & Hall: International Thomson Pub. (1997).

Haykin, S., *Digital Communications*. New York, USA: John Wiley and Sons (1988).

HDTV.NET. Technical report, www.hdtv.net, acess in February, 2006 (2006).

He, N., Kuhlmann, F., and Buzo, A., Multiloop sigma-delta quantization. *IEEE Transactions on Information Theory*, **38(3)** (1992).

Heegard, C. and Wicker, S. B., *Turbo Coding*. Boston, USA: Kluwer Academic Publishers (1999).

Herbster, A. F., Carvalho, F. B. S., and Alencar, M. S., Use of the IEEE 802.16 standard in ad hoc networks to transmit the return channel for interactive digital TV (Portuguese). In *XI Brazilian Symposium on Multimedia and Web Systems (WebMedia 2005), Poços de Caldas, Brazil*, pp. 187–194 (2005).

Herrero, C., Cesar, P., and Vuorimaa, P., Delivering MHP applications into a real DVB-T network, OTADIGI. In *Proceedings of the 6th IEEE International Conference on Telecommunications in Modern Satellite, Cable, and Broadcasting Services*, pp. 231–234 (2003).

Hocquenghem, A., Codes correcteurs d'erreurs. *Chiffres*, **(2)**:147–156 (1959).

Hong-pang, Z. and Shi-bao, Z., An extensible digital television middleware architecture based on hardware abstraction layer. In *2004 IEEE International Conference on Multimedia and Expo (ICME)*, pp. 711–714 (2004).

Hsu, H. P., *Fourier Analysis* (in Portuguese). Rio de Janeiro, Brazil: Livros Técnicos e Científicos Publishers Ltd. (1973).

Immink, K. A. S., *Coding Techniques for Digital Recorders*. Englewood Cliffs, USA: Prentice Hall (1991).

ISO/IEC, *ISO/IEC JTC1/SC29: Information Technology – Coding of Moving Pictures and Associated Audio for Digital Storage Media at up to about 1.5 Mbit/s – IS 11172 (Part 3, Audio)* (1992).

ISO/IEC, *ISO/IEC JTC1/SC29: Information Technology – Coding of Moving Pictures and Associated Audio Information – IS 13818 (Part 3, Audio)* (1994).

ISO/IEC, *ISO/IEC 13818-7: Information Technology – Generic Coding of Moving Pictures and Associated Audio Information – Part 7: Advanced Audio Coding (AAC)*. Number BSI ISO/IEC (1997).

ISO/IEC, *Coding of Audio-Visual Objects – Part 2: Visual. ISO/IEC 14496-2 (MPEG-4 visual version 1)*, ISO/IEC JTC1 MPEG and ITU-T VCEG (1999).

ISO/IEC, *ISO/IEC 14496-3: Information Technology – Coding of Audio-Visual Objects – Part 3: Audio*. Number BSI ISO/IEC (2005).

Itoh, N. and Tsuchida, K., HDTV Mobile Reception in Automobiles. *IEEE Proceedings*, **94**: 274–280 (2006).

ITU, *ISO/IEC 13818: Generic coding of moving pictures and associated audio (MPEG-2)* (1998).

ITU, *Recommendation ITU-T H.264: Advanced Video Coding for Generic Audiovisual Services*, (2003).

ITU-T, *A Guide to Digital Terrestrial Television Broadcasting in the VHF/UMF Bands* (1996).

ITU-T, *Recommendation J.83: Digital Multi-Programme Systems for Television, Sound and Data Services for Cable Distribution* (1998).

ITU-T, *Video coding for low bit rate communication*. Technical report, ITU-T (2000a).

ITU-T, *Recommendation H.222 Information Technology – Generic Coding of a Moving Picture and Associated Audio Information* (2000b).

ITU-T, *Recommendation J.200: Worldwide Common Core Application Environment for Digital Interactive Television Services* (2001).

ITU-T, *Recommendation J.202: Harmonization of Procedural Content Formats for Interactive TV Applications* (2003).

ITU-T, *Recommendation J.201: Harmonization of Declarative Content Format for Interactive Television Applications* (2004).

Johannesson, R. and Zigangirov, K. S., *Fundamentals of Convolutional Coding*. IEEE Series on Digital & Mobile Communication. Piscataway, USA: IEEE (1999).

Joint Video Team of ITU-T and ISO/IEC JTC 1, Draft ITU-T Recommendation and Final Draft International Standard of Joint Video Specification (ITU-T Rec. H.264 ISO/IEC 14496-10 AVC). Jvt-g050, ISO/IEC MPEG and ITU-T VCEG (2003).

Joint Video Team of ITU-T and ISO/IEC JTC 1, Draft ITU-T Recommendation and Final Draft International Standard of Joint Video Specification (ITU-T Rec. H.264 ISO/IEC 14496-10 AVC) – Fidelity Range Extensions documents JVT-L047. Jvt-l047 (non-integrated form) and jvt-l050(integrated form), ISO/IEC MPEG and ITU-T VCEG (2004).

Jones, G. A., Defilippis, J. M., Hoffmann, H., and Williams, E. A., Digital television station and network implementation. *Proceedings of the IEEE*, **94(1)**:22–36 (2006).

Jurgen, R. K., *Digital Consumer Electronics Handbook*. New York, USA: McGraw-Hill (1997).

Kou, Y., Lin, S., and Fossorier, M., Low density parity check codes based on finite geometries: a rediscovery. In *Proceedings of the IEEE International Symposium on Information Theory, Sorrento, Italy*. Piscataway, USA: IEEE (2000).

Ladebusch, U. and Liss, C., Terrestrial DVB (DVB-T): a broadcast technology for stationary portable and mobile use. In *IEEE Proceedings*. Piscataway, USA: IEEE (2006).

Lambert, P., de Neve, W., deNeve, P., Moerman, I., Demeester, P., and de Walle, R. V., Rate-distortion performance of H.264/AVC compared to state-of-the-art video codecs. *Circuits and Systems for Video Technology, IEEE Transactions on Circuits and Systems for Video Technology*, **16(1)**:134–140 (2006).

Laroia, R. and Fravardin, N., A Structured fixed-rate vector quantizer derived from a variable-length scalar quantizer: Part I – Memoryless sources. *IEEE Transactions on Information Theory*, **39(3)**:851–867 (1993).

Lathi, B. P., *Modern Digital and Analog Communication Systems*. Philadelphia, USA: Holt, Rinehart and Winston, Inc. (1989).

Lee, W. C. Y., *Mobile Cellular Telecommunications Systems*. New York, USA: McGraw-Hill Book Company (1989).

Lévine, B., *Fondements Théoriques de la Radiotechnique Statistique*. Moscow, USSR: Éditions de Moscou (1973).

Lin, T. -Y. and Palais, J. C., Laser diode linewidth measurement techniques. In *Proceedings of the IEEE Instrumentation and Measurement Technology Conference – IMTC'86*, pp. 226–228 Piscataway, USA: IEEE (1986).

Lloyd, S. P., Least Squares Quantization in PCM. *IEEE Transactions on Information Theory*, **28(2)**:129–137 (1982).

MacKay, D. and Neal, R., Near Shannon limit performance of low density parity check codes. *Electronics Letters*, **32(18)**:1645 (1996).

MacKay, D. J. C. and Neal, R. M., Near Shannon limit performance of low density parity check codes. *Electronics Letters*, **33(6)**:457–458 (1997).

MacKay, D. J. C. and Neal, R. M., Good error-correcting codes based on very sparse matrices. *IEEE Transactions on Information Theory*, **45(2)**:399–432 (1999).

MacWilliams, F. and Sloane, N., *The Theory of Error-Correcting Codes*. Amsterdam: North-Holland (1977).

Madhow, U., *Fundamentals of Digital Communication*. Cambridge: Cambridge University Press (2008).

Manhaes, M. A. R. and Shieh, P. J., Interactive channel: concepts, potential and commitments. In *Workshop on Digital and Interactive TV – Sibgrapi 2005, Natal, Brazil* (2005).

Massey, J. L., Shift-register synthesis and BCH decoding. *IEEE Transactions on Information Theory*, **15(1)**:122–127 (1969).

Massey, J. L., Cryptography: Fundamentals and Applications. Class notes. Zurich, Switzerland: ETH-Zurich (1998).

Max, J., Quantizing for Minimum Distortion. *IEEE Transactions on Information Theory*, **6(1)**: 7–12 (1960).

MC/MCT/FINEP/FUNTTEL, Descriptive Form for Deliverable Request – Middleware (2004).

Meloni, L. G. P., Return channel for the Brazilian digital television system-terrestrial. *Journal of the Brazilian Computer Society*, **13**:83–94 (2007).

Mendes, L. L. and Fasolo, S. A., Introduction to Digital TV. Technical report, Inatel (2002).

MHP. Technical report, available at www.mhp.org, access February (2006).

Mitianoudis, N., Management of Telecommunication Systems - Essay 3: Digital television (2004).

MobileWord, Mobileword's glossary. www.mobileworld.org (2001).

Moreira, J. C. and Farrell, P. G., *Essentials of Error-Control Codes*. New York, USA: John Wiley & Sons Ltd. (2006).

Moreira, R. A., Digital Television and Interativity. Technical report, Available at www.teleco.com. br/tutoriais/tutorialinteratividade/default.asp (2006).

Morello, A. and Mignone, V., DVB-S2: The second generation standard for satellite broadband services. *Proceedings of the IEEE*, **14(1)** (2006).

MPEG, MPEG-4 frequently asked questions. http://www.chiariglione.org/mpeg/faq/mp4-vid/mp4-vid.htm (2000).

MPEG, Overview of the MPEG-4 Standard. http://www.chiariglione.org/mpeg/standards/mpeg-4/mpeg-4.htm (2002).

Muck, M., de Courville, M., Debbah, M., and Duhamel, P., A pseudo random postfix OFDM modulator and inherent channel estimation techniques. In *Proceedings of the IEEE Global Telecommunications Conference (GLOBECOM)*, **4**, pp. 2380–2384 (2003).

Muquet, B., Wang, Z., Giannakis, G., de Courville, M., and Duhamel, P., Cyclic prefixing or zero padding for wireless multicarrier transmissions? *IEEE Transactions on Communications*, **50(12)**:2136–2148 (2002).

Nebeker, F., July and August in EE History. *Institute*, 9 (2001).

NHK, Super hi-vision standard. Internet webpage, Japan Broadcasting Corporation, www.nhk.or.jp/digital/en/super hi/index.html (2008).

Oberhettinger, F., *Tables of Fourier Transforms and Fourier Transforms of Distributions*. Berlin, Germany: Springer-Verlag (1990).

Oppenheim, A. V. and Schafer, R. W., *Discrete-Time Signal Processing*. Englewoods Cliffs, USA: Prentice-Hall (1989).

Paez, M. D. and Glisson, T. H., Minimum mean-squared-error quantization in speech PCM and DPCM systems. *IEEE Transactions on Communications*, **20(2)**: 225–230 (1972).

Papoulis, A., *Probability, Random Variables, and Stochastic Processes*. Tokyo, Japan: McGraw-Hill (1981).

Papoulis, A., *Signal Analysis*. Tokyo, Japan: McGraw-Hill (1983).

Peterson, W. W. and Weldon, E. J., *Error-Correcting Codes*. Second Edition. Cambridge, USA: MIT Press (1972).

Phamdo, N., Farvardin, N., and Moriya, T., A unified approach to tree-structured and multistage vector quantization for noisy channels. *IEEE Transactions on Information Theory*, **39(3)**: 835–850 (1993).

Piesing, J., The DVB multimedia home platform (MHP) and related specifications. *IEEE Proceedings*, **94**:237–247 (2006).

Poynton, C., *Digital Video and HDTV Algorithms and Interfaces*. San Francisco, USA: Morgan Kaufman Publishers (2003).

Pyndiah, R., Glavieux, A., Picart, A., and Jacq, S., Near optimum decoding of product codes. In *Proceedings Global Telecommunications Conference, 1994, (GLOBECOM'94)*, **1**, pp. 339–343 (1994).

Ramsey, J. L., Realization of optimum interleavers. *IEEE Transactions on Information Theory*, **16**:338–345 (1970).

Reed, I. S. and Solomon, G., Polynomial codes over certain finite fields. *Journal of the Society of Industrial and Applied*, **8**:300–304 (1960).

Reeves, A. H., US Patent 2,272,070. Technical report, International Standard Electric Coorporation (1942).

Reimers, U., DVB-T: the COFDM-based system for terrestrial television. *Electronics Communication Engineering Journal* (1997).

Reimers, U., *DVB – The Family of International Standards for Digital Video Broadcasting*. New York, USA: Springer-Verlag (2004).

Reimers, U., *DVB – The Family of International Standards for Digital Video*. Berlin, Germany: Springer (2005a).

Reimers, U., *DVB – The Family of International Standards for Digital Video Broadcasting*. Second edition. Berlin, Germany: Springer (2005b).

Reimers, U., DVB – The Family of International Standards for Digital Video Broadcasting. *IEEE Proceedings*, **94**:173–192 (2006).

Resende, L. E. A., *Development of a performance analysis platform for the ISDB-T television standard*. Master's dissertation. PUC-RJ, Rio de Janeiro, Brazil (2004).

Richardson, I. E., *H.264 and MPEG-4 Video Compression*. New York, USA: John Wiley & Sons (2003a).

Richardson, I. E. G., *H.264 and MPEG-4 Video Compression: Video Coding for Next Generation Multimedia*. New York, USA: Addison Wesley (2003b).

Richardson, I. E. G., H.264/MPEG-4 Part 10 White Paper – Reconstruction Filter. http://www.vcodex.com (2003c).

Richer, M. S., Reitmeier, G., Gurley, T., Jones, G. A., Whitaker, J., and Rast, R., The ATSC Digital Television System. *IEEE Proceedings*, **94**:37–43 (2006).

Santos, D. F. S., Silva, E. F., Carvalho, F. B. S., and Alencar, M. S., RF-Intraband Solution for the Interactivity Channel of the Brazilian Digital TV System. *Proceedings of the Workshop on Digital TV, Brazilian Symposium on Graphic Computing – Sibigrapi'2005* (2005).

Scaglione, A., Giannakis, G., and Barbarossa, S., Redundant filterbank precoders and equalizers – Part I: Unification and optimal designs. *IEEE Transactions on Signal Processing*, **47**:1988–2006 (1999a).

Scaglione, A., Giannakis, G., and Barbarossa, S., Redundant filterbank precoders and equalizers – Part II: Blind channel estimation, synchronization and direct equalization. *IEEE Transactions on Signal Processing*, 47:2007–2022 (1999b).

Schäfer, R. Wiegand, T., and Schwarz, H., The emerging H.264/AVC standard. In EBU Technical Review, Special Issue on Best of 2003 (2003).

Shannon, C. E., A Mathematical Theory of Communication. *The Bell System Technical Journal*, 27:379–423 (1948a).

Shannon, C. E., The Philosophy of PCM. *Proceedings of the IRE*, 36:1324–1331 (1948b).

Silva, H. V. O., Rodrigues, R. F., Soares, L. F. G., and Saade, D. C. M., NCL 2.0: Integrating new concepts to XML modular languages. In *DocEng'04: Proceedings of the 2004 ACM Symposium on Document Engineering*, pp. 188–197, New York, USA: ACM (2004).

Simonetta, J., *Televisión Digital Avanzada – Handbook*. Buenos Aires, Argentina: Intertel (2002).

Skycell, Glossary of satellite terminology. www.satellitetelephone.com (2001).

Soares, L. F. G., Rodrigues, R. F., and Moreno, M. F., Ginga-NCL: the declarative environment of the Brazilian digital TV system. *Journal of the Brazilian Computer Society*, 13:37–46 (2007).

Song, B., Gui, L., Guan, Y., and Zhang, W., On channel estimation and equalization in TDS-OFDM based terrestrial HDTV broadcasting system. *IEEE Transactions on Consumer Electronics*, 51(3):790–797 (2005).

Song, J., Yang, Z., Yang, L., Gong, K., Pan, C., Wang, J., and Wu, Y., Technical review on Chinese digital terrestrial television broadcasting standard and measurements on some working modes. *IEEE Transactions on Broadcasting*, 53(1):1–7 (2007).

Spiegel, M. R., *Fourier Analysis* (in Portuguese). São Paulo, Brazil: McGraw-Hill, Brazil, Ltd. (1976).

Sripad, A. B. and Snyder, D. L., A necessary and sufficient condition for quantization errors to be uniform and white. *IEEE Transactions on Accoustics, Speech and Signal Processing*, 25(5):442–448 (1977).

Stienstra, A. J., Technologies for DVB Services on the Internet. In *Proceedings of the IEEE*, 94(1): (2006).

Stremler, F. G., *Introduction to Communication Systems*. Reading, USA: Addison-Wesley Publishing Company (1982).

Sullivan, G. J. and Wiegand, T., Rate-distortion optimization for video compression. *IEEE Signal Processing Magazine*, 15:74–79 (1998).

Sun, H. Chiang, T. and Chen X., *Digital Video Transcoding for Transmission and Storage*. Boca Raton, USA: CRC Press (2005).

Takada, M. and Saito, M., Transmission System for ISDB-T. *Proceedings of the IEEE*, 94:251–256 (2006).

Tanner, R. M., A recursive approach to low complexity codes. *IEEE Transactions on Information Theory*, 27:533–547 (1981).

Taub, H. and Schilling, D. L., *Principles of Communication Systems*. McGraw-Hill, Tokyo, Japan (1971).

TELECO. Technical report. Available at http://www.teleco.com.br (2006).

TIAB2B.com, Everything communications. www.tiab2b.com/glossary (2001).

Trushkin, A. V., On the design of an optimal quantizer. *IEEE Transactions on Information Theory*, 39(4):1180–1194 (1993).

Tude, E., UMTS Tutorial (in Portuguese). Technical report. Available at www.teleco.com.br/tutoriais /tutorialwcdma (2006).

Tufvesson, F., Edfors, O., and Faulkner, M., Time and frequency synchronization for OFDM using PN-sequence preambles. In *Proceedings of the 50th IEEE Vehicular Technology Conference*, Vol. 4, pp. 2203–2207, Amsterdam, The Netherlands (1999).

UIT ITU-R, Document 11-3/3-E: A Guide to Digital Terrestrial Television Broadcasting in the VHF/UHF Bands. Geneva, Switzerland (1996).

Valdestilhas, A., Almeida, F. A., Carvalho, F. B. S., and Alencar, M. S., An approach on the development of the interactive digital television: new concepts and tools (Portuguese). In *XI Brazilian Symposium on Multimedia and Web Systems (WebMedia 2005), Poços de Caldas, Brazil*, pp. 313–318 (2005).

Views, Fazendo vídeo. Views Imagem & Comunicação Ltda., www.fazendovideo.com.br (2006).

Viterbi, A. J., When not to spread spectrum – a sequel. *IEEE Communications Magazine*, **23(4)**: 12–17 (1985).

Wagdy, M. F., Effect of ADC quantization errors on some periodic signal measurements. *IEEE Transactions on Instrumentation and Measurement*, **36(4)**:983–989 (1987).

Wang, J., Yang, Z., Pan, C., Han, M., and Yang, L., A combined code acquisition and symbol timing recovery method for TDS-OFDM. *IEEE Transactions on Broadcasting*, **49(3)**:304–308 (2003).

Wang, J., Yang, Z., Pan, C., Song, J., and Yang, L., Iterative padding subtraction of the PN sequence for the TDS-OFDM over broadcast channels. *IEEE Transactions on Consumer Electronics*, **51(4)**:1148–1152 (2005).

Whitaker, J. C., *DTV Handbook: The Revolution in Digital Video*. New York, USA: McGraw-Hill Professional (2001).

Wiberg, N., Loeliger, N., and Kotter, R., Codes and iterative decoding on general graphs. *European Transactions on Telecommunications*, **6(5)**:513–525 (1995).

Wicker, S. and Bhargava, V., *Reed Solomon Codes and Their Applications*, chapter RS codes and the compact disk, by K.A.S. Immink. Piscataway, USA: IEEE Press (1994).

Wicker, S. B., *Error Control Systems for Digital Communication and Storage*. New Jersey, USA: Prentice-Hall (1995).

Wiegand, T. and Andrews, B. D., An improved H.263 coder using rate-distortion optimization. *ITU-T/SG16/Q15-D-13* (1998).

Wikipedia, DLP. Wikipedia Encyclopedia, pt.wikipedia.org/wiki/DLP (2007a).

Wikipedia, MPEG-4. http://en.wikipedia.org/wiki/MPEG-4 (2007b).

Woodward, P. M., The spectrum of random frequency modulation. Memo. 666, Telecommunications Research Establishment, Great Malvern, England (1952).

Wozencraft, J. M. and Reiffen, B., *Sequential Decoding*. Cambridge, USA: MIT Press (1961).

Wu, X. and Zhang, K., Quantizer monoticities and globally optimal scalar quantizer design. *IEEE Transactions on Information Theory*, **39(3)**:1049–1052 (1993).

Yamada, F., Sukys, F., Jr., G. B., Akamine, C., Raunheitte, L. T. M., and Dantas, C. E., Digital TV system, fifth edition (in Portuguese). *Mackenzie Computing and Engineering Magazine*. Mackenzie Presbyterian University (2004).

Yang, Z., Tong, L., and Yang, L., Outage probability comparison of CP-OFDM and TDS-OFDM for broadcast channels. In *Proceedings of the IEEE Global Telecommunications Conference (GLOBECOM)*, Vol. 1, pp. 594–598 Piscataway, USA: IEEE Press (2002).

Zamir, R. and Feder, M., On universal quantization by randomized uniform/lattice quantizers. *IEEE Transactions on Information Theory*, **38(2)**:428–436 (1992).

Zeger, K. and Manzella, V., Asymptotic bounds on optimal noisy channel quantization via random coding. *IEEE Transactions on Information Theory*, **40(6)**:1926–1938 (1994).

Zhang, J., Perkis, A., and Georganas, N., H.264/AVC and transcoding for multimedia adaptation. In *Proceedings 6th COST 276 Workshop*. Greece: European Cooperation in the Field of Scientific and Technical Research (2004).

Zheng, Z., Yang, Z., Pan, C., and Zhu, Y., Synchronization and channel estimation for TDS-OFDM systems. In *Proceedings of the 58th IEEE Vehicular Technology Conference (VTC)*, Vol. 2, pp. 1229–1233. Piscataway: IEEE Press (2003).

Zheng, Z., Yang, Z., Pan, C., and Zhu, Y., Novel synchronization for TDS-OFDM-based digital television terrestrial broadcast systems. *IEEE Transactions on Broadcasting*, **50(2)**:148–153 (2004).

Zuffo, M. K., Open Digital TV in Brazil – Structuring Policies for a National Model. Technical report. Available at www.lsi.usp.br/mkzuffo (2006).

Index